Motors and Controls

James T. Humphries
Southern Illinois University

Merrill Publishing Company
A Bell & Howell Information Company
Columbus • Toronto • London • Melbourne

Cover Photo: An exploded view of a moving-coil motor, courtesy of PMI Motion Technologies

Published by Merrill Publishing Company
A Bell & Howell Information Company
Columbus, Ohio 43216

This book was set in Century Schoolbook.

Administrative Editor: Stephen Helba
Text Designer and Production Coordinator: Jeffrey Putnam
Art Coordinator: James Hubbard
Cover Designer: Jolie Muren
Library of Congress Catalog Card Number: 87-63486
International Standard Book Number: 0-675-20235-3
Printed in the United States of America
1 2 3 4 5 6 7 8 9—92 91 90 89 88

MERRILL'S INTERNATIONAL SERIES IN ELECTRICAL AND ELECTRONICS TECHNOLOGY

HUMPHRIES	*Motors and Controls*, 20235-3
KULATHINAL	*Transform Analysis and Electronic Networks with Applications*, 20765-7
LAMIT/LLOYD	*Drafting for Electronics*, 20200-0
LAMIT/WAHLER/HIGGINS	*Workbook in Drafting for Electronics*, 20417-8
MARUGGI	*Technical Graphics: Electronics Worktext*, 20161-6
MILLER	*The 68000 Microprocessor: Architecture, Programming, and Applications*, 20522-0
ROSENBLATT/FRIEDMAN	*Direct and Alternating Current Machinery, Second Edition*, 20160-8
SCHOENBECK	*Electronic Communication: Modulation and Transmission*, 20473-9
SCHWARTZ	*Survey of Electronics, Third Edition*, 20162-4
STANLEY, B. H.	*Experiments in Electric Circuits, Second Edition*, 20403-8
STANLEY, W. D.	*Operational Amplifiers with Linear Integrated Circuits*, 20090-3
TOCCI	*Fundamentals of Electronic Devices, Third Edition*, 9887-4
	Electronic Devices, Third Edition, Conventional Flow Version, 20063-6
	Fundamentals of Pulse and Digital Circuits, Third Edition, 20033-4
	Introduction to Electric Circuit Analysis, Second Edition, 20002-4
WEBB	*Programmable Controllers: Principles and Applications*, 20452-6
YOUNG	*Electronic Communication Techniques*, 20202-7

MERRILL'S SERIES IN MECHANICAL AND CIVIL TECHNOLOGY

BATESON	*Introduction to Control System Technology, Second Edition*, 8255-2
HOPEMAN	*Production and Operations Management: Planning, Analysis, and Control, Fourth Edition*, 8140-8
HUMPHRIES	*Motors and Controls*, 20235-3
KEYSER	*Materials Science in Engineering, Fourth Edition*, 20401-1
LAMIT/PAIGE	*Computer-Aided Design and Drafting*, 20475-5
MARUGGI	*Technical Graphics: Electronics Worktext*, 20311-2
MOTT	*Applied Fluid Mechanics, Second Edition*, 8305-2
	Machine Elements in Mechanical Design, 20326-0
ROLLE	*Introduction to Thermodynamics, Second Edition*, 8268-4
ROSENBLATT/ FRIEDMAN	*Direct and Alternating Current Machinery, Second Edition*, 20160-8
WEBB	*Programmable Controllers: Principles and Applications*, 20452-6
WOLF	*Statics and Strength of Materials*, 20622-7

CONTENTS

Preface

This is a text on AC and DC machines and the circuits that control them. The electric motor is the principal means of converting electrical energy into mechanical energy in industry. In fact, most of the energy used in industry today is used to drive motors of some kind. Because motors are so widely used in industry, personnel responsible for maintaining electrical systems in industry need to have a good basic knowledge of the subject of motors. This book is designed to meet that need. All the major motors used in industry today are covered in such a way that technical personnel can understand their basic operational principles. In an effort to be as up-to-date as possible, the most recent developments in motors have been included, such as brushless DC motors and ironless disc stepper motors. The older motors, such as the induction and synchronous motors, have received equal attention. Design problems and questions are avoided.

Every motor used in industry must be controlled. As a result, seven chapters in this text deal with motor control devices and circuits, from relay control to digital, microprocessor, and programmable controllers. Again, every effort has been made to present the basic information technical personnel need to understand basic motor control devices, circuits, and systems. A book that trains people to take a useful place in industry must be practical. Maintenance principles and practices are discussed throughout this text, including subjects such as strategies for replacing semiconductors in control equipment. This practical emphasis is often overlooked in many texts.

This book is intended for use in electronics technology programs offered in two-year community colleges, technical institutes, and vocational/technical programs. Students taking this course should have a basic course in algebra as a prerequisite. Some motor systems can best be described and understood by working with the equations that describe their characteristics. Understanding these motors and controls has been enhanced by presenting many charts, diagrams, and photographs, in addition to the mathematical descriptions, where necessary. The higher level mathematics often seen in engineering texts on motors and controls has been avoided, since the emphasis of this text has a technical rather than an engineering emphasis. Students should also have a course in basic DC and AC (including magnetic fields) and semiconductor theory. A course in digital and microprocessor circuits would be helpful but is not necessary.

Learning aids have been built into this text. First, learning objectives appear at the beginning of the chapter. Where concepts involve mathematics, examples are given. Summaries of the most important points in the chapter appear at the end of each chapter. Since many of the concepts relating to motors and controls are difficult to understand, hundreds of photographs, line drawings, and charts have been included. Finally, problems and exercises appear at the end of each chapter. The appendices also contain much valuable information, such as a bibliography for further reference and data sheets for motors and controls.

I would like to acknowledge the assistance of the reviewers who undertook the task of reading and suggesting improvements to this text in its preliminary stages: Russ Bowker, Norfolk Community College; Ted Brown, Southwest Missouri State University; Kenneth Edwards, International Brotherhood of Electrical Workers; Charles Hahn, Aiken Technical College; and Homer Solesbee, Augusta Area Technical College. In particular, I would like to thank Russ Bowker for his comprehensive review of the final manuscript. His efforts were absolutely invaluable. I would also like to thank the companies that provided many of the photographs and line drawings. In particular, I would like to thank Mr. Don Wismann of PMI Motion Technologies, Mr. Fenimore Fisher of IMC Magnetics, Mr. John Mazurkiewicz of Pacific Scientific, and Ms. Yvonne Shaffer of Inland Motor.

I would also like to take this opportunity to thank my colleagues at Santa Fe Community College—Dr. Tony Blalock, Mr. George Purdue, Mr. Jerry Rosenberg, Mr. Don Dekold, Mrs. Joanne Mott, Dr. Alton Stevens, and Mr. Bruce Bush—for their support and encouragement during the final stages of this project. I would also like to thank the patient staff at Merrill Publishing Company for their professional and enthusiastic support. I would especially like to thank Steve Helba.

Last, but certainly not least, I want to thank my family, and especially my wife Margaret, for their sacrifices and understanding during the preparation of this manuscript.

To my mother

SAFETY NOTE

Installing, testing, and operating rotating machinery can be dangerous. Beyond the obvious mechanical dangers, the voltages used for motors and controls are frequently high, well above the 120 VAC used in our homes. Both students and maintenance personnel need to observe standard industrial safety practices when working on this type of equipment.

This text is intended to be a general introduction to the concepts behind motors and controls, generators, and transformers and is NOT intended to be used as a laboratory manual. In some cases, the topics have been simplified to ensure better understanding. Because of the widely varying types and complexities of motors and controls, this simplified treatment may not cover the safe operation of every type of motor and control found in industry or in the classroom.

If you do not know how to operate or troubleshoot a particular piece of equipment safely, do not proceed until you do. Your instructor or supervisor is the best source of this information. Also, most manufacturers of rotating machinery and drives provide information about safe operation and testing of their equipment. Look in the manufacturer's data for these safety notes.

NEVER operate or make measurements on rotating machinery without your instructor's or supervisor's knowledge and approval. ALWAYS follow standard safety practices when operating or testing rotating machinery.

1

MOTOR FUNDAMENTALS

At the end of this chapter, you should be able to

- define and give an example of each of the following:
  - *work* (in the scientific sense)
  - *torque*
  - *power*.
- contrast the terms *kilowatt-hour* and *kilowatt*.
- explain how electric current is generated electromagnetically.
- list the three things necessary to generate electric current by electromagnetic means.
- explain and apply the left-hand rule for generators.
- define *power factor*.
- draw the two configurations for three-phase generators.
- explain and apply the right-hand rule for motors.
- define *CEMF* and *motor effect*.

1–1 INTRODUCTION

This chapter examines many concepts crucial to the study of motors. This chapter will be divided into two general sections: mechanical and electrical concepts. In the mechanical section, we will discuss torque, work, and power. In the electrical section, we will discuss how motors and generators produce torque and power. We will also consider single-phase as well as three-phase generators.

1

1–2 MECHANICAL CONCEPTS

1–2.1 Torque

A *force* is a quantity that produces motion. *Torque* is a particular type of force, the twisting kind that causes an object to rotate. It is exerted, for example, at a tangent to a motor shaft. This force causes the motor shaft to rotate. Whether the object moves or not, torque is still exerted. This concept is illustrated in Figure 1–1. Note that the hand is pushing down on the wrench with a force, F. The force is exerted at a distance, d. This distance is measured between the pivot point (the center of the bolt) and the place on the wrench where the force is exerted. The amount of torque produced is calculated by multiplying the force, F, and the distance, d. (Force is measured in newtons (N), where 1 newton is the force required to accelerate 1 kilogram (kg) of mass at a rate of 1 meter per second (m/s). One N is approximately equal to two-tenths of a pound.) For example, let us say that the force exerted is 2 N and the distance is 0.5 m.

$$T = F \times d = (2 \text{ N}) (0.5 \text{ m}) = 1 \text{ N} \cdot \text{m} \qquad \textbf{(eq. 1–1)}$$

The torque T produced in this system is 1 newton-meter (N · m). The newton-meter is the SI unit for torque. The English system torque units of pound-feet (lb · ft) and ounce-inches (oz · in) are still widely used in the United States.

The diagram in Figure 1–2 shows this concept by showing the calculation of two torques, each exerted in opposite directions. The motor produces torque that is in the direction of F_1, called the *clockwise* (CW) direction. A torque

FIGURE 1–1 A force is exerted on a wrench at distance d

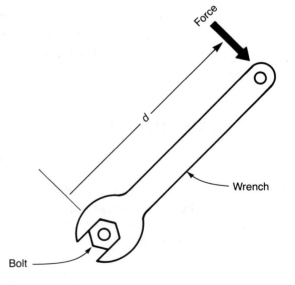

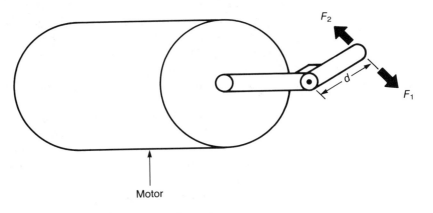

FIGURE 1–2 A motor produces torque

caused by F_2 is exerted in the opposite, or counterclockwise (CCW), direction. This ***counter torque*** can be the result of a mechanical load on the motor, such as a conveyor belt. The total resultant torque (T_r) can be calculated by equation 1–2:

$$T_r = T_1 - T_2 = (F_1 \times d) - (F_2 \times d) \qquad \text{(eq. 1–2)}$$

If $F_1 = 10$ N, $F_2 = 2$ N, and $d = 3$ m, the resultant torque would be

$$T_r = T_1 - T_2 = (F_1 \times d) - (F_2 \times d)$$
$$= (10 \text{ N} \times 3 \text{ m}) - (2 \text{ N} \times 3 \text{ m}) = 30 - 6 = 24 \text{ N} \cdot \text{m}$$

It should be stressed at this point that torque is produced whether the motor shaft is turning or not. If the torque produced by the motor is greater than the counter torque of the resisting load, the motor shaft will turn. If the load resists with a counter torque greater than the torque produced in the motor shaft, the shaft will not turn.

1–2.2 Work

Closely allied to the concept of torque is the idea of ***work.*** Work is accomplished when a force acting on an object causes it to move through a distance. The force acting on the object must overcome some resisting force. For example, work is done when a box is pushed across a floor. The pushing force must overcome the friction (resisting force) of the box on the floor. The unit for work in the SI system is the joule. One joule is a force of 1 N moving through a distance of 1 m (1 J · m). An example of work being done can be illustrated by a person lifting a box. If the box weighs 10 N (about 2.2 lb) and the person lifts it 1 m, the amount of work done is 10 joules (J). In the English system, torque is represented by the pound-foot (lb · ft), while the unit for work is the foot-pound (ft · lb). The definition for *foot-pound* is very similar to the definition of *joule*. One foot-pound is the work done in moving one pound one foot. One foot-pound is equal to approximately 1.36 J. One joule is equal to 0.737 ft · lb.

In motor behavior, work is produced by the motor torque acting through a distance. As the previous definition of *work* indicates, motion must be present for work to be done.

In our earlier example, we used a wrench that was 0.5 ft long. A person exerted a force of 2 lb at the end of the wrench. We know that the person made a complete circle with the wrench. How much work did this person do? We know how much force is being exerted, i.e., 2 lb. Since work involves the distance covered, what is the value of the distance traveled by the end of the wrench? We know that the circumference of a circle is

$$C = 2\pi r = (2)(3.14)(0.5 \text{ ft}) = 3.14 \text{ ft} \qquad \textbf{(eq. 1–3)}$$

where C is the circumference in feet (ft) and r is the radius of the circle traveled by the wrench in feet. We can see from this calculation that the distance traveled by the end of the wrench is 3.14 ft. If we multiply this distance by the force in pounds, we will know the amount of work (W) done:

$$W = F \times d = (2 \text{ lb})(3.14 \text{ ft}) = 6.28 \text{ ft} \cdot \text{lb} \qquad \textbf{(eq. 1–4)}$$

where W is the amount of work done in foot-pounds, F is the force applied in pounds, and d is the distance traveled in feet.

The difference between torque and work cannot be stressed enough. Torque is the turning force exerted by a motor shaft. Work, in a motor system, is the energy developed when the motor shaft turns as a result of the torque generated. Torque can be produced whether the motor shaft rotates or not. Work, however, can be done only when movement is involved. A motor shaft producing torque—without moving that shaft—does no work.

1–2.3   Power

Another useful concept in the study of motors is the concept of ***power*** **(P).** We can define *power* as work per unit of time or as the rate of doing work. Power is normally expressed in watts in the SI system. One watt (W) is 1 J/s or 0.737 ft · lb per second. You are probably familiar with the electrical definition of *one watt* as the energy released when one ampere of current causes a one-volt potential. In the English system, one ***horsepower*** **(HP)** is equal to 550 ft · lb per second or 746 J/s (746 W). Until recently, motor power ratings were given exclusively in HP. Today, more manufacturers are giving motor ratings in watts and kilowatts (kW). As previously indicated, we can change from watts to horsepower, or horsepower to watts, by using the relation 1 HP = 746 W. For example, the output of a 2-HP motor would be

$$P \text{ (in W)} = P \text{ (in HP)} \times 746 \qquad \textbf{(eq. 1–5)}$$

$$= (2)(746) = 1492 \text{ W, or } 1.492 \text{ kW}$$

This 2-HP motor (when operated at 2 HP) produces 1492 W, or 1.492 kW.

While we are discussing the topic of work, we should mention how power consumption is measured. Power consumption in our homes is measured in **kilowatt-hours.** The kilowatt-hour (kWh), as its name suggests, is the number of kilowatts consumed in one hour. For example, let's say we used a 100-W light bulb for two hours. If we multiply the power we use by the time we use it, we get the rate of energy consumption:

$$\text{rate of energy usage} = \text{power} \times \text{time}$$

$$= 1000 \text{ W} \times 2 \text{ h} = 2000 \text{ Wh, or } 2 \text{ kWh}$$

The kilowatt-hour is widely used in the English system of measurement. The SI equivalent is the megajoule (MJ). One kWh is equal to 3.6 MJ.

When a motor shaft rotates, we can calculate the amount of power the motor produces by knowing two things: the motor's speed in revolutions per minute (r/min) and the torque produced by the motor either in N · m or ft · lb. If we need to know the motor's power output in watts, the relation is

$$P \text{ (in W)} = 0.105 N_a T \qquad \textbf{(eq. 1–6)}$$

where N_a is the motor's speed and T is the torque the motor produces in N · m. Using equation 1–6, let us say that a motor is producing 10 N · m of torque at a speed of 900 r/min. The power output of the machine is

$$\text{power output in watts} = 0.105 N_a T$$

$$= (0.105)(900 \text{ r/min})(10 \text{ N} \cdot \text{m})$$

$$= 945 \text{ W}$$

In the English system, the power output in HP can be found by the equation

$$P \text{ (in HP)} = \frac{T N_a}{5252} \qquad \textbf{(eq. 1–7)}$$

where N_a is the motor's speed in r/min and T is the torque the motor produces in lb · ft.

This equation is derived from knowing that

$$1 \text{ HP} = \frac{550 \text{ ft} \cdot \text{lb}}{\text{sec}} \times \frac{60 \text{ sec}}{\text{min}} = \frac{33,000 \text{ ft} \cdot \text{lb}}{\text{min}}$$

To summarize, torque is a twisting force, such as that produced by a motor shaft. Torque is normally associated with rotating machinery like motors, since torque produces a tendency to rotate. Work differs from torque in that work must involve movement. An object must be moved for work to be done. Torque, however, can be produced with or without motion. *Power* refers to the rate at which work is done. Let us suppose two people lift the same weight the same distance. Both are doing the same *amount* of work. But, let's suppose one person lifts the weight twice as fast as the other. The person who works faster is developing twice as much power as the other. Finally, the *kilowatt-hour* or the *megajoule* refers to the amount of power consumed.

1–3 ELECTRICAL CONCEPTS

1–3.1 Generated Voltage—The Generator

Before discussing how electricity is produced by a generator, let us review magnetic theory. A magnet possesses both a north and a south pole. Between north and south poles lies a magnetic field of force. This force field is made up of lines of force, or flux, that are assumed to emerge from the magnetic north pole and enter the south pole. The lines of flux in a magnetic field are collectively called the *magnetic flux*. The Greek letter phi (ϕ) is used to represent magnetic flux. The unit of measurement of magnetic flux is the weber (wb). An important property of flux lines is their density within a certain area. This concept, called *magnetic flux density,* is symbolized by the upper case letter B. Flux density is measured in webers per square meter (wb/m^2). With this short review behind us, let us examine how a generator produces electricity.

A study of electromagnetism shows that three things are needed to produce electric potential electromagnetically: (1) a conductor, (2) a magnetic field, and (3) relative motion between the field and the conductor, where magnetic flux lines are cut. This concept was discovered by Michael Faraday in the early 1830s. It is illustrated in Figure 1–3. Note in Figure 1–3a that we have all three things we need to generate voltage, i.e., the conductor (copper), a field from a permanent magnet, and motion in the direction shown. Voltage polarity is in the direction shown. Figure 1–3b shows what happens when one requirement is taken away. Here motion is present, but no flux lines are cut. There is, therefore, no *electromotive force* (EMF) produced. What would happen if we were to move the conductor in the opposite direction from that shown in Figure 1–3a? The result is shown in Figure 1–3c. The direction of the EMF is reversed. We also know that a complete circuit must be present for current to flow. This concept is illustrated by Figure 1–3d.

A relationship obviously exists in this example between the direction of the magnetic field, the direction of motion, and the direction of current flow (or induced EMF). A good memory aid to help you in recalling these relationships is found in the left-hand rule for generator action (Figure 1–4). If the motion of

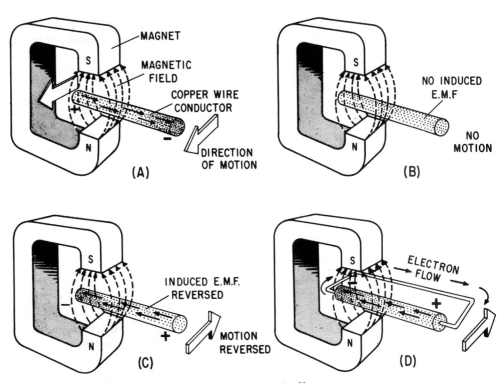

FIGURE 1–3 Voltage produced electromagnetically

the conductor is in the direction of the thumb (*TH*umb–thrust), and the index finger (*F*orefinger–flux) points toward the south pole of the magnet, the direction of current flow (or induced EMF) is in the direction of the middle finger (*C*enter finger–current). Please note that this rule applies only to electron flow. For conventional current flow, the hand rule is known as the right-hand rule for generators. The convention is the same, but current flow is in the opposite direction. (*Conventional current flow*, which is in the opposite direction from electron flow, is a term often used in engineering texts.)

The amount of EMF produced by the generator depends on several factors. These factors are expressed in the equation

$$e = \frac{\phi l N}{1 \times 10^8}$$ (eq. 1–8)

where e is the voltage produced at any instant, ϕ is the flux density in lines per square centimeter, l is the effective length of the conductor in centimeters, N the speed of the conductor in centimeters per minute, and 1×10^8 is a constant. We are assuming the conductor is moving at right angles to the field.

FIGURE 1–4 Left-hand rule for generators

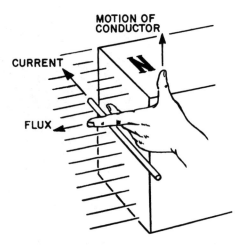

Alternating Current (AC) Generation In real generators, we seldom see
only one conductor moving through a field. Usually the conductors are coils of
wire. Each turn of the coil is seen by the generator as one conductor. Since all
of these conductors are connected in series, the effect is additive. The principle
of voltage creation, however, remains the same. Voltage is produced in the
same way as previously described. It is possible, with multiple coils, to get
higher voltages and currents than we could with the single wire example presented earlier. A single coil alternating current (AC) voltage generator is illustrated in Figure 1–5. If we were to plot the relationship between the output
voltage of the generator (on the X-axis) and time (on the Y-axis), we would see
a sine wave. Typically, the unit for voltage is the volt, where time can be either
in seconds, degrees, or radians. As the coil rotates CCW, we can see that current flows in the direction shown. Try the left-hand rule for generators on this
system to see if you could have predicted the direction of current flow. We call
this generator on *AC generator* because current flows first in one direction and
then in the opposite direction. In direct current (DC), current flows in one
direction at all times.

The drawing in Figure 1–6 shows the basic parts of an AC generator, or
alternator, as it is usually called. We show only one coil of wire in this picture.
Actually, there will be many coils, which are usually wound around a form.
The entire assembly is called the *armature,* or *rotor.* Voltage is passed from the
coil to the external circuit through slip rings and brushes. The slip rings and
the armature loop rotate CW or CCW. The brushes do not move. They ride on
the slip rings as the slip rings rotate. The brushes and slip rings connect the
generator to the load.

Recall that when we started this discussion on generators, it was pointed
out that three things were necessary for a generator: a magnetic field, a conductor, and relative motion between the two. In our previous discussion, we
assumed that the field would be stationary and the coils would be rotating.

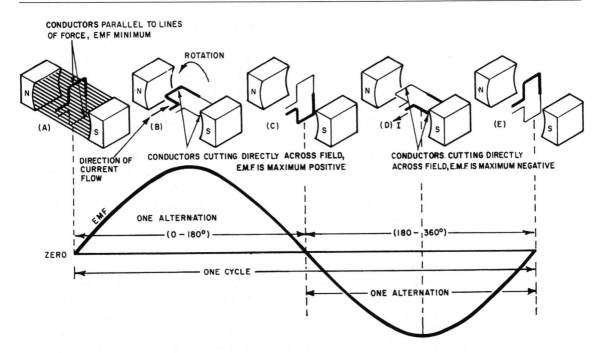

FIGURE 1–5 Basic alternating current (AC) generator

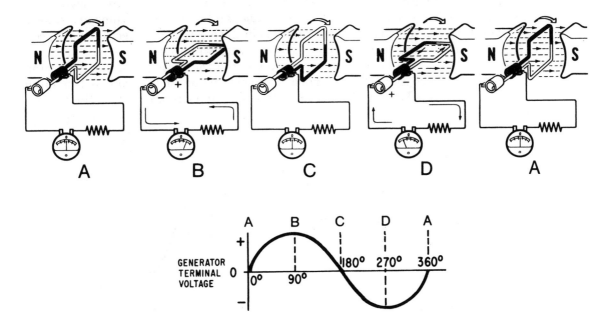

FIGURE 1–6 Output voltage of an elementary AC generator *(alternator)* during one revolution

This situation is not always the case. You can see from Figure 1–7 that the opposite can be true. The field (here a permanent magnet) rotates while the coils remain stationary. Note also the four-pole field (called the *rotor* since it rotates) and the four coil windings (called the *stator* since they are stationary). As the rotor turns, AC voltages are induced into the stator windings. The voltages induced in each winding will be equal because all stator windings cut the same number of flux lines at any given time. The four stator windings are connected so that the AC voltages are in phase or series-aiding. All four voltages induced in the stator windings add together. The result is a total voltage that is four times the voltage in any one winding. Note the output wave form is a sine wave. This type of output is called a *single-phase AC output*. The device that produced this output is, therefore, called a *single-phase AC generator*.

A very common type of AC generation involves creating more than one AC wave form at the same time. We call such a generator a *multiphase*, or *polyphase, AC generator*. These generators usually have three single-phase windings symmetrically spaced around the stator. In the three-phase generator, there are three single-phase windings physically spaced 120 degrees apart. The three voltages induced are then 120 electrical degrees apart. They are similar to the voltages that would be generated by three single-phase AC generators whose voltages are out of phase by 120°. A schematic diagram of a three-phase stator showing all coils becomes complex since it is difficult to see what is happening. A simplified schematic diagram, showing all the windings of a single phase lumped together as one winding, is illustrated in Figure 1–8. The rotor is omitted for simplicity. The wave forms of voltage are shown to the right of the schematic. Note that both **wye** and **delta** connections are shown. The difference between wye and delta is in the electrical connection, not necessarily in the physical placement of the stator windings.

Rather than have six leads coming out of the three-phase AC generator, one lead can be connected to form a common junction. In Figure 1–9, the center of the Y-shaped stator connection is the neutral, or common. It is the connec-

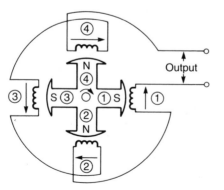
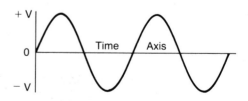

FIGURE 1–7 Single-phase alternating current generator

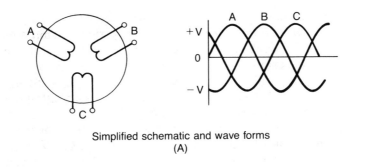

Simplified schematic and wave forms
(A)

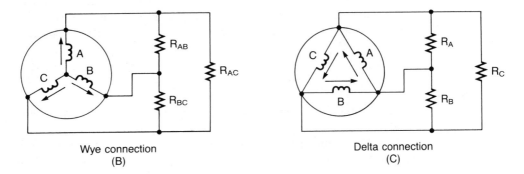

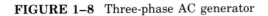

Wye connection
(B)

Delta connection
(C)

FIGURE 1–8   Three-phase AC generator

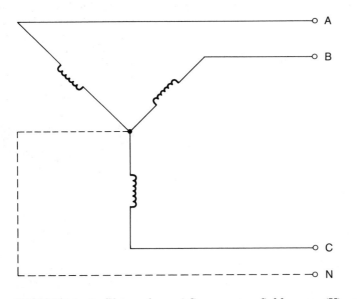

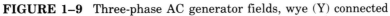

FIGURE 1–9   Three-phase AC generator fields, wye (Y) connected

tion with the dotted line attached. This stator is called *wye,* or *star-connected.* The common lead may or may not be brought out of the machine. If the common lead is brought out of the generator housing, it is called the *neutral.* When the neutral is brought out, this system is called a *three-phase, four-wire generator.* It is the most common type of generation system in industry. It can be used to power three-phase motors. It also has single-phase power available between any of the three outputs and the neutral. The voltage available between any of the three leads appears to be the sum of the two individual voltages if the voltage from point A to the neutral is 120 V. The total voltage is not, however, 240 V. Voltages can be added this way only when they are in phase. The total voltage between any two of the three outputs is equal to the square root of three times either of the voltages. For example, if we wanted to know the voltage between leads A and B, we would multiply the square root of three (1.73) by 120 V:

$$V_{AB} = (1.73)(120 \text{ V}) = 208 \text{ V} \qquad \text{(eq. 1–9)}$$

We then have 208 V across any of the three output phases, as shown in Figure 1–10.

Not all generators are connected in the wye configuration. Some are connected in the configuration shown in Figure 1–11. This is called a *delta connection.* In the delta connection, the voltage between phases is the same as the voltage across the coil. No neutral exists in the delta configuration. Because of its versatility, the wye-connected generator is more popular than the delta in most applications.

Power Factor   We have assumed that the current and voltage produced by the AC generator will be in phase in the load. If the load is a pure resistance,

FIGURE 1–10 Voltage distribution in a three-phase, wye-connected AC generator

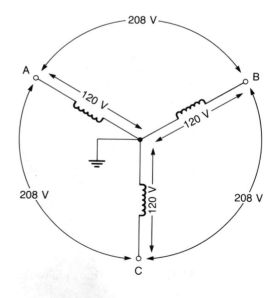

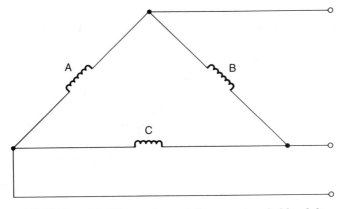

FIGURE 1–11 Three-phase AC generator fields, delta connected

this is true. In a system with a resistive load, all the power produced by the generator can be consumed by the load. To find out the power consumed by a resistive load, we measure voltage and current in the load and multiply the two together. No actual load is purely resistive (even a straight piece of wire has some inductance). All loads have a mixture of resistance and reactance (stored). Since this situation is true, we can then say that a load that is made up of resistance and reactance will usually have a phase difference between current and voltage in the load. If we were to measure current and voltage in this type of load and multiply them together, we would not have the real power consumed by the load. We would have only the power that appears to be consumed by the load. (Remember that a reactive load consumes no power.) This power that appears to be consumed by a load is called ***apparent power.*** It is measured in volt-amperes or volt-amps (VA) to separate it from real power, which is measured in watts. In a purely resistive load, the real power and the apparent power are equal. In a purely reactive load (a perfect capacitor or inductor), the real power consumed would be zero. Actually, all loads fall somewhere between these two extremes. Current through the reactances in the circuit produces reactive power, sometimes symbolized by the letter Q. Reactive power can be useful and necessary in a system. Reactive power is power that is stored in a capacitor or inductor. The inductor or capacitor ultimately returns it to the circuit, as in the motor armature coils. Power I^2R losses increase because current increases with reactive increases. Since we normally regulate voltages, the voltage stays constant and the power delivered stays the same.

A convenient way to represent these two parameters, real and apparent power, is through the concept of the ***power factor.*** The power factor is the ratio of real power to apparent power.

$$\text{power factor} = \frac{\text{real power}}{\text{apparent power}} \qquad \text{(eq. 1–10)}$$

The real power or true power, as it is sometimes called, is the power used by the load, as measured by a wattmeter. The apparent power, as mentioned earlier, is the product of the current and voltage produced by the generator. In electrical terms, the power factor is the cosine of the angle between the current and voltage. In a purely resistive circuit, the current and voltage are exactly in phase. The phase angle difference is zero electrical degrees. The cosine of zero is one. This circuit then has a power factor of one. This agrees with what we said earlier about resistive circuits, i.e., their power factor is one. In a purely reactive circuit, the difference between current and voltage is 90°. The cosine of 90° is zero; thus, the power factor is zero. This agrees with what we said earlier about reactive loads not consuming power.

There are two kinds of reactive components: capacitors and inductors. When a load is capacitive, the current leads the voltage. In this situation, the circuit has a leading power factor. When a load appears inductive, the current lags the voltage. This situation is called a lagging power factor. Most loads in industry are inductive because they are made up of coils of wire. These coils are found in motors, transformers, solenoids, relays, etc. Since the loads in industry are primarily inductive, industrial power consumption has a low, lagging power factor. There are many reasons a low, lagging power factor is undesirable. One important reason is the power cost issue. The power companies supplying energy to industry have established penalties for low, lagging power factors in their rates. This means that power users pay more for consumption with a low, lagging power factor. When power is purchased under a power factor clause, the user benefits when the power factor is kept high. A low power factor is also undesirable because most loads require a certain amount of true power. The apparent power will be higher than the true power, causing current to be higher. This extra current adds to the heating of conductors which, in turn, limits their current-carrying ability. Power factors usually range from 0.7 to 1.0.

To summarize, equation 1–10 stated that the power factor is the real power divided by the apparent power.

$$\text{power factor} = \frac{\text{real power}}{\text{apparent power}} = \begin{array}{c}\text{cosine of phase angle} \\ \text{between voltage and} \\ \text{current}\end{array}$$

In electrical circuits, the power factor is equal to the cosine of the angle between current and voltage. Apparent power (measured in VA) is due to the applied voltage and current, while true power (measured in W) is the useful power extracted by devices. (Motors, for instance, have little resistance but draw and convert real power, as well as provide a reactive load.) Since true power is usually less than the apparent power, the power factor will be a percentage or a decimal number less than one. A low, lagging power factor is generally unwanted by industry, since it makes utility bills higher and can limit circuit performance.

DC Generation The generator system described above is an AC generator. This means that current flows alternately through a circuit, first in one direction, then in the other. Another type of generator, using the same principle, is the DC generator.

This generator, shown in Figure 1–12, has many similarities to the AC generator. The DC generator has a coil of wire moving through a permanent magnet field. It also has brushes but not slip rings. Instead of slip rings, the DC generator uses a ***commutator.*** The commutator is a cylinder cut in half lengthwise with the two sections separated by an insulator. The commutator functions as a ***rectifier,*** changing AC into DC. Without the commutator, the DC generator would be an AC generator. The commutator ensures that current flows through the load in only one direction. To demonstrate the commutator function, look at Figure 1–13. In position *A,* the coil cuts no flux lines; therefore, it produces no voltage. In position *B* (90 mechanical degrees), current flows in the direction shown, through the load from left to right. In position *D,* the coil is in the opposite position from *B.* In the AC generator, the current would be flowing in the opposite direction, right to left. The commutator keeps current flowing in the same direction through the load. You can prove this to yourself by using the left-hand rule we learned earlier. Figure 1–13 also shows the output voltage wave form we would expect to see from this kind of generator. Note that it is pulsating DC, not a sine wave. All the voltage is above zero volts. Since the output voltage never goes negative, we can conclude that this is, indeed, a DC generator.

Note also the polarity of the voltage. Within the generator itself, current flow is into the negative brush. In the external circuit, current is from the negative brush. In the DC motor and generator, the brushes make electrical connection between a moving part (a rotor) and a stationary part (the load).

Actual generators seldom have only one coil of wire. Note what happens in Figure 1–14 when we add another coil. The first thing to observe is the difference in the commutator. It is divided into four parts instead of two. Figure 1–14 shows the lessened ripple of the voltage when two armature coils are used. Since there are four commutator segments in the commutator and only two brushes, the voltage cannot fall any lower than point *A.* The ripple, there-

FIGURE 1–12 DC generator—single coil

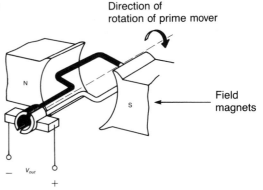

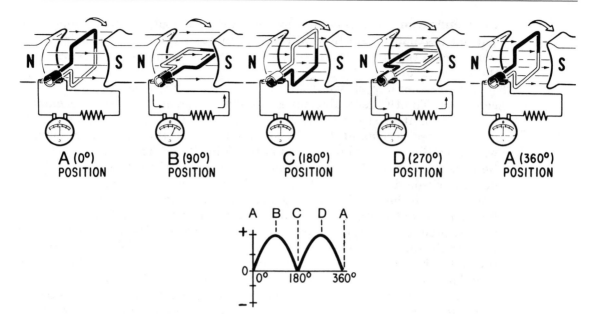

A (0°) POSITION   B (90°) POSITION   C (180°) POSITION   D (270°) POSITION   A (360°) POSITION

FIGURE 1–13   Effects of commutation

fore, is limited to the points between *A* and *B*. Compare this wave form with the single coil generator in the previous figure. Notice that the ripple has been reduced. By adding more coils, we can further reduce this ripple effect. (Ripple can also be reduced by using filter components, such as capacitors and inductors.) Decreasing the ripple gives a higher average voltage out. Practical generators use many armature coils. They also use more than one set of magnetic

FIGURE 1–14   Effects of additional coils

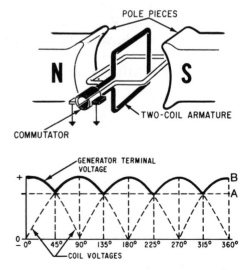

poles. To simplify our discussion, we have depicted only one set of poles. Adding more sets of magnetic poles further reduces the ripple, as did the added armature coils. If we knew all the factors that contributed to the generator's output voltage, we could calculate the output voltage produced. Equation 1–11 allows us to make this calculation:

$$V = \frac{\phi NPZ}{60A(10^8)} \qquad \text{(eq. 1–11)}$$

where V is the output voltage in volts, ϕ is the flux from one pole, N is the speed in r/min, P is the number of poles, Z is the number of conductors, and A is the number of parallel paths through the armature. Everything in this equation stays relatively constant except ϕ and N. We can simplify this equation by gathering all the constants into one grand constant. Let us call this grand constant K_v. Our equation then becomes

$$V = \phi N K_v \qquad \text{(eq. 1–12)}$$

Equation 1–12 is an important one and will be used again in the chapters on DC machines. Apart from that, we can get a clue to the performance of a DC generator from this equation. We can see immediately that the voltage produced by this generator is directly proportional to the strength of the magnetic field and to the speed at which the armature coils turn. Thus, increasing the voltage produced would simply be a matter of increasing the strength of the field or increasing the armature rotational speed.

1–4 GENERATED TORQUE—THE MOTOR

A generator is a device that converts mechanical energy (torque) to electrical energy. In most machines the opposite conversion can also take place. Many textbooks begin with a general discussion of a machine that can do both conversions. Such a machine is called a **dynamo.** We have spent some time discussing the conversion of mechanical energy to electrical energy. Let us now turn to the opposite conversion, i.e., electrical energy to mechanical. A device that converts electrical energy to mechanical motion is called a **motor.** If we supply a source of power to a motor, the motor armature should produce torque, whether it turns or not.

We explained the operation of the DC generator by first discussing the left-hand rule for generators. The rule that governs the operation of motors is the right-hand rule for motors. This rule is illustrated in Figure 1–15. When a conductor carrying current is placed in a magnetic field, the conductor will try to move. If you know the direction of current flow in the conductor and the magnetic field direction, the right-hand rule can tell you the direction of mo-

FIGURE 1–15 Right-hand rule for motors

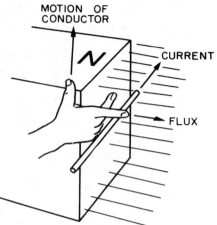

tion. In this rule, the conductor will move in the direction of the thumb. The index finger points in the direction of the flux while the middle finger indicates the direction of electron current flow.

To explain this phenomenon, we need to talk about the motor effect. The motor effect can be defined as the tendency of a current-carrying conductor to move in an external magnetic field. Figure 1–16 will help you to understand this concept.

This illustration shows a cutaway view of a conductor carrying current in a magnetic field produced by a permanent magnet. Here, we will use the dot-cross convention to indicate the direction of current flow in the conductor. A dot in the center of the conductor indicates the electron flow is out of the page. If the conductor had a cross instead of a dot, current would flow into the page. The cutaway diagram (Figure 1–16) shows a cross-section of a current-carry-

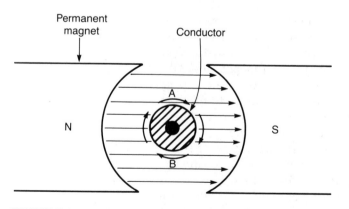

FIGURE 1–16 Interacting fields around conductor and magnet

ing conductor. Conductors carrying current produce a magnetic field. Another left-hand rule will help us with the direction of the magnetic field (Figure 1–17). Hold your left hand as if grasping the wire in such a way that your thumb points in the direction of the current flow. The fingers of your left hand will then point in the direction of the field. You will see that this rule fits in with the conductor in Figure 1–16. Using our dot convention, we can see that current is flowing out of the page. Using the left-hand rule, we can see the field is clockwise as shown. Now note that we have *two* magnetic fields present in this diagram, one from the permanent magnet and one from the conductor. Note also that these two force fields add at point *A* and subtract or cancel each other at point *B*. The conductor will try to equalize this distribution by moving downward, in the direction shown. This tendency to move is called the motor effect. This is consistent with the right-hand rule for motors.

The same motor effect would work if we used a coil of wire instead of a straight conductor. The illustration in Figure 1–18 shows motor action on a single-turn coil in a magnetic field. Current is flowing out of the left-hand conductor and into the right-hand conductor. The force moves the right-hand conductor up and the left-hand conductor down. Both forces acting together on the coil produce a turning effect. This force that causes rotation is called torque. When the coil reaches the position shown in Figure 1–18b, the forces spread the coil apart. The motor produces no torque in this position, which is called *dead center.* If the current in the coil reverses at this point and the coil is

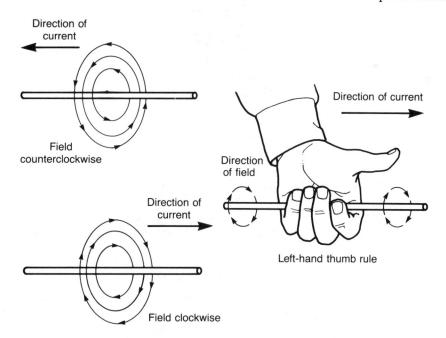

FIGURE 1–17 Fields around current-carrying wires

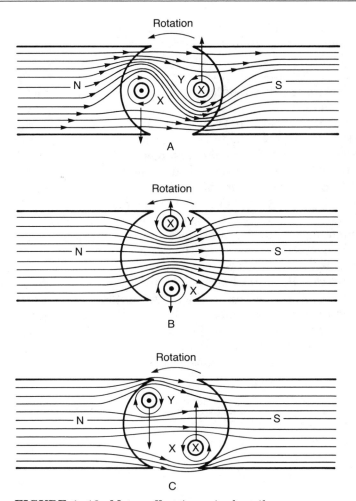

FIGURE 1–18 Motor effect in a single coil

carried beyond dead center, forces will be developed as shown in Figure 1–18c. The torque continues turning the coil in the same direction as before. If the direction of current flow is not reversed, torque would be exerted in the *opposite* direction. We would have torque reversing twice each cycle; the armature would not move.

If the direction of current is always reversed at the right time, the motor will develop a continuous rotating force. The device that allows us to do this is called the commutator. The commutator serves the same purpose in the DC motor as in the DC generator; i.e., it changes the direction of current flow at the right time to make sure the device operates properly.

In our discussion of generators, we stated that we could predict the output voltage of a generator if we knew all the factors that went into producing the voltage. We can do the same thing with the motor effect. If we are dealing with a single conductor in a field, such as that shown in Figure 1–16, the equation used to calculate the force exerted by the conductor is

$$F = BLI \qquad \text{(eq. 1–13)}$$

where F is the force in newtons, B is the number of magnetic lines of force, and I is the current flow in amperes. We can convert this force equation to a torque equation by multiplying by the distance at which the force was exerted. The distance would be from the center point of the armature to the conductor itself. We can further modify this equation to apply to motors which have more than one coil of wire, as most motors do. The modified equation will be

$$T = \phi K_t I_a \qquad \text{(eq. 1–14)}$$

where T is the torque in newtons, ϕ is the field flux, and K_t is the grand constant that is made up of all those constant factors in the motor, such as the length of conductors, distance of the conductors from the center of the armature, etc. This equation tells us that field strength and the armature current are directly proportional to the amount of torque the motor produces.

1–4.1 Motor Reaction and Counter Electromotive Force (CEMF)

Motor Reaction When a generator delivers current (illustrated in Figure 1–19b), the current flow induced in the conductor creates a magnetic force that opposes the driving action on the conductor. This counterforce is called *motor reaction* or *countertorque*. A single conductor of a generator armature is pictured in Figure 1–19.

An earlier equation (1–14) related the torque a motor produced to several factors:

$$T = K_t \phi I_a$$

where T is the torque produced by the motor in newtons, ϕ is the field flux, and K_t is the grand constant made up of all those constant factors in the motor, such as the length of conductors, distance of the conductors from the center of the armature, etc. We can modify this equation by replacing T with CT

$$CT = K_t \phi I_a \qquad \text{(eq. 1–15)}$$

This equation tells us that the amount of countertorque produced by a generator armature is proportional to the strength of the field through which the conductor is moving. It is also proportional to the amount of armature current in the conductor coils.

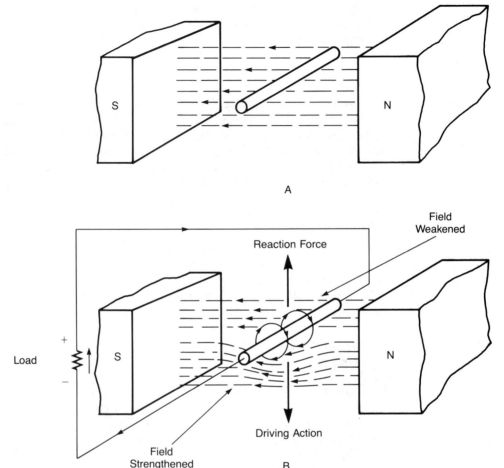

FIGURE 1–19 Motor reaction in a generator

With no armature current, no motor reaction exists. The force needed to turn the armature is very low. As armature current increases, the reaction of each armature conductor increases. The actual force in a DC generator is multiplied by the number of conductors in the armature. This factor is contained in the K_t term in equation 1–15. The driving force needed to keep the generator armature at a certain speed must be increased to overcome the motor reaction. The device that provides the turning force applied to the generator armature is called the *prime mover.* The prime mover can be an electric motor, a gasoline

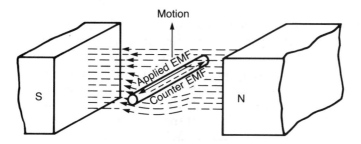

FIGURE 1–20 CEMF opposes the applied EMF

or diesel engine, a steam engine, or any other mechanical device that provides turning force.

Counter EMF (CEMF) *Counter electromotive force* (CEMF) is that EMF generated by the armature that opposes the potential supplied to the armature. Observe the diagram in Figure 1–20. As the conductor is being forced down by the motor effect (remember the right-hand rule), it is also cutting flux and generating an EMF. By applying the left-hand rule for generators, we can see that the potential of this induced EMF is in the opposite direction to that of the potential applied to the armature. This induced EMF is called CEMF. The effect of the CEMF is to reduce the total armature current.

Instead of a straight conductor, Figure 1–21 shows a motor armature consisting of a coil of wire in a field. For the moment, disregard the current flowing through the loop. As the loop sides cut the magnetic field, a voltage is induced in them. The voltage induced is the same as would be induced in a DC generator. Now, consider the direction of this induced voltage compared to the applied voltage from the battery. First, check the direction of the EMF induced

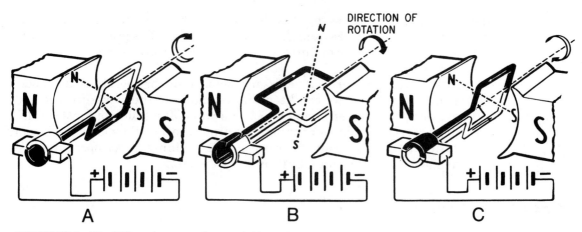

FIGURE 1–21 DC motor armature rotation

by generator action. Use the left-hand rule to do this. Using the left hand, hold it so that the index finger points in the direction of the magnetic field, towards the south. Point the thumb in the direction that the black side of the armature is moving (up). Your middle finger then points out of the paper, showing that the current caused by the generator action flows in the same direction in the black half of the armature. Note that this is opposite to the direction of the current caused by the battery. We call the voltage which produces this opposing current flow the CEMF, or *back EMF*.

At the beginning of this discussion, we disregarded the current caused by the battery when finding the direction of the CEMF in the armature. Although there are two EMFs (the applied EMF and the CEMF), only one current flows. The CEMF can never be as high as the applied EMF. Current in a DC motor, therefore, will flow in the direction of the applied EMF. Because the two EMFs oppose one another, the CEMF effectively cancels part of the applied EMF. Thus the amount of current flowing in the armature is restricted by the CEMF. In other words, the current that flows in the armature is far less with CEMF than it would be without it. The CEMF greatly reduces the size of the armature current.

CEMF is always developed in a DC motor with a turning armature. The CEMF cannot be equal to or greater than the applied battery voltage. If the CEMF were greater than the applied voltage, the motor would not run. Actually, in a running DC motor, the CEMF is a little less than the applied voltage. Just as we did in the motor reaction section, we can develop an equation by which we can predict how much CEMF we will have in any DC motor. Since the motor operates as a generator when producing CEMF, you should not be surprised by the generator formula given earlier as equation 1–12, $V = \phi N K_e$. We can adapt this equation into a CEMF equation by substituting V_{CEMF} for V

$$V_{\text{CEMF}} = \phi N K_e \qquad \qquad \textbf{(eq. 1–16)}$$

where ϕ is the field strength, N is the speed of the armature, and K_v is our grand constant. We can see from this formula that CEMF is proportional to the field strength and the armature speed. If the armature is not turning, no CEMF is produced.

The CEMF, however, is a necessary factor in motor operation. By opposing the applied voltage, we can keep the resistance of the armature low and still have armature current limited to a safe value. If we did not have CEMF, much more current would flow through the armature, and the motor would run much faster. CEMF is a valuable concept in explaining motor operation. We will have more to say about CEMF in later chapters.

CHAPTER SUMMARY

- Torque (measured in N · m) is the twisting force produced by the rotor of a motor.

- Work (measured in joules) is accomplished when torque is produced by a motor resulting in shaft rotation.
- Power is the rate of doing work and is measured in watts (1 horsepower = 746 W).
- The power produced by a DC motor is directly proportional to the speed of rotation and the torque produced.
- Voltage is generated when a conductor cuts magnetic lines of force.
- Power factor is the ratio of real power to apparent power and is equal to the cosine of the angle between circuit current and applied voltage.
- AC and DC generators produce counter emf (CEMF). CEMF is exerted in the opposite direction to applied voltage and is directly proportional to speed and field current.

QUESTIONS AND PROBLEMS

1. Define the following terms:
   a. *torque*
   b. *work*
   c. *power*
2. A 5-N force is exerted at a distance of 2 m. What is the torque exerted? in N · m? Do we need to have motion to exert torque? Why?
3. A weight of 2 N is moved upward 3 m. How much work was done?
4. What are the units of torque in the SI system? In the English system? What are the units of work in the SI and English systems?
5. A 10-HP motor produces how much power in watts?
6. Differentiate between a kilowatt and a kilowatt-hour.
7. A motor shaft is turning at 1500 r/min, producing a torque of 30 N · m. What is the power produced by the motor in watts and HP?
8. Name the three requirements for generating EMF with an electromagnet.
9. Draw the schematic diagrams for delta and wye generator connections.
10. A delta generator connection has 240 V produced across each phase. What is the voltage from phase to phase? With the same voltage across the wye phase windings, what is the phase-to-phase voltage?
11. Define power factor. Why is it important to reduce power factor for the industrial power consumer?
12. What is motor reaction and how does it affect generator behavior?
13. What is CEMF and how does it affect motor behavior? What factors in a motor determine CEMF?

2

WOUND-FIELD DC MOTORS

At the end of this chapter, you should be able to

- define a DC motor
- describe DC motor construction by identifying the major parts.
- describe feedback in a DC motor and how it regulates speed.
- define and calculate *speed regulation*.
- describe at least two methods of DC motor classification.
- calculate various performance characteristics of a series, shunt, and separately excited DC motor.
- list the major power losses in a DC motor.

2-1 INTRODUCTION

Until recently, the DC motor was considered the workhorse of modern industry. Its flexibility made it the motor of choice in many industrial drives. The DC motor can be accelerated and decelerated quickly and smoothly, and speed can be controlled accurately over a wide range. As with any machine, the DC motor has some disadvantages. It is costly when compared to its rival, the AC induction motor. It is also larger and more expensive to maintain. Since industry is supplied with AC power, a conversion to DC is necessary for running DC motors. Industry has accomplished this conversion by generating its own power. AC power is used to drive DC generators, which, in turn, are used to supply DC motors. Solid state rectifier diodes and thyristors do this job today when DC motors are used.

Chapter 1 introduced general concepts crucial to the study of DC motors. This chapter will examine the construction and operation of the DC motor. Although some types of DC motors are becoming less popular, others are increasing in popularity.

2-2 WOUND-FIELD MOTORS

A DC motor converts DC electrical energy to torque. Electrical energy, in the form of direct current, is put into the DC motor. The mechanical energy output from the motor is in the form of torque. You will recall from Chapter 1 that every motor operates because of the interaction between two magnetic fields. Every conductor carrying current produces a magnetic field. The strength of this field is directly related to the amount of current flowing in the conductor. The stronger the current flow, the stronger the magnetic field. When this current-carrying conductor is placed within a magnetic field, a force is exerted on the conductor. The interaction of these two fields produces a force on the conductor. If the conductor is in the form of a coil and the coil is placed in a magnetic field, the coil produces a force, called torque, which causes it to rotate. Although we have discussed a single coil up to this point, modern DC motors use more than one coil to generate torque.

In the motor armature, an induced voltage always *opposes* the current due to the applied voltage. As defined previously, this induced voltage is called the counter-electromotive force (CEMF) or back EMF. The total EMF acting on the armature windings is equal to the applied voltage (usually called the line voltage) minus the CEMF.

$$V_A = V_{app} - V_{CEMF} \qquad \text{(eq. 2–1)}$$

In a DC motor, CEMF is developed only when the motor armature is turning. When the motor armature is stopped, no flux lines are cut. No CEMF is then developed. CEMF cannot be equal to or greater than the applied voltage, or the motor will not turn.

An equation can be developed to calculate the amount of CEMF in a motor armature. Only a few factors actually determine the amount of CEMF a motor armature generates. We already know that the speed that the armature moves through the field influences CEMF. Recall that when the coil was not moving through the field, no flux lines were cut. Therefore, no CEMF was developed. We also know that the number and orientation of the flux lines will affect the induced EMF. These factors are related in equation 2–2:

$$V_{CEMF} = \frac{N_a P Z}{60\, P'\, 10^8} \qquad \text{(eq. 2–2)}$$

where ϕ is the number of flux lines from one pole, N_a is the speed of the armature in rpm, P is the number of poles, Z is the number of conductors, and P' is the number of parallel paths through the armature. (We will not attempt to derive this equation as it is outside the scope of this text. P, Z, and P' are all constant for a specified motor. These three constant values can be gathered into one large constant K_e:

$$V_{\text{CEMF}} = \phi N_a K_e$$

where K is the constant made up of P, Z, and P'. Rearranging this equation, we can solve for speed:

$$Na = \frac{V_{\text{CEMF}}}{K_e \phi} \qquad \textbf{(eq. 2--3)}$$

This equation demonstrates that the speed of a motor is directly proportional to the CEMF and inversely proportional to field flux strength.

2–3 TORQUE

We can develop an equation to calculate the average (not instantaneous) torque developed by a running DC motor in much the same way as we did for the CEMF. As we noted in Chapter 1, the force exerted by a conductor in a field is related by the equation

$$F = \phi I L \qquad \textbf{(eq. 2--4)}$$

where F is the force exerted, ϕ is the field strength, I is the current, and L is the length of the conductor. Since the force in a coil is exerted at a distance, the torque equation can be derived from this force equation (equation 1–14) into the form

$$T = \phi I_a K_t$$

where T is torque, I_a is the armature current, and K_t is the torque constant. The torque constant, like the constant in the CEMF equation, is a collection of things that remain constant in the motor system, such as the number of poles, parallel paths, length of conductors, etc. Please note at this point that the torque constant and the CEMF constant are different, since they came from different equations.

These two equations are fundamental to the understanding of motor operation. We will use them in the remainder of this chapter to describe DC motor operation.

2–4 PRACTICAL MOTOR CONSTRUCTION

A popular DC motor used today in industry is the ***wound-field DC motor.*** This motor gets its name from the way its field is generated. In the wound-field motor, the field is generated by an electromagnet. Figure 2–1 shows the basic components of the wound-field DC motor. The motor creates magnetic flux

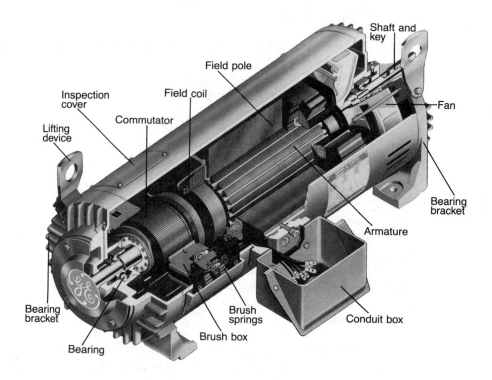

FIGURE 2–1 Cutaway view of a wound-field DC motor (courtesy of General Electric)

when current flows through the field windings. The field windings, or coils, are connected so that they produce alternate north and south poles. The field windings in the wound-field motor form an electromagnet that creates the motor field flux. These field windings either get their current from an external source or they get voltage from the same potential that supplies the armature.

The field windings are wound around the pole pieces, or cores, that are usually made of sheet steel laminations riveted together. The laminated sheets are insulated from each other so that current does not flow in the pole pieces. These currents are called *eddy currents,* and they rob power from the motor. The pole faces are shaped to fit the curvature of the armature, which is circular (Figure 2–2a). The diagram in Figure 2–2b shows a field coil. The field coil is made of many turns, or windings, of copper wire. In larger motors, the coils can be made of solid copper bars.

The armature also has several parts. The core, made of sheet steel laminations in much the same way as the pole pieces, reduces eddy currents in the armature. The outer surface of this cylindrical core is slotted to provide a place for the armature coils that are wound around the core (Figure 2–3). Each winding is connected in series to form a field coil. The windings are usually fastened by fiber wedges driven into the tops of the slots. The ends of the armature coils are then connected to the commutator by soldering.

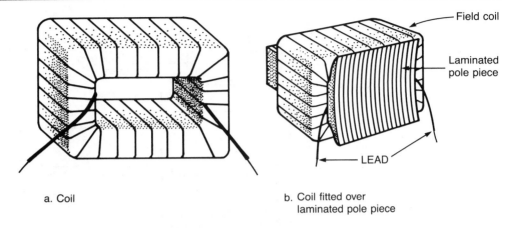

a. Coil

b. Coil fitted over
laminated pole piece

FIGURE 2–2 Wound-field DC motor field pole construction

You will recall that the purpose of the commutator is to keep current flowing in the proper direction through the armature coil. The commutator is made of several copper wedges or bars arranged in a cylinder (Figure 2–4). Each commutator segment is insulated from its neighbor by mica sheets. The voltage to be applied across the armature and the number of poles in the motor determine the number of commutator segments needed. The entire armature assembly is shown in Figure 2–5.

The brushes ride on the commutator and carry the current from the source to the armature coils. The brushes are usually made of a mixture of carbon and graphite or carbon and metallic powder and are held in place by the brush holder. They are free to slide in their holders so that they can follow any unevenness in the commutator. The brushes are pressed against the commutator by springs. Springs are necessary to keep the brushes pressed against the commutator as the brushes wear down. The usual pressure of the brushes

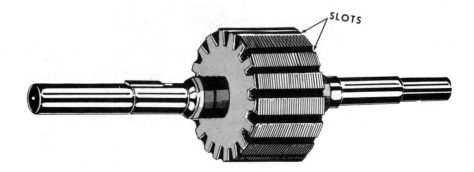

FIGURE 2–3 Slotted rotor construction

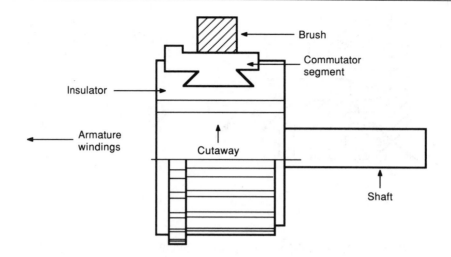

FIGURE 2–4 Side view of a DC motor commutator

against the commutator is from one to two pounds per square inch of brush contact area.

The frame, or yoke, is usually made of annealed steel, either cast or rolled. It provides mechanical support for the pole pieces and is also part of the magnetic circuit of the motor.

2–5 MOTOR OPERATION

The motor is a feedback system. Feedback is that action where part of the output of a circuit or a device is fed back to the input. If the feedback increases the output, it is called positive feedback. If the feedback decreases the output,

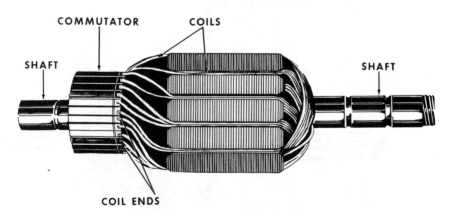

FIGURE 2–5 Entire armature assembly

it is called negative feedback. An illustration might help at this point. Let us suppose that we have a DC motor which has no power applied to it. The motor armature will not be turning, since no torque is created. Let us then apply power to both armature and field (assuming a wound-rotor motor). Since the motor armature is at rest, CEMF will be 0 and armature current will be correspondingly high. Remember that the major factor limiting or controlling armature current is CEMF. The motor armature will start to turn since current will flow in both armature and field, creating magnetic flux. As the motor armature picks up speed, CEMF will increase, causing less armature current to flow. Armature current continues to decrease until enough current flows to produce enough torque to handle the load on the motor.

It is apparent from the CEMF equation that the field strength is directly related to the motor armature speed. As field strength decreases, CEMF decreases. CEMF changes in this way because it is directly proportional to field strength. A drop in CEMF causes greater armature current to flow; therefore, the armature speed increases due to increased torque production. Likewise, if you were to increase the load on the motor (by increasing the torque demand), the motor would slow down, CEMF would decrease, armature current would increase, and more torque would be produced. It is obvious from this discussion that the motor really is a feedback system, and the key is the CEMF.

2–5.1 Speed Regulation

Speed regulation is the ability of a motor armature to maintain its speed when a changing load is applied. It is an inherent characteristic of a motor and remains the same if the applied voltages do not vary and a physical or mechanical change is not made in the machine. The speed regulation of a motor is a comparison of its **no-load speed** to its **full-load speed.** It is usually expressed as a percentage of full-load speed. Thus,

$$\text{percent speed regulation} = \frac{N_{nl} - N_{fl}}{N_{fl}} \times 100 \qquad \textbf{(eq. 2–5)}$$

where N_{fl} is the full-load speed and N_{nl} is the no-load speed. For example, if the no-load speed of a motor is 1,600 rpm and the full-load speed is 1,500 rpm, the speed regulation is

$$\frac{1,600 - 1,500}{1,500} \times 100 = 6.6\%$$

The lower the speed regulation percentage figure of a motor, the more constant the speed will be under varying load conditions and the better the speed regulation. The higher the speed regulation figure is, the poorer the speed regulation.

2–6 DC MOTOR CLASSIFICATION

There are many ways of classifying DC motors. One way is by the type of load they drive. ***Continuous duty*** motors, found in fans or blowers, have continuous steady-running loads over long periods. ***Intermittent duty*** motors are designed to run only for short periods of time. Examples of intermittent duty motors are those found in compressors or pumps. Another way of classifying DC motors is by their horsepower (HP) rating. Motors rated 1 HP and above are called ***integral HP*** machines. Motors below 1 HP are called ***fractional HP*** motors. DC motors are most frequently classified by their internal construction. Generally, DC motors can be broken down into three different categories: 1) wound field, 2) permanent magnet and 3) electronic commutation (see Figure 2–6). The wound-field motor differs from the permanent-magnet motor in the method used to create the field. The permanent-magnet motor has a field generated by a permanent magnet. There are no field windings in this motor. The wound-field motor, as its name implies, uses the electromagnet principle to create its field. The motor shown in Figure 2–1 is a wound-field motor. Both of these motors have brushes and commutators, unlike the electronically-commutated motor. The electronically-commutated motor has no brushes; it is commutated by semiconductor devices. Each of these motors will be discussed in later chapters. In this chapter, we will examine wound-field motors.

The wound-field motor is one of the oldest and most popular DC motors. In size, wound-field motors range from fractional HP to integral HP up to a high of 8000 HP (6000 kW). Wound-field motors can be broken down into three basic configurations, depending on how the field is connected to the armature.

2–6.1 Series-Connected Field DC Motors

A schematic diagram of a series-connected field is shown in Figure 2–7. Note that the field is connected in series with the motor armature. A DC motor designed to run with a series-connected field can be identified by the gauge and thickness of the wire used in the field. The field coils are usually a few turns of heavy gauge, low-resistance wire. The field coil wire must have low resistance because all the armature current must flow through it. Since the field and the armature are in series, the strength of the field depends on the armature current. Thus, the torque of the series motor is proportional to $I_A{}^2$.

FIGURE 2–6 Motor classification chart

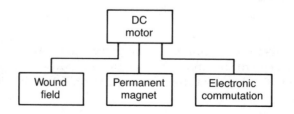

FIGURE 2–7 Series-connected field DC motor
schematic diagram

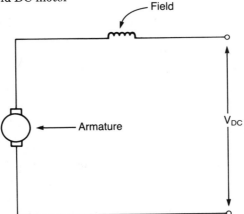

We can see this relationship from the torque equation developed earlier. Since

$$T = \phi I_a K_t \qquad \text{(eq. 2–6)}$$

and ϕ is directly proportional to field current (I_f), we can write a new torque equation for the wound-field motor:

$$T = I_f I_a K_{tw} \qquad \text{(eq. 2–7)}$$

In the series motor, the equation can be further simplified by realizing that $I_f = I_a$.

$$T = I_a{}^2 K_{tw} \qquad \text{(eq. 2–8)}$$

Note that the torque constant in this equation is different from the previous one. In this equation, the units of K_{tw} are $N \cdot m/A^2$. An example may help to clarify the usefulness of this equation.

Let us say that we have a motor in which we have measured the torque produced as 4 N $\cdot$ m and the armature current as 5.0 A. We can now calculate the value of K_t:

$$K_{tw} = \frac{T}{I_a{}^2} = \frac{4\ N \cdot m}{(5.0\ A)(5.0\ A)} = 0.16\ N \cdot m/A2 \qquad \text{(eq. 2–9)}$$

If we increase the load on the motor, the motor armature will slow down, causing less CEMF and, therefore, more armature current. More armature current will produce more torque. To determine how much more torque, we measure the armature and find it equal to 10.0 A. Assuming the torque constant stays constant, the new torque can be calculated from equation 2–8.

$$T = I_a{}^2 K_{tw} =$$

$$= (10.0 \text{ A})(10.0 \text{ A})(0.16 \text{ N} \cdot \text{m/A}^2) = 16 \text{ N} \cdot \text{m}$$

In this motor system, reducing speed by increasing the load increases armature current to 10.0 A and torque to 16 N · m.

It should be mentioned at this point that the torque constant stays relatively constant if temperature remains constant. If temperature increases, the resistance of the wire in the armature and field will increase, reducing current flow. If the motor reduces armature current, torque is also diminished. The torque of a hot motor will then be less than that of a cold motor.

In this example, we have held the applied or line voltage constant while changing the torque. We have seen that increasing the torque demand on the motor causes the motor armature to slow down and produce more torque. If we were to continue increasing the torque demanded, the motor armature would eventually stop because the motor can produce only so much torque. The torque produced when the shaft stops moving is called the **stall torque.** It is the amount of torque the motor armature produces when power is applied to the armature and the torque demanded is greater than the torque produced. How can we calculate a DC motor's stall torque? Figure 2–8a shows the electrical equivalent circuit for a series DC motor. Note that a CEMF is produced in the

FIGURE 2–8 Electrical equivalent of the series DC motor

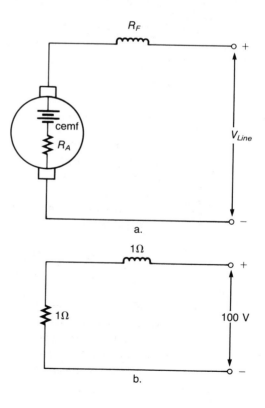

armature section of the motor, opposing the current produced by the supply, V_{line}. The resistance R_f is the series field winding resistance. The resistance R_a is the armature winding resistance as well as the brush resistance. If the motor armature stopped turning, CEMF would be 0 V. The equivalent circuit would then look like the one in Figure 2–8b. The current flow is limited only by R_f and R_a since CEMF is zero. By using Ohm's law, we can calculate the armature current flowing with 100 V applied:

$$I_a = \frac{V}{R} = \frac{100 \text{ V}}{(1 \text{ } \Omega) + (1 \text{ } \Omega)} = 50 \text{ A} \qquad \text{(eq. 2–10)}$$

If we were to use the torque constant from the previous problem, the approximate torque produced would be:

$$T = I_a{}^2 K_t = (50 \text{ A})^2 (0.16 \text{ N} \cdot \text{m/A}^2) = 400 \text{ N} \cdot \text{m}$$

This value is the stall torque, the maximum torque the motor can produce.

The relationship between speed and torque is shown in Figure 2–9. Note that as torque increases, the speed decreases to zero. The value on the torque axis where speed equals zero is the stall torque value. Going in the opposite direction on the torque axis, we notice an interesting but dangerous event. Decreasing the load torque increases the speed. The line never touches the speed (N_a) axis. The reason is simple. As you decrease the load, speed increases dramatically. If the load torque is decreased too far, the motor can come apart

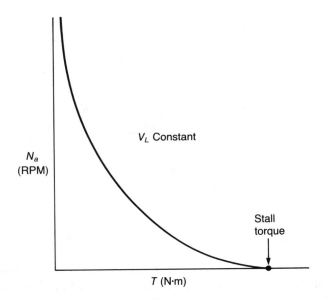

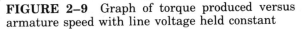

FIGURE 2–9 Graph of torque produced versus armature speed with line voltage held constant

due to overspeed forces. Specifically, the bearings can be damaged or the windings can fly out of the slots in the armature. In both instances, considerable damage can occur to both equipment and personnel. To prevent this, you must ensure that a load is ALWAYS connected to the series motor before turning it on. This precaution is primarily for large DC motors. Small motors, such as those found in electric hand drills, usually have enough internal friction to load themselves and prevent damage. It is best, however, to avoid any possibility of damage.

Series Motor Speed Regulation   In the previous discussion, we saw that the torque developed by the series motor armature is proportional to the current squared. With a relatively low flux density in the field pole pieces, the field strength is proportional to armature current. Examine the circuit in Figure 2–10 and the torque equation. If the supply voltage and the load torque remain constant, the armature current will remain constant. If there is no load on the motor, the armature will speed up to such an extent that the windings may be thrown from the armature slots and the commutator destroyed from excessive centrifugal forces.

To explain this phenomenon, we must go back to equation 1–16.

$$V_{\text{CEMF}} = \phi N_a K_e$$

Since flux is proportional to field current, we can rewrite the equation:

$$V_{\text{CEMF}} = I_f N_a K_e \qquad \textbf{(eq. 2–11)}$$

Solving this equation for N_a, we get

$$N_a = \frac{V_{\text{CEMF}}}{I_f K_e} \qquad \textbf{(eq. 2–12)}$$

Since CEMF is not easily measured, we can say that CEMF equals the applied voltage V_{line} minus the drop across field and armature resistances.

$$V_{\text{CEMF}} = V_{line} - [I_f(R_a + R_f)] \qquad \textbf{(eq. 2–13)}$$

Substituting into our CEMF equation, we get

$$N_a = \frac{V_{line} - [I_f(R_a + R_f)]}{I_f K_e} \qquad \textbf{(eq. 2–14)}$$

We can estimate the speed of this motor if we know the line voltage applied, the resistance of the armature and field, the amount of field current, and K_e. All these factors are easily measured. To use this equation, we must first calculate K_e. Solving this equation in terms of K_e, we get

$$K_e = \frac{V_{line} - [I_f(R_a + R_f)]}{I_f N_a}$$ **(eq. 2–15)**

If we use our previous example with a 100-V series motor turning at 900 rpm, an armature and field resistance of 1 Ω each, and 5 A of current flowing, we get

$$K_e = \frac{100 \text{ V} - [5 \text{ A}(1 \text{ Ω} + 1 \text{ Ω})]}{(5 \text{ A})(900 \text{ rpm})} = 0.02 \text{ V/(A} \cdot \text{rpm)}$$

Note that this constant differs from the one in the torque equation.

We can use this equation to calculate the speed of this motor. Let's suppose we decrease the torque on our series motor system. If we measure 20 A of current flow after this torque change, what will the new torque be? Using the torque equation (2–8), the new torque will be

$$T = I_a^2 K_t = (20 \text{ A})^2 (0.16 \text{ N} \cdot \text{m/A}^2) = 64 \text{ N} \cdot \text{m}$$

Using the CEMF equation (2–14), we can find the new speed

$$N_a = \frac{V_{line} - [I_f(R_a + R_f)]}{I_f K_e}$$

$$= \frac{100 \text{ V}[20 \text{ A}(1 \text{ Ω} + 1 \text{ Ω})]}{(20 \text{ A})(0.02 \text{ V/A} \cdot \text{rpm})} = 150 \text{ rpm}$$

When the armature current is 5 A, the armature and field drops are 10 V. The CEMF is then 90 V. At this load, the speed is 900 rpm and the torque is 4 N · m. Using equation 1–6, we find the output power in watts is

$$P_{out} = 0.105 \, N_a T = (0.105)(900 \text{ rpm})(4 \text{ N} \cdot \text{m}) = 378 \text{ W}$$

This output power can be converted to horsepower by dividing by 746 (1 HP = 746 W). Input power can be found by multiplying input current (5 A) by the input voltage (500 V).

The series DC motor is described as having poor speed regulation. It is classified as a variable speed device with changing load. An increase in load torque causes a decrease in speed. How much the speed will decrease with a load increase depends on the design of the motor.

Calculations of power at five different torques for a series motor with line voltage held constant at 100 V are listed in Table 2–1 for your reference.

In both of these examples, we have kept the line voltage constant and changed the torque produced by varying the torque demanded from the motor. This relationship is graphed in Figure 2–8. You can see from this graph and the speed equation that the armature always tends to rotate at such a speed

Table 2–1 Operational features of a typical series motor

I Armature (amps)	Counter EMF (V)	Speed (rpm)	Torque (lb · ft)	(N · m)	Power (W)
5	90	900	3	4.00	378
10	80	400	12	16.27	683
15	70	233	27	36.6	895
20	60	150	48	65.1	1025
25	50	100	75	101.68	1068

that the sum of the CEMF and the drops across R_a and R_f equal the applied voltage. If the load is removed from the motor, the armature will speed up and a higher CEMF will be induced into the armature. This reduces the current through the armature and field, weakening the field. The weakened field causes the armature to turn faster. The CEMF is, therefore, increased, and the speed of the machine is further increased. This regenerative process increases until the machine is destroyed. For this reason, series motors are never connected to their loads by belts or chains. The belt or chain may break and the motor may overspeed and destroy itself. Series motors are connected to their loads by direct drives or through gears, or they may have overspeed protection built into the motor.

Constant Torque Applications In some applications, the torque demanded of the motor will remain relatively constant. If the torque demand stays constant, the torque produced stays constant. An examination of the torque equation reveals that if torque stays constant, so does the armature current. In this situation, a change in the motor's speed will have to be produced by a method other than changing the torque. Our speed equation (developed from the basic CEMF equation) shows us that the speed can be changed by varying the line voltage V_{line}.

Let us, for example, take the series motor system we have used previously. The motor armature is turning at 900 rpm, drawing 5 A of armature current with 100 V_{line} applied. We increase the line voltage to 125 V. Using a K_e of 0.02, we can calculate the new speed

$$N_a = \frac{V_{line} - [I_f(R_a + R_f)]}{I_f K_e}$$

$$= \frac{125 - [5 \text{ A}(1 \text{ } \Omega + 1 \text{ } \Omega)}{(5 \text{ A})(0.02 \text{ V/A} \cdot \text{rpm})} = 1150 \text{ rpm}$$

We have assumed that the currents have not changed. The speed has increased to 1150 rpm. Note that the CEMF has increased to 115 V. If we were to do many calculations of the speed versus line voltage, we would see a relationship such as the one shown in Figure 2–10. As line voltage increases, so does the armature speed. We can then change the motor speed by increasing or decreasing the voltage supplied.

Series Motor Application Industry uses series DC motors in applications that require high starting torque, high accelerating torque, and high speed with light loads. Since the torque at any time is proportional to the square of the armature current, torque at starting is particularly high. Cranes and hoists are applications that need high starting torque. These devices lift stationary objects by overcoming their inertia. When a crane is unloaded, we want it to move quickly. But when it is loaded with a heavy weight, we want it to move slowly. The series motor has both of these characteristics. Traction equipment also depends heavily on series DC motors in such devices as locomotive train drives and forklifts. As in the crane or hoist, the locomotive must overcome the considerable inertia of the motionless cars. To accomplish this, one series motor is mounted on each axle. The diesel engine powers a DC generator that, in turn, provides the power to the series motors. The series motor also adjusts its speed automatically, depending on the grade. The motor armature turns fast when running on level ground. When going up a hill, it automatically slows down. An added benefit of the series motor is *regenerative braking*. We will discuss braking in greater detail later in this chapter. For now, we can point out that when the engine goes down a grade, it has a tendency to accelerate, due to the force of gravity on the mass of the engine. At some point, the armature will turn faster than it would normally turn with power applied. From that point on, the motor will act as a generator. The countertorque produced will then tend to slow it down.

FIGURE 2–10 Graph of speed versus line voltage with torque held constant

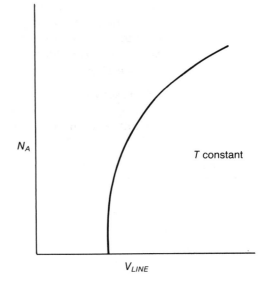

The series-wound DC motor also has one of the highest power/dollar, power/pound, and speed of any motor. For example, a 1/10 HP series DC motor that can run at 10,000 r/min weighs only four pounds. A comparable AC induction motor rated at 1/10 HP will weigh about 15 pounds and will run at 1725 r/min. As we will see in Chapter 5, AC induction motors cannot run at speeds greater than 3600 r/min from a 60-Hz AC supply. It is common to see a small series DC motor running at speeds above 15,000 r/min. Higher speeds, however, are not attained without a price. Higher speeds give greater brush and bearing wear. The brushes in residential appliances can have a brush life as low as 200 hours. The balance of the armature also becomes more important at high speeds. A slight unbalance at high speeds can cause serious vibration damage to the motor. Series motors are cost effective compared to shunt and permanent magnet DC motors at sizes up to approximately five inches in diameter. In larger sizes, the shunt and permanent-magnet motor types will have a higher cost/performance tradeoff.

One important use of the series motor is in the automobile starter. The starter motor must overcome the large inertia of a cold and unlubricated engine. Owing to its widespread use in automotive applications, the series DC motor is probably the most widely used motor in the world.

2–6.2 Shunt-Connected Field DC Motors

Construction In the shunt-connected field DC motor (hereafter called the shunt motor), the armature and field coils are connected in parallel with the applied voltage (Figure 2–11). ("Shunt" is a synonym for parallel.)

Like the series motor, the shunt motor can be recognized by its field windings. The series motor has field windings consisting of a few turns of heavy gauge wire. Such construction is necessary because heavy armature current flows through the field. In the shunt motor, the field windings are usually made of many turns of fine wire. The shunt field windings then have a much greater resistance than their series field counterparts. The actual field resistance varies widely from motor to motor and can be found in manufacturers' data sheets on series motors.

Speed Regulation A study of the circuit diagram in Figure 2–11 gives a clue to the operation of the shunt motor. Note that if we hold the supply voltage constant, the current through the field will be constant. Consequently, the

FIGURE 2–11 Shunt-connected field DC motor schematic diagram

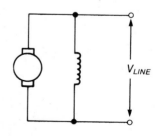

field flux remains relatively constant. When no load is applied to the motor, the only torque produced is that necessary to overcome bearing friction and the wind resistance of the armature (called *windage*). The rotation of the armature coils through the field produces a CEMF, as in the series motor. The CEMF limits the armature current to the relatively small value needed to produce the small amount of torque necessary to run the motor on no load.

When a load is applied to the motor, the speed of the motor starts to decrease. Using the CEMF formula from equation 2–11, you can see that a decrease in speed causes a decrease in CEMF.

$$V_{\text{CEMF}} = I_f N_a K_e$$

A decrease in CEMF causes armature current to increase. Equation 2–7 shows the torque produced by the shunt motor to be proportional to the armature and field current.

$$T = I_f I_a K_{tw}$$

Since torque depends on armature current as well as field current, this increase in armature current increases the torque produced by the motor. The CEMF decreases and armature current increases until the developed torque is nearly equal to the applied torque.

The speed equation is similar to the one used for the series motor. The difference is that the CEMF is equal to the line voltage minus the voltage developed across the armature resistance. The field resistance is not in series with the line voltage; it is parallel or shunt. The speed equation (2–15) tells us something else about the motor's operation.

$$N_a = \frac{V_{line} - [I_a(R_a)]}{I_f K_e}$$

Note that when the field current and line voltage remain constant, the motor's speed remains constant. As we have shown, the armature current increases when the load on the motor increases. However, because of armature reaction, the flux field will decrease slightly as armature current increases. This slight decrease in flux, added to the increase in armature current, causes the speed of the motor to remain relatively constant. The shunt motor, therefore, is a constant speed device with changing torque demand. When the load on the motor increases, the motor armature, in reality, slows down slightly. Conversely, if the load on the shunt motor is reduced, the motor speeds up slightly. A 1/4 HP shunt motor will change about 15% from no-load to full-load. If speed regulation techniques are used, the shunt motor can be adjusted to within 1% of its set speed from no-load to full-load. **National Electrical Manufacturer's Association (NEMA)** standard base speed ratings for shunt motors are 1140,

1725, 2500, and 3450 r/min. (The **base speed** of a motor is that speed which a motor develops at rated voltage with the rated load applied.) You may, however, find shunt motors operating at other base speeds. Since they are not standard, they must be specially wound.

A graphic example may be helpful at this point. Figure 2–12 shows a **synchrogram,** which is a diagram showing the simultaneous effects on different parameters with respect to time. The synchrogram shows the effects of a load change on the CEMF, armature speed, torque, and armature current in a shunt motor. In this example, we will assume that the flux stays constant.

During the interval between time t_0 and t_1, the motor is operating at equilibrium. This means all the values have stabilized and remain constant. At time t_1, an increased load is applied to the motor. Instantly, the load torque curve rises, showing the increased value. Note that all the other values change gradually due to the inertia of the armature. Because the load torque is greater than the torque generated in the motor armature, the motor speed is reduced until a balance is again attained. When the torque developed or produced is again equal to the load torque demand, the motor deceleration stops and the motor operates at a constant reduced speed.

At time t_2, the motor reaches a new equilibrium state, corresponding to the new load demand. At time t_3, the load torque demand is suddenly decreased. Again, the inertia of the armature keeps the generated torque from following instantly with it. The armature's inertia also prevents the armature speed from changing instantly. Because the generated torque exceeds the load torque, the motor accelerates. As the speed increases, the CEMF decreases. The increased CEMF opposes the applied EMF, reducing armature current. Armature current is reduced until the generated torque and load torque are again at equilibrium. The speed levels off at a constant value once this equilibrium point is reached.

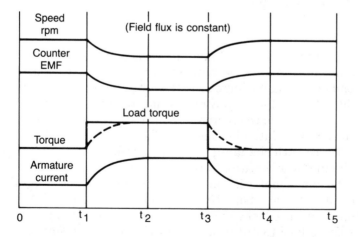

FIGURE 2–12 Shunt motor load-speed-torque-EMF relationships with respect to time

Table 2–2 Operational features of a typical shunt motor

I Line (amps)	I Armature (amps)	Counter EMF (V)	Speed (rpm)	Torque	
				(lb · ft)	(N · m)
2	1	99	990	0.7	0.95
4	3	97	970	2.1	2.85
6	5	95	950	3.5	4.75
8	7	93	930	4.9	6.64
10	9	91	910	6.3	8.54
21	20	80	800	14.0	18.98

Constant Line Voltage As we did in the series motor, we shall calculate the operational features of a shunt motor. The variation of line current, armature current, CEMF, speed, and torque are shown in Table 2–2.

Assume a shunt motor with 100-V line voltage, armature resistance of 1 Ω, field resistance of 100 Ω, an armature current of 1 A, an armature speed of 990 rpm, and a torque of 0.949 N · m. With an armature current of 1 A and an armature resistance of 1 Ω, the drop across the armature resistance is 1 V, by Ohm's law. We can see from the schematic diagram in Figure 2–13 that the CEMF is equal to the line voltage minus the drop across the armature resistance:

$$V_{\mathrm{CEMF}} = V_{line} - (I_a R_a) = 100 \text{ V} - 1 \text{ V} = 99 \text{ V} \qquad \textbf{(eq. 2–16)}$$

The power output in watts can be found using equation 1–6 from Chapter 1.

$$P_{out} = 0.105 \, N_a T = (0.1048)(990 \text{ rpm})(0.949 \text{ N} \cdot \text{m})$$

$$= 98.5 \text{ W or } 0.132 \text{ HP}$$

The total input current to the system is equal to the sum of the field and armature currents, or 2 A. The input power is equal to the input current multiplied by the input voltage of 100 V.

$$P_{in} = I_{in} V_{line} = (2 \text{ A})(100 \text{ V}) = 200 \text{ W} \qquad \textbf{(eq. 2–17)}$$

FIGURE 2–13 Shunt DC motor with applied voltage

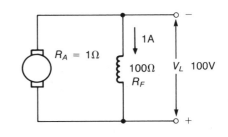

The torque calculations for other values of load are based on the proportionality between armature current and torque. For example, if torque is 0.949 N · m when the armature current is 1 A, the torque for an armature current of 2 A will be 2 times 0.949 or 1.898 N · m.

The speed calculations are based on the proportionality between speed and CEMF. For example, when the armature current is 2 A, the drop across the armature resistance is 2 V and the CEMF is 100 V − 2 V or 98 V. If the speed is 990 rpm when the CEMF is 99 V, the speed is 980 rpm when CEMF is 98 V.

Curves showing speed versus torque are shown in Figure 2–14. Note how this speed torque curve differs from the series motor curve.

Constant Torque Operation Speed of a shunt motor is often achieved by placing a potentiometer or rheostat in series with the field of the motor. Changing the amount of resistance in series with the field changes the amount of field current. Analysis of our speed equation (2–15) shows that speed is inversely related to field current.

$$N_a = \frac{V_{line} - [(I_a)(R_a)]}{I_f K_e}$$

If the resistance increases, field current will decrease, thus increasing speed. This condition creates a potential runaway problem. Recall that the series motor can overspeed if the load is lost. In the shunt motor, an identical problem can result if the field is lost. Some shunt motors use relays to sense the decreasing strength of the field. Power is disconnected when field current reaches a certain minimum value.

FIGURE 2–14 Speed-torque curve for the shunt DC motor with line voltage held constant

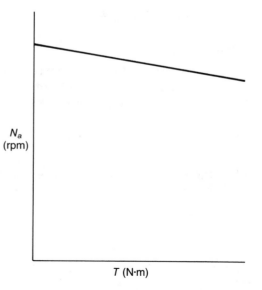

N_a
(rpm)

T (N·m)

If the torque were to remain constant, the field flux strength could be changed by varying either the resistance in the field or the voltage across the field. Usually it is easier to vary the field resistance than to vary the line voltage. Increasing or decreasing field resistance can be accomplished by placing a potentiometer or rheostat in series with the field winding. With a field resistance of 125 Ω, for example, the field current is

$$I_f = \frac{V_{line}}{R_f} = \frac{100 \text{ V}}{125 \text{ Ω}} = 0.8 \text{ A} \qquad\qquad \textbf{(eq. 2–18)}$$

Since the torque remains at 0.949 N · m, the armature current can be solved from the torque equation (Equation 2–7):

$$I_a = \frac{T}{I_f K_{tw}} = \frac{0.949 \text{ N} \cdot \text{m}}{(0.8 \text{ A})(0.949 \text{ N} \cdot \text{m/A}^2)} = 1.25 \text{ A}$$

With 1.25 A of armature current, the drop across the armature resistance is 1.25 V. This gives a CEMF of 98.75 V. Finally, the speed can be worked out with the speed equation (Equation 2–15):

$$N_a = \frac{V_{line} - [I_a(R_a)]}{I_f K_e} = 1234 \text{ rpm}$$

The same effect can be accomplished by varying the voltage instead of the resistance. A curve showing the relationship between speed and field resistance (or voltage) appears in Figure 2–15.

FIGURE 2–15 Speed-line voltage curve for the shunt DC motor

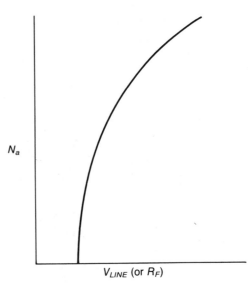

N_a

V_{LINE} (or R_F)

Application As we have seen from the preceding discussion, the shunt motor can be considered a constant-speed device. Although the speed can be varied by changing the field current (usually by a rheostat), the speed remains constant for a given field current. Machine tools, elevators, and other devices that need constant speed with changing loads use the shunt motor. A good example is a conveyor belt, with parts being loaded randomly. The belt speed should be constant. Random fluctuations will occur when parts are placed on the conveyor. A shunt-connected DC motor is a good choice for this application. When a heavy load is thrown on a shunt motor, it tries to take on the new load at only slightly less speed. The flux stays constant, and the increase in torque is followed by an increase in armature current. With a very heavy load, however, the armature current can become too high. The motor temperature can then increase to a very high value. Unlike the series motor, the shunt motor cannot slow down appreciably with a heavy load. Thus, the shunt motor is more vulnerable to overloads. To illustrate, let us assume that a DC motor is used to move an electric vehicle up an incline. If it is a shunt motor, it will try to keep the same speed as it goes up the grade. As a result, too much armature current can flow. The self-adjusting nature of the series motor protects it from excessive overloads.

2–6.3 Compound Motor

Both series and shunt motors have unique properties. The high starting torque of the series motor is not found in the shunt motor. By the same token, the constant-speed characteristic of the shunt motor is not found in the series motor. Some applications demand both of these characteristics. If we give a DC motor both shunt and series fields, we will have a motor that shares the features of both types of configurations. We call this type of motor a **compound** motor. In most cases, the series winding is connected so that its field aids the shunt field. This is done by winding both shunt and series fields in the same

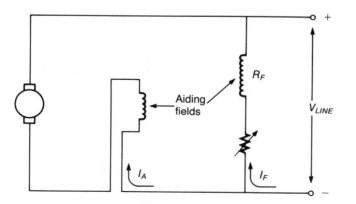

FIGURE 2–16 Cumulative compound DC motor

direction. Motors connected in this way are given the name *cumulative compound* motors. A schematic diagram of this configuration is shown in Figure 2–16. Cumulative compound motor characteristics lie between those of shunt and series motors. As the load is increased, the increase in current increases the flux, due to the series winding. This increases the torque faster than the increase for a straight shunt motor. The increase in flux decreases the speed more rapidly than the speed decrease in a shunt motor. The applied and developed torques balance with less speed decrease than in a series motor but more than in a shunt motor. The speed of the cumulative compound motor depends, therefore, on the sum of the series and shunt fields. In a condition where the load is small, only a very small amount of current flows through the series field. In the low-load condition, the speed depends on the shunt winding, not the series winding. This overcomes the disadvantage of the series motor, which is likely to run away at low loads. As always, an advantage gained in one area is offset by a disadvantage in another. The cumulative compound motor has poorer speed regulation than the shunt motor.

Because of the series field winding, the cumulative compound motor has a higher starting torque than a shunt motor. Cumulative compound motors are used to drive machines that are subject to sudden changes in load. Sometimes, after the motor gets up to full-load speed, the series winding is shorted out. The motor then runs as a shunt motor, with its better speed regulation.

If the series winding is connected so that its field opposes the shunt winding field, the motor is called a *differential compound* motor. This effect is achieved by winding series and shunt coils in opposite directions. Since the series field opposes the flux of the shunt field, an increase in load causes an increase in current and a *decrease* in total flux. These two flux fields tend to subtract from each other. A decrease in load causes an increase in current and a *decrease* in total flux. These two flux fields tend to cancel each other out, and the result is a practically constant speed. Under heavy loads, the speed of the differential compound motor is unstable. This motor possesses very low starting torque and very good speed regulation. However, due to the instability at heavy loads and the fact that the AC motor performs equally well, this motor configuration is rarely used. Improvements in permanent-magnet motors and electronic controls have done much to displace the compound motor in industry.

2–6.4 Separately-Excited DC Motor

The separately-excited motor configuration, shown in Figure 2–17, is very similar to the shunt motor in operation. Electrically, it differs from the shunt motor in that the field voltage can be controlled independently of the armature voltage. This configuration has the advantage of better speed regulation and wider speed range than the shunt motor. In all other respects, the separately-excited motor has the same characteristics as the shunt motor. The torque and speed curves are identical.

FIGURE 2–17 Separately-excited DC motor schematic diagram

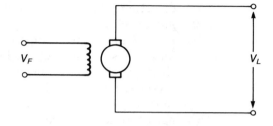

2–6.5 Speed Regulation

If a separate source excites the field, two different areas of motor operation are possible, as shown in Figure 2–18. In armature voltage speed control, the field voltage is held constant and the armature voltage varied. The power output of the motor is directly proportional to motor speed in this range. As speed increases, the motor power output increases by the same amount. This region is called the *constant torque region* because the motor torque output stays the same. Although the speed of a DC motor can theoretically be decreased to zero, heat becomes a problem at low speeds. At speeds below approximately 1/10 base speed, the fan on the armature turns too slowly, allowing heat to build up in the armature.

Speed is increased above the base speed by holding armature voltage at the maximum level and by decreasing the strength of the field. As you can see from Figure 2–18, speed control above the base speed gives a constant power out. Unfortunately, the torque output decreases as speed increases in this area of the curve. It appears from this curve that we may be able to increase the speed without limit by weakening the field. This is not true for several reasons. First, the flux generated by the current in the armature opposes the flux in the field. If the armature current is held constant and the field weakened, a point is reached where the total motor flux is reduced to a low value. At this point, the motor speed becomes unstable, which can result in damage to the motor. Second, the motor armature has speed limitations imposed by the commutator. If speeds become too high, the commutator bars can be thrown out of the commutator. A machine designed to be used in a variable-speed application must be specially constructed to withstand speeds over the base speed. Such motors are larger, heavier, and more expensive than motors that are designed to be operated at speeds below the base speed. Third, motors operated over the base speed have greater problems with armature reaction.

DC motors today use field current speed control widely. For manual control, speed can be changed by a variable resistance in the field circuit. Automatic control can be achieved by transistors or power MOSFETs controlling field current.

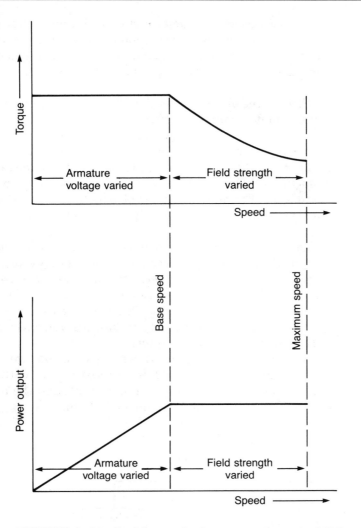

FIGURE 2–18 Speed control of a separately-excited DC motor

2–7 DC MOTOR POWER OUTPUT AND EFFICIENCY

Electric motor *efficiency* has been an increasingly important topic in industry in the last few years. In response to industry interest, motor manufacturers have tried to produce more efficient electrical machines.

We need to keep three things in mind before we look at the efficiency of a DC motor. First, a motor's load, horsepower rating, and speed are related to efficiency. If a motor is operated at a load other than rated load, a change in efficiency can take place. Second, a general rule of thumb is that motor efficiency increases with power rating. A large motor is generally more efficient than a small motor. Third, motors with higher base speeds generally have higher efficiencies than motors with lower base speeds with the same power rating.

A glance at these three observations should convince you that the proper choice of motor helps to determine its efficiency when operating in a specific application. Loads, for example, can vary significantly from one application to another. Some applications require a motor to run either continuously or for long periods of time. Examples of such applications might include fans, pumps, and processing machinery. On the other hand, many motors are operated for very short periods of time. Examples include valve motors, dam gates, and door openers. Choosing the right type of motor will increase efficiency and decrease cost.

One way of looking at the efficiency of a motor is to consider the motor an energy conversion device. A motor converts electrical energy to mechanical energy. No energy conversion device is perfect; some energy must be lost in this transition. Most of this energy is lost as heat and cannot be recovered. A diagram showing this distribution of losses is found in Figure 2–19. The losses are divided into two basic classifications: mechanical and electrical losses, which will be discussed later in this chapter.

For the purposes of this discussion, we will assume that we are using a series DC motor, with armature and field resistances of 1 Ω, with 100 V applied to the motor. The motor is drawing 5 A of current, and the armature is turning at 900 rpm. Torque produced equals 4.07 N · m.

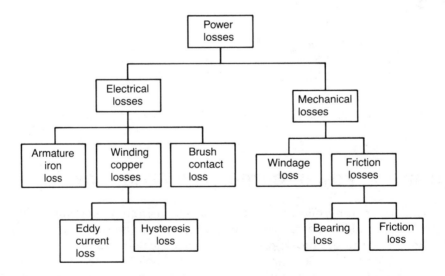

FIGURE 2–19 Losses in a DC motor

When the armature current is 5 A, the armature and field drops are each 5 V, for a total drop of 10 V. The CEMF is then 90 V. At this load, the speed is 900 rpm and the torque is 4.07 N · m. The output power in watts is

$$P_{out} = 0.105 \, N_a T = (0.105)(900 \text{ rpm})(4.07 \text{ N} \cdot \text{m}) = 384 \text{ W}$$

This corresponds to 0.516 HP, since 1 HP = 746 W. We can now calculate the efficiency of the machine. To do this, we must first find the input power. The efficiency of a system is found by dividing the output power by the input power and multiplying by 100 (for a percentage). Efficiency can also be expressed

$$\text{efficiency} = \frac{\text{output}}{\text{output} + \text{losses}} \times 100 \qquad \text{(eq. 2–19)}$$

$$= \frac{\text{input} - \text{losses}}{\text{input}} \times 100 \qquad \text{(eq. 2–20)}$$

These equations demonstrate that the efficiency will never be 100% because all machines have some losses. The first equation is most often seen in calculations of efficiency involving generators. The second is more useful in motor applications. In these equations, it is important that both input and output be expressed in the same units. For example, if the output power is expressed in horsepower (HP) and the input is expressed in watts (W), one should be changed so that the units correspond. The input power of this motor system is

$$P_{in} = I_{in} V_{in} = (5 \text{ A})(100 \text{ V}) = 500 \text{ W} \qquad \text{(eq. 2–21)}$$

The output power in watts is

$$P_{out} = 0.105 \, N_a T = (0.105)(900 \text{ rpm})(4.07 \text{ N} \cdot \text{m}) = 385 \text{ W}$$

The percent efficiency is then

$$\% \text{ efficiency} = \frac{P_{out}}{P_{in}} = \frac{385 \text{ W}}{500 \text{ W}} = 76.9\% \qquad \text{(eq. 2–22)}$$

This efficiency figure tells us that the motor system provides approximately 77% of the power we put in to turn a mechanical load. The rest of the power (13% or 115 W) is lost in the motor.

The cost of operating two motors with different efficiencies but the same load can be easily calculated. Such a calculation shows the cost of energy savings over a year, a significant factor in the choice of a motor. Equation 2–22 gives the dollars saved in a year:

$$S = (0.746)(HP)(C)(N)\left(\frac{100}{E_A} - \frac{100}{E_B}\right)$$ **(eq. 2–23)**

where S is the dollars saved in a year, HP is the power of the motors in horse-power, C is the energy cost in per kilowatt hour, N is the time the motors will run in hours per year, E_A is the efficiency of motor A in percent, and E_B is the efficiency of motor B. This equation applies only to motors operated at a constant, specific load.

Let's use this equation to see the cost savings between two motors operated on a 10-HP load 8,760 hours per year. If motor A is 85% efficient, motor B is 86% efficient, and 1 kWh costs \$0.10, what is the savings in dollars per year? Using our equation, we get

$$S = (0.746)(HP)(C)(N)\left(\frac{100}{E_A} - \frac{100}{E_B}\right)$$

$$S = (0.746)(10 \text{ HP})(\$0.10)(8760 \text{ hr.})\left(\frac{100}{85} - \frac{100}{86}\right)$$

$$= \$89.40/\text{year savings}$$

Various types of power losses, which affect the efficiency of the motor, occur in the DC motor. They can be grouped generally into two categories: electrical and mechanical, as mentioned previously (see Figure 2–19).

The important electrical losses are:

1. Armature iron losses—The armature *iron losses* can be further divided into two categories: *eddy current* losses and *hysteresis* losses. Eddy current losses arise from the current induced into the armature's steel core. These losses are reduced, but not eliminated, by making the core out of laminated steel sheets. Generally speaking, the thinner the laminations, the less the eddy current effect. The eddy current produced in the iron has a significant effect on the high speed operation of the DC motor. Silicon steel with a high electrical resistance also reduces eddy currents. Hysteresis loss comes from the rotation of the magnetic domains in the armature. The resistance to this magnetic domain shift increases with speed, causing heat to be generated. Hysteresis loss is affected by the type of steel used in the armature.

2. Winding copper losses—These losses include the I^2R losses from the copper conductors in the motor windings. These losses increase as the current increases. They are normally computed when the motor is hot, 75°C for class A, B, and E insulation and 115°C for class F and H insulation. Because R increases with temperature and current must be kept constant to keep torque constant, the I^2R losses will increase as motor temperature increases. In our example, if we assume that the

armature and field resistance were all copper loss (1 Ω each), the power loss will be

$$P = I^2R = (5 \text{ A})^2(1 \text{ } \Omega) = 25 \text{ W} \qquad \text{(eq. 2–24)}$$

for each winding, for a total loss of 50 W.

3. Brush contact loss—As the brush slides over the commutator, a film is formed on the surface of the commutator. This film has the properties of resistance. It is very difficult to calculate brush resistance, since it is affected by many factors. We can, however, estimate it from measurements of the voltage drop across the brushes. Generally, the brush drop is 2 V. With 5 A of current flow, the resistance will be 0.4 Ω. We have been including this resistance in with the armature resistance, since it is difficult to separate the brush drop from the drop across the copper in the armature windings. In our example, the actual winding resistance will be 0.6 Ω and the brush resistance will be 0.4 Ω. The total loss will still be 25 W in the armature.

The important mechanical losses are

1. Friction losses—These losses can further be broken down into two categories: bearing friction and brush friction. Bearing friction is the loss caused by the heat generated by one metal surface rubbing against another. This type of loss can be reduced by proper lubrication. Ordinarily, roller bearings have less friction loss than sleeve bearings. The other type of friction loss, brush friction, results from the brushes rubbing against the commutator. Brush friction can be reduced by a well-polished commutator and properly-fitted brushes.
2. Windage losses—This loss is caused by the wind resistance presented to a moving armature.

Motor losses can also be grouped by the two factors affecting them: torque (or load) and speed. The winding losses are considered load sensitive losses since changes in load determine the amount of current that flows in a motor. More load usually means more current. More current causes more I^2R losses. The other losses are classified as speed sensitive losses. These losses increase with increasing speed, except for brush resistance, which is unpredictable.

Calculations of power and efficiency of a 1-HP 100-V shunt motor at six different torques with line voltage at 100 V are listed in Table 2–3.

Several things about the DC motor system are clear. The power a motor produces is dependent on the speed at which the armature is turning and the torque produced by the motor at that speed. As the motor is loaded down, more torque is produced, but torque increases more than speed decreases. This action produces an overall increase in the total power produced. The increase

Table 2–3 Operational features of a typical shunt motor with efficiency added

I Line (amps)	I Arm. (amps)	CEMF (V)	Speed (r/min)	Torque (lb · ft)	Power Output (W)	Input Power (W)	Efficiency (%)
2	1	99	990	0.7	98.4	200	49.2
4	3	97	970	2.1	289	400	72.2
6	5	95	950	3.5	472	600	78.5
8	7	93	930	4.9	646	800	80.8
10	9	91	910	6.3	815	1,000	81.5
21	20	80	800	14.0	1,590	2,100	76

in power produced comes with a price tag, a decrease in efficiency. Efficiency of motor systems is an important topic. Greater motor efficiency means bigger profits for a manufacturer using that motor in a process. Greater motor efficiency also means less wasted energy. In 1977, almost half of the total electrical energy used in the United States was used to power electric motors! Efficiency is important.

2–8 DC MOTOR MAINTENANCE, TESTING, AND TROUBLESHOOTING

2–8.1 Maintenance

The best maintenance programs stress preventing problems before they occur, rather than repairing them after they occur. Many problems with motors can be prevented or, at least minimized, with a program of preventive maintenance. A good preventive maintenance program includes regularly scheduled equipment inspection and repair, and documentation of problems. The heart of preventive maintenance is routine equipment inspections. Such regular inspections can uncover minor problems that can be fixed before they become major problems.

A good system of record keeping is vital to a preventive maintenance program. Good records can show deterioration of a piece of equipment over time. If a certain repair (for example, a brush replacement) is done often, good record keeping can show this. Further investigation can then uncover why brushes are being replaced so frequently. Although every system of record keeping is different, four areas seem to be important. First, every piece of equipment should have an *equipment record* containing the following information: equipment location, serial number, and date of installation. Second, a checklist is necessary to perform regular preventive maintenance. The checklist should contain a list of tools needed, how long it will take to complete the job, and precise instructions on what to do. These instructions should include a checklist to indicate when parts of the task are completed and safety notes. This document should be part of a system that lets maintenance personnel know when preventive maintenance should be done on a particular piece of equipment. Third, a record of costs incurred in maintaining a piece of equipment should be kept. Last, an inventory control system is needed to keep track of what parts are used in maintenance procedures, when to order more parts, and whether enough of the right parts are on hand.

When discussing preventive maintenance on motors, we should realize that damage due to environmental conditions causes the majority of motor failures. A good preventive maintenance program will keep motor equipment clean, dry, tight, and frictionless.

Of the environmental factors mentioned, dirt is the most important cause of motor failure. Dust and dirt cover equipment and restrict proper motor ventilation. This causes the motor to overheat, since the motor cannot dissipate heat effectively. When a motor continuously overheats, the insulation breaks down and windings short out. Dirt can also cause wear in moving parts, such as bearings. Since dirt can be conductive, dirt buildup on moving electrical parts can cause arcing and burning. A regularly scheduled cleaning of motors is essential for proper operation. If a motor cannot be kept clean with frequent cleanings, a totally enclosed motor should be considered.

Moisture, which tends to soften insulation, also causes problems in motors. Soft insulation can flow, leaving thin spots that have lower dielectric strength. This increases the possibility of arcing and short circuits. Moisture also increases the speed at which parts of a motor collect dust and dirt. Moisture can also cause corrosion of copper and iron parts, causing higher conductor resistances. Guarding against moisture is best done by selecting the right motor enclosure. All motor manufacturers have an enclosure for a motor that will run in an environment where moisture is a problem. Moisture damage can be prevented by other procedures. First, if motors are run infrequently, make sure they are turned on and run for a short time at least once a week. This prevents any moisture buildup due to condensation. Second, in very moist locations, install heaters that turn on when the motor turns off.

Friction is another cause of mechanical damage in a motor. Friction damage can be caused by unlubricated or damaged bearings, a belt or unbalanced motor shaft, loose mechanical connections, or a misalignment between motor and load. Preventive maintenance should involve checking the motor to see if it rotates smoothly without drag or vibration. The motor shaft usually sits in bearings or sleeves that, in turn, are placed in the end bell of the motor. The bearings or sleeves must be lubricated properly with the correct amount of the right lubricant.

Operating motors in places where ambient temperatures are high will damage motors. The area of the motor likely to be damaged is the winding insulation. Insulation damage can occur at temperatures above the standard reference ambient temperature of 40°C (104°F). In general, DC motors are designed to operate at temperatures between 0°C and 40°C at altitudes below

3300 ft. Motor winding insulation comes in four different classes. Classification depends on the temperature at which each type of insulation will give normal life. Classifications of insulation are A, B, F, and H. Class A insulation is found on many fractional HP motors and some older integral HP motors. Most fractional HP motors have class B insulation. Classes F and H are reserved for special applications. Insulation is one of the most important factors that determine the useful life of the motor and how much maintenance will be required. The insulation protects the windings from contaminants as well as short circuits.

Motor insulation can be destroyed by internal, as well as external, temperature rises. An operating condition where the motor windings draw more than normal amounts of currents is called an **overload.** It is the primary cause of internal temperature rise. Overloads can be caused by improper line voltages or by excessive mechanical loads. Motor currents and voltages should be monitored from time to time to see if they are within limits.

The quality of the insulation can be checked during preventive maintenance. Insulation checks are done with a special ohmmeter used to measure high values of electrical resistance, called a **megohmmeter** or **"megger."** Resistance is checked from winding to ground and from winding to winding. A lower-than-normal resistance reading could be caused by deteriorating insulation.

2–8.2 Troubleshooting Motors

Although the motor is a relatively simple device, troubleshooting a motor system can be complicated. A motor system can be broken down into four basic parts: 1) the DC motor itself, 2) the drive circuitry, 3) the power supply, and 4) the load. If a motor malfunctions, the problem can be in any one of these areas. The technician should be familiar with all four of these areas.

One common problem with DC motors is overheating. If ambient temperature has not increased, there can be any number of causes for motor overheating. The most common cause is an overload. Other causes might include incorrect armature or field voltages or poor ventilation. The technician will need to narrow the choices by testing possibilities, throwing out unlikely causes. For example, incorrect armature or field voltages can be eliminated by voltage measurements.

Another problem sometimes exhibited by DC motors is noise and vibration. Vibration should be carefully measured in a motor. Excessive vibration can cause wearing of insulation, loosened windings, uneven wear on bearings, and general structural failure in the motor. Ultimately, vibration can lead to electrical failure by causing short or open circuits. Vibration can also cause sparking at the brushes of the motor. Excessive sparking at the brushes causes the commutator to burn. The causes of noise can be broken down into two categories: mechanical and electrical. A good way to see if the source of noise is electrical or mechanical is to bring the motor up to speed and then disconnect it from the power source and let it coast. If the noise is present after the motor is disconnected, the problem is mechanical and not electrical. If the noise goes away when the power is disconnected, the source is electrical.

Noise or vibration in a motor can have several mechanical causes: a bad bearing, a bent or unbalanced armature shaft, or loose mounting hardware. Sometimes a measurement of the frequency of the vibration can give a clue to its origin. Noise can come from an obstruction on the armature or coupling to the load, or, the load and motor may have become misaligned. In each of these cases, the technician should investigate each problem and take corrective action. Part of a good preventive maintenance program should include checking for loose parts and hardware. Loose parts should be tightened or secured with cement. A bent shaft is usually caused by a blow to the shaft. A forklift, for example, may accidentally run into a motor shaft and bend it. It may be possible to straighten the shaft. If the shaft cannot be straightened, it will have to be discarded. Never heat a motor shaft to straighten it. Heating removes the temper from the steel, weakening the shaft.

Noise and vibration can also have an electrical cause. The mechanical causes of vibration and noise are usually apparent and, with some investigation, can be located and isolated. Electrical causes of noise are not so readily apparent. One frequent cause of noise in a motor is oversaturation of the magnetic poles. Most magnetic paths in a motor are designed to carry a maximum amount of flux. If the amount of flux is increased, stresses occur on the inside of the motor, causing parts of the motor to vibrate. Improper commutation can also be a source of noise. Brushes must ride smoothly and quietly on the commutator and hold sparking to a minimum. Brush tension should be measured periodically, as shown in Figure 2–20. Another cause of vibration is a short in an armature or field winding. A short circuit in either of these places reduces

FIGURE 2–20 Measuring brush tension

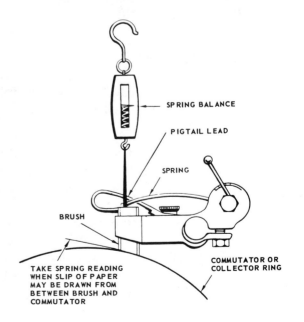

the magnetic field in one area of the motor and causes the motor shaft to deflect. An open armature or field winding can also cause vibration, for the same reason.

CHAPTER SUMMARY

- A DC motor is a device that converts electrical energy, in the form of DC voltages and currents, to torque.
- A wound-field DC motor gets its field from an electromagnet.
- The CEMF generated in the armature of a DC motor is directly proportional to the speed of the armature and to the strength of the field.
- The torque produced by a wound-field DC motor is directly proportional to the field and armature currents.
- The speed regulation of a DC motor is the ability of the motor armature to keep the same speed, regardless of the load applied.
- Wound-field DC motors are usually classified by the field connection: series, shunt, compound, and separately-excited.
- Series DC motors have poor speed regulation and are most often used in hoist and crane applications.
- Shunt/separately-excited DC motors have good speed regulation and are used where a closely-controlled speed is desired.
- The efficiency of a motor is the power the motor produces compared to the power consumed.

QUESTIONS AND PROBLEMS

1. What energy conversion takes place in a DC motor?
2. To what is CEMF proportional in a DC motor?
3. Identify the major parts of a practical DC motor.
4. In a DC motor, what purpose does the commutator serve? How does the commutator accomplish this?
5. In the DC motor, what is the motor CEMF when power is applied to the armature and the armature is stalled?
6. Discuss the motor as a feedback system. From where does the feedback come? How is the speed regulated?
7. What is armature reaction? How does it affect a DC motor's operation? Discuss two ways to reduce the effects of armature reaction.
8. What are interpoles and compensating windings? What purpose do they serve in a DC motor?

9. Define speed regulation in a motor. How is speed regulation expressed?

10. If the no-load speed of a motor is 3000 r/min and the full-load speed is 2750 r/min, what is the motor's speed regulation?

11. What is the difference between continuous duty and intermittent duty motors?

12. Draw a series DC motor circuit schematic diagram.

13. A series DC motor produces a torque of 10 N · m with an armature current of 3 A. Find the torque constant (K_t) for this motor. We applied a heavier load to this motor and measured a current of 6 A. What is the torque produced for this load?

14. Under what conditions does the torque constant (K_t) change?

15. A series DC motor has armature and field resistances of 2 Ω each. If the applied voltage is 50 V, what is the stall torque?

16. In the preceding problem, with an armature current of 6 A, an applied voltage of 50 V, and a speed of 1000 r/min, what is the CEMF? If we loaded the motor down to where it was drawing 10 A, what would the new torque demand be?, The new speed and CEMF?

17. How much power is the motor producing when it is drawing 6 A of armature current? At 10 A?

18. In the motor system above, if we kept the torque constant with 6 A of armature current and increased the line voltage to 100 V, what would be the new speed?

19. Draw a circuit that uses a shunt-connected field DC motor.

20. A shunt DC motor has an armature resistance of 1 Ω, a field resistance of 500 Ω, and an applied voltage of 100 V. How much CEMF is produced?

21. In the problem above, the armature speed is 1500 r/min. If we decreased the armature resistance to 400 Ω, what would the new speed be? The new CEMF?

22. What condition could cause a runaway condition in a shunt DC motor? In a series DC motor?

23. In what kinds of application would you find a shunt DC motor? A series DC motor?

24. Describe the characteristics of a compound DC motor. Differentiate between a differential compound and a cumulative compound motor.

25. Describe the characteristics of a separately-excited DC motor.

26. A shunt motor has an armature current of 2 A with an armature voltage of 50 V and an efficiency of 85%. What is the motor's output power in W and HP?

27. List the losses likely to occur in a DC motor and tell where they come from.

3

PERMANENT MAGNET, STEPPER, AND BRUSHLESS DC MOTORS

At the end of this chapter, you should be able to

- explain the difference between permanent-magnet motors and wound-field motors.
- predict and calculate permanent-magnet motor performance.
- explain the difference between conventional permanent-magnet motors, moving coil motors, and torque motors.
- name the two types of stepper motors.
- describe the construction and operation of the two types of stepper motors.
- describe the construction and operation of the brushless DC motor.

3–1 INTRODUCTION

Chapter 1 introduced the subject of the DC motor and examined the wound-rotor DC motor in depth. This chapter will discuss other popular types of DC motors, such as *stepper motors, brushless DC motors (BDCMs),* and *permanent-magnet (DM) motors.* These motors have power ratings to approximately 4 kW. Some of these motors are used as servo motors. A *servo motor* is any motor that is employed in a remote control application and has identical characteristics in both directions of rotation. The word *servo motor* came into use around the turn of the century, when the motor was used to turn a ship's rudder.

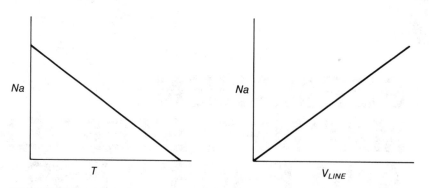

FIGURE 3–1 Speed-torque and speed line voltage curves for the PM motor

3–2 PERMANENT-MAGNET MOTORS

The PM motor differs from the wound-rotor DC motor in one respect. The field
is supplied by a PM motor, as its name implies. In operation, the PM motor
stator flux is always constant. In practice, this means that the speed-torque
and speed-armature voltage curves are linear, as shown in Figure 3–1a,b.

3–2.1 Measuring and Predicting PM Motor Performance

PM motor performance can be predicted from curves, just as in the wound-rotor
DC motor. In Figure 3–2, a composite curve is used to show the relationships
between torque, speed, input current, and output power. Recall from Chapter 2
that the output power of a DC motor is proportional to the product of the torque
and speed. Note that at the no-load speed (N_a), the output torque and power
are both zero. This is not surprising, since the load is not demanding any
torque. Note also that some armature current is flowing. Some armature current must flow to overcome mechanical and electrical losses at no load. Mechanical losses are usually friction related. Electrical losses are primarily eddy
current and hysteresis losses. As the motor is loaded down, it produces useful
torque. At this point, the output power and armature current increase. The
motor's efficiency rises quickly to a peak and then falls off. Note that the
motor's maximum output power occurs at a higher torque. As the stall point is
reached, both efficiency and output power fall off. As the load is increased
further, a point is reached where the motor cannot produce any more torque.
The motor then stalls. At this stall point, the output torque reaches its maximum value, called the **stall torque.** The output power and efficiency have decreased to zero, and armature current has gone to a maximum value.

Data sheets often do not give adequate information to predict maximum
efficiency and the best operating point of the motor. A simple measurement

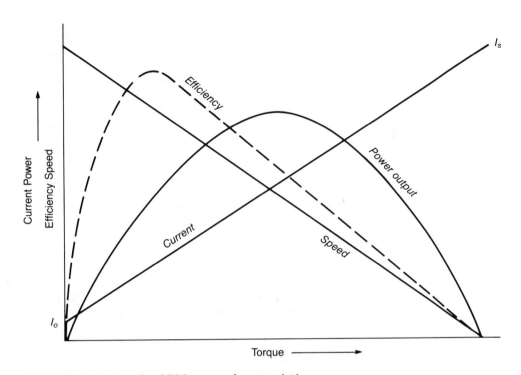

FIGURE 3–2 Graph of PM motor characteristics

and several calculations will allow us to predict the motor's behavior with some degree of accuracy.

After the motor has warmed up with no load connected, read the applied voltage (V_l), the armature current (I_a), and the armature speed (see Figure 3–3 and note the schematic symbol for the PM motor). Next, decrease the motor's speed by loading it down. You may want to actually connect a load to the motor for this test. If no load is available and the motor is small, holding a rag or wooden blocks on the shaft will suffice. The motor should be loaded to approximately 80% of its no-load speed. For purposes of illustration, we will measure an applied voltage of 12 V, a current of 0.5 A, and a speed of 3000 rpm, all measured without a load. After loading, we measure a voltage of 12 V, a current of 1.5 A, and a speed of 2500 rpm. We will use these values to calculate a parameter called the *factor of merit* (M) that will, in turn, be used to calculate other values. The factor of merit can be calculated by using the following equation:

$$M = \sqrt{\left(\frac{N_{nl}}{I_{nl}} \times \frac{\Delta I}{\Delta S}\right) + 1} \qquad \textbf{(eq. 3–1)}$$

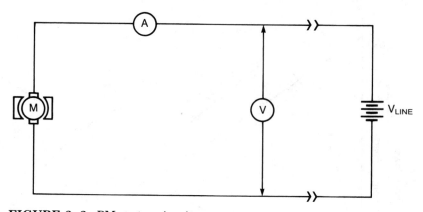

FIGURE 3–3 PM motor circuit

where N_{nl} is the no-load speed, I_{nl} is the no-load current, ΔS is the difference between the loaded speed and the no-load speed, and ΔI is the difference between the loaded current and the no-load current. Substituting into this equation, we get

$$M = \sqrt{\left(\frac{3000 \text{ rpm}}{0.5 \text{ A}} \times \frac{1.0 \text{ A}}{500 \text{ rpm}}\right) + 1} = 3.6$$

The maximum motor efficiency, shown in Figure 3–2, can be calculated from the factor of merit

$$Em = \left[1 - \left(\frac{1}{M}\right)\right]^2 = 52.2\% \qquad \textbf{(eq. 3–2)}$$

The input current (I_m) at the point of maximum efficiency is

$$I_m = MI_{nl} = (3.6)(0.5 \text{ A}) = 1.8 \text{ A} \qquad \textbf{(eq. 3–3)}$$

The power output (P_{out}) at maximum efficiency is

$$P_{out} = V_l I_{nl} E_m = (12 \text{ V})(0.5 \text{ A})(0.522) = 3.132\text{W} \qquad \textbf{(eq. 3–4)}$$

The speed (N_m) at maximum efficiency is

$$N_m = Nnl\left(\frac{M}{M+1}\right) = 3000\left(\frac{3.6}{3.6+1}\right) = 2348 \qquad \textbf{(eq. 3–5)}$$

The output torque (T_m) at maximum efficiency can also be calculated.

$$T_m = \frac{9.458\, P_{out}}{N_m} = 0.0126 \text{ N} \cdot \text{m (1.8 oz} \cdot \text{in)} \qquad \textbf{(eq. 3–6)}$$

Output torque can be estimated by use of the nomogram in Figure 3–4. For example, a motor turning at 100 rpm at 400 oz · in (approximately 2.8 N · m) torque provides slightly over 30 W of output power. (The left side of Figure 3–4 will be used in our discussion of stepper motors later in this chapter.)

None of these calculations are particularly difficult, and they can yield valuable information about motor performance.

3–2.2 PM Motor Classification and Characteristics

Permanent-magnet motors can be classified into three types: 1) the conventional PM motor, 2) the moving-coil PM motor, and 3) the torque motor. These motors use the same basic operating principle. Torque is developed by the interaction of two fields. These motors are usually more reliable and sturdy than their wound-rotor counterparts due to the lack of field windings, which also cuts down on copper costs.

PM motors are available in sizes up to the low integral HP range. They are efficient, easy to control, and have linear performance characteristics. Only one power supply is needed, since there is no field to be excited. You will recall that losing the field in a shunt motor can result in a runaway condition. This is not possible in the PM motor. Efficiency is improved since there is no power dissipated in the field winding. Behavior in changing temperature conditions depends on the type of magnet used.

Generally, three types of magnets are used: the Alnico type, the ferrite or ceramic type, and the rare earth type. The Alnico magnet has a very high flux density but is easily demagnetized. (Resistance to demagnetization is called *coercivity*. The Alnico magnets have low coercivity.) Alnico magnets are less affected by temperature because they have lower temperature coefficients. The flux density of the ceramic magnet is low compared to the Alnico magnets but has high coercivity. Ceramic magnets are inexpensive, both in materials and production costs.

Recently, another type of magnetic material has become cost-competitive, the rare earth-cobalt magnet. This new magnet, usually made of samarium cobalt, cuts the weight of the motor by 30 to 50% and reduces the diameter significantly without increasing cost or decreasing performance. Another advantage of the rare earth-cobalt magnet is its resistance to demagnetization. One of the disadvantages of the PM motor is that overloads can cause the magnet to demagnetize or lose some of its strength, which can cause a significant change in the operating characteristics. The samarium-cobalt magnet has a higher coercivity or resistance to demagnetization than the other types of magnets. Costs of the rare earth-cobalt magnet have decreased recently with the introduction of the rare earth material neodymium. Magnets based on neodymium can take advantage of inexpensive fabrication techniques. The

Load Operating at a Radius

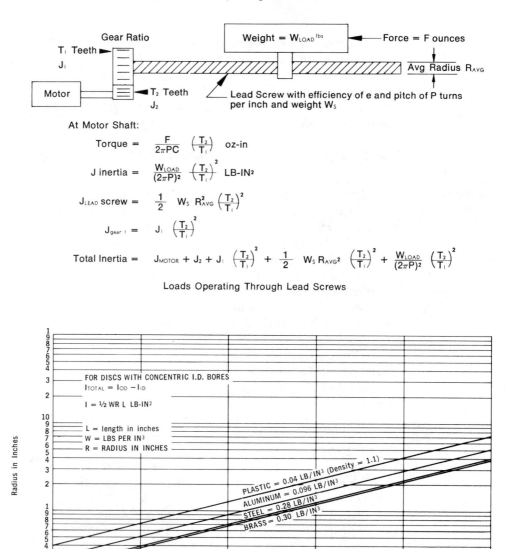

At Motor Shaft:

$$\text{Torque} = \frac{F}{2\pi PC} \left(\frac{T_2}{T_1}\right) \quad \text{oz-in}$$

$$J \text{ inertia} = \frac{W_{LOAD}}{(2\pi P)^2} \left(\frac{T_2}{T_1}\right)^2 \quad \text{LB-IN}^2$$

$$J_{LEAD} \text{ screw} = \frac{1}{2} W_S R_{AVG}^2 \left(\frac{T_2}{T_1}\right)^2$$

$$J_{gear\ 1} = J_1 \left(\frac{T_2}{T_1}\right)^2$$

$$\text{Total Inertia} = J_{MOTOR} + J_2 + J_1 \left(\frac{T_2}{T_1}\right)^2 + \frac{1}{2} W_S R_{AVG}^2 \left(\frac{T_2}{T_1}\right)^2 + \frac{W_{LOAD}}{(2\pi P)^2} \left(\frac{T_2}{T_1}\right)^2$$

Loads Operating Through Lead Screws

FIGURE 3–4 Motor nomogram

new neodymium-cobalt magnet also has high flux-to-volume ratios. A magnet cobalt can be 1/8 the size of a ferrite magnet and provide flux.

Motors The PM motor armature is similar in construc- tor motor. It has an iron core with wound coils placed in in construction lies in the field. Each of the different types ts has slightly different field constructions.

used as a magnet, the field has a structure similar to that . Because they have low coercivity, Alnico magnets must hwise, as shown in Figure 3–5. Magnets can be placed in ructure or the four-pole structure. Although two-pole and nets are popular, six or more magnets have been used in [motor.

agnets are used, a structure such as that shown in Figure ese magnets have a low flux density, the magnet is made r. This construction produces a higher flux density in the r can have a very small *air gap,* which increases the effi-

gnet motors made with rare earth-cobalt magnets often n shown in Figure 3–7. These magnets can be magnetized s in the ferrite magnet. This helps keep the cost of the agnets can be made very thin. A samarium-cobalt magnet of the same length and width has a flux density about twice that of the ferrite magnet.

The conventional PM motor is the motor of choice for electric vehicle applications. It is chosen for its low weight and space requirements, a key feature for efficient vehicle operation. PM magnets are used to power wheel-

FIGURE 3–5 Permanent-magnet field struc- ture with Alnico magnet

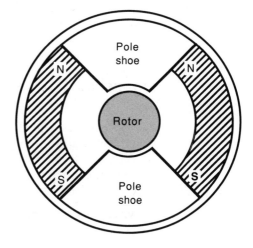

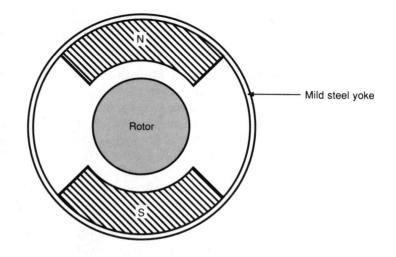

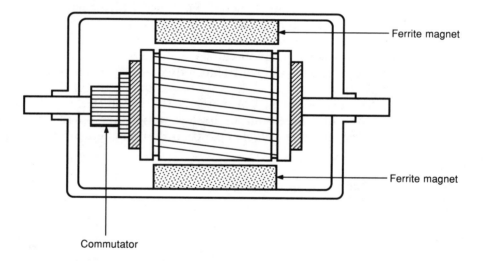

FIGURE 3–6 Ferrite magnet field structure

chairs, forklifts, and electric mopeds. In general, they are preferred for applications demanding high efficiency, high peak power, and fast response.

Figure 3–8 shows a conventional PM motor. This motor is attached to gears that change the rotational speed of the motor shaft. Note the large ferrite magnet encircling the armature.

Moving-coil Motors (MCM) The MCM, although classified as a PM motor, differs in several ways from the conventional PM motor. The primary differ-

FIGURE 3–7 Permanent-magnet field structure using rare earth magnet

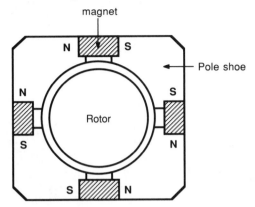

ence is in the armature. The MCM motor is a result of an engineering requirement that motors give high torque and have low inertia. These requirements have been met in the MCM.

The illustrations in Figure 3–9 show the two types of MCMs. As the figure indicates, MCMs are classified by the type of armature construction, the

FIGURE 3–8 PM motor using ceramic magnets (courtesy of Bodine Electric Company)

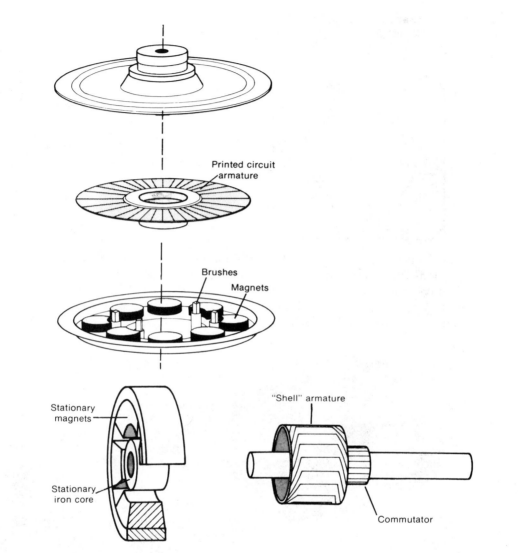

Printed circuit
armature

Brushes

Magnets

Stationary
magnets

Stationary
iron core

"Shell" armature

Commutator

FIGURE 3-9 Two types of moving-coil motors

disk armature and the shell armature. Both of these armatures use the same concept to generate torque; many conductors move in a magnetic field. The main difference between the MCMs and the conventional PM motor is in the armature. The conventional PM motor uses the same type of armature as the wound-rotor motor. The armature is a series of coils wound on an iron base. The iron base makes the PM and wound-rotor motor armatures heavy, slow to respond to changes in applied voltage. The MCM armature does not have any iron. The lack of iron does two things. First, it makes the armature light and gives it a low inertia. Second, lack of iron decreases the electrical time constant by lowering the armature inductance.

The disk armature, shown in Figure 3–9a, has conductors that are flat as if they were placed on a plate. Two types of disc armatures are used in this type of construction. The printed-motor armature has conductors that are stamped from a sheet of copper, welded together, and placed on a disc. The conductor segments are then joined with a commutator at the center of the disc. The figure shows the conductors radiating from the center of the disc. These conductors serve the same function as the armature coils in a conventional DC motor. When current from a supply is passed through them, they generate a field. The field interacts with the permanent-magnet field (from Alnico or ferrite magnets) to produce torque needed to turn the armature. The field, in this case, is generated by two sets of eight permanent magnets arranged on each side of the armature.

The second type of disc motor is called the pancake motor. It uses wire conductors embedded in a disc made of epoxy resin. The function is the same as the printed disk motor.

The second type of rotor, the shell, is pictured in Figure 3–9b. It is a hollow rotor in the shape of a cylinder. The shell is made by bonding copper or aluminum wires together using polymer resins and fiberglass. A closeup of a printed circuit armature is shown in Figure 3–10. Figure 3–11 shows a photo of an MCM. Note the two sets of permanent magnets in the stator; these magnets provide the field in this DC motor. The ironless rotor is a printed circuit type. A view of this motor, assembled, is shown in Figure 3–12.

The low inertia of the armature gives the MCM a higher acceleration capability than any other type of motor. In some cases, the acceleration can be as high as 10,000,000 rpm/s^2. In addition to high acceleration, the MCM runs well at speeds as low as 1 rpm. This low speed is due to the high number of conductors in the armature. The larger the number of conductors, the

FIGURE 3–10

FIGURE 3–11 Exploded view of an MCM (courtesy of PMI Motion Technologies, division of Kollmorgen Corp.)

smoother the torque produced by the armature. You may recall an earlier comparison between the torque produced by a single-coil motor and a double-coil motor. In the disk armature, the number of conductors can run as high as 200. Another advantage of the MCM over other DC motors is in the area of losses. We discussed in Chapter 2 that part of the losses in the DC motor were due to eddy current and hysteresis losses. An armature that has no iron, like the MCM, will not have these losses. This improves the efficiency of the MCM over conventional DC motors, especially small motors. Finally, the MCM has a lower inductance than conventional DC motors. High inductances generate large inductive kicks when the motor commutator disconnects a coil. These large inductive kicks cause arcing, destroying both the commutator and the brushes.

MCMs are gaining popularity because of their advantages over conventional DC motors. Because of their smooth output torque and fast acceleration, they are being used in computer peripheral devices and tape transport systems.

Torque Motors A case can be made that all motors produce torque. All motors could, therefore, be called torque motors. A *torque motor* is different from most other DC motors in that it is designed to run for long periods in a stalled or a low rpm condition. Not all DC motors are designed to operate in this manner. Recall from Chapter 2 that CEMF is directly proportional to armature speed. A low CEMF means that a large amount of armature current will flow. Thus, most conventional DC motors will overheat at low speeds or in a

FIGURE 3–12

stalled condition. Torque motors are designed to be run under a low torque or stall condition, found in applications such as spooling and tape drives. In spooling applications, the tension is often controlled by a torque motor.

Torque motors are found in three major application areas in industry. First, they are found in applications, like spooling, where the motor is stalled with no rotation required. In this case, the torque motor will operate much like a spring that exerts tension or pressure. It can easily be changed to control the amount of tension and the direction of the tension. Second, the torque motor is found in those applications requiring only a few revolutions or degrees of revolution. Examples of this type of application involve opening and closing switches, valves, and clamping devices. Third, torque motors are found in low-speed applications where they must run constantly at a low speed. Reel drives in tape transport mechanisms are a good example of this type of application.

Almost all types of motors, including DC motors, can be specially designed as torque motors. Torque motors are rated by the torque they produce under stalled conditions and by duty cycle. Some torque motors may be allowed to run in a stalled condition only 50% of the time. If the motor is designed to run at a low rpm, the power or HP rating will also be specified. A photograph of a torque motor with a samarium-cobalt permanent magnet is shown in Figure 3–13.

3–3   BRUSHLESS DC MOTORS

Chapter 1 discussed briefly the subject of commutation. We learned that the commutator mechanically switches the current direction in the armature coils. The wound-rotor motor field is divided into several sections. Each field section has an associated commutator segment. By dividing the field windings, the total torque output is smoother. Depending on the design, fractional HP DC motors can have from 7 to 32 commutator bars or segments. Integral HP motors can have over 100 commutator bars. The purpose of these commutator bars is to keep current flowing in the same direction in the field windings. Torque is then produced in one direction only. Commutation essentially converts an AC generator or motor into a DC generator or motor. The commutation speed occurs at all armature speeds.

One of the major problems with wound-rotor DC motors lies in the commutation process. Mechanical commutation places limits on motor performance. If the armature speed is too high, the centrifugal force developed will throw the commutator bars out of the armature. If it overheats, the commutator can bend and change shape. The commutator must also be protected from moisture and dirt particles. The commutator bars are separated by mica washers. To prevent arcing, the voltage across these bars must be kept low. The brushes riding on the commutator must be checked and adjusted occasionally. Commutator wear produces brush dust that can affect the motor's bearings and create a voltage leakage path.

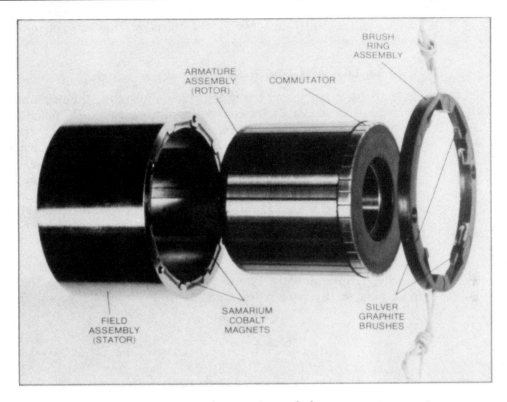

FIGURE 3–13 Torque motor with samarium-cobalt permanent magnet

In addition to these limitations, mechanical commutation has many other disadvantages. Since arcing is always present in the mechanical commutator, the wound-rotor DC motors generate radio frequency noise. Arcing and friction can also limit brush life to less than 2000 hours. These factors increase the maintenance requirements and decrease the reliability of wound-rotor motors.

Since commutation is switching, we should be able to use electronic switching instead of mechanical switching in wound-rotor motors. (***Electronic commutation*** is the use of electronic switches instead of mechanical switches.) We would, however, need to use two transistors for each commutator bar. This would result in a large number of semiconductors, as many as 64 in a fractional HP motor. Such a system would be complex, inefficient, and cost ineffective. An efficient, cost effective design would reduce the number of semiconductor switches and still produce the minimum operating characteristics. Motors using electronic commutation are freed from many of the restrictions of mechanical commutation. Since electronically commutated motors operate without brushes, these motors are called *brushless DC motors (BDCMs)*. They are motors specially designed for this purpose. They are not converted wound-field motors.

3–3.1 BDCM Construction

BDCMs have four basic parts: the rotor, the stator, the commutator, and the rotor position sensor. In the conventional wound-field DC machine, the field is stationary, generated by a permanent magnet or an electromagnet. This field would then be called a stator, since it is stationary. In the normal DC motor, the DC power is fed to the armature, which is free to rotate. In the BDCM, the field rotates, and we call the field the rotor. It is usually a permanent magnet. DC power is supplied to the armature, or stator, which is stationary. These two parts are illustrated in Figure 3–14. The stator coils are labeled A through D; the rotor is a permanent magnet with the polarities shown.

3–3.2 BDCM Operation

Blocks A through D in Figure 3–14 contain the switches that are usually transistor or thyristor pairs. Current from the power supply flows in either

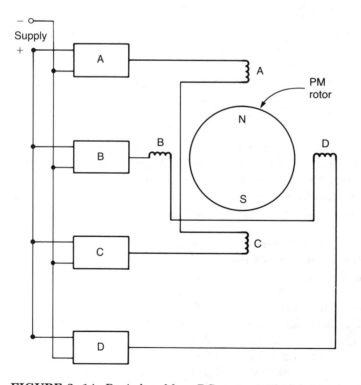

FIGURE 3–14 Basic brushless DC motor with driving circuits

direction through the coils. Let us suppose that current flows up through coils
A and C, placing a north pole at the bottom of C and a south pole at the bottom
of A. The rotor positions itself as in Figure 3–15a. Next, keeping the current
flow in A and C, we switch the current in B and D in the direction shown in
Figure 3–15b. The motor rotor moves to align itself with the total resultant
field, shown by the dotted line. Next, the current in A and C coils is shut off.
The rotor proceeds to the position shown in Figure 3–15c. Notice how the rotor

FIGURE 3–15 Brushless DC motor
turns through 90°

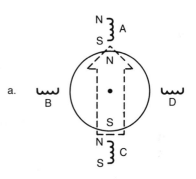

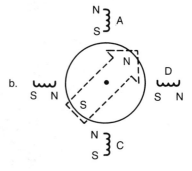

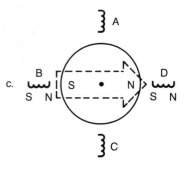

has moved 90°. No other parts in the motor have moved. This principle will be used in our discussion of the AC induction motor. The stepper motor, discussed later in this chapter, works in the same manner.

There is one major difference between the stepper motor and the BDCM. The BDCM uses feedback to switch the current through the right coils at the right time. The thyristors or transistors need firing pulses at the right time depending on the rotor position.

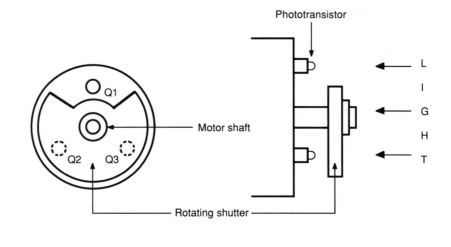

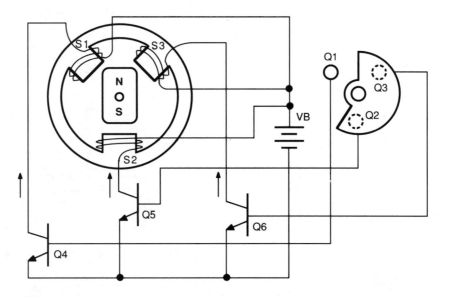

FIGURE 3–16 Three-phase BDCM operation with photoelectric switching

Two types of sensors are used to detect the rotor's position: magnetic and optic. The magnetic sensors use either a coil or a ***Hall effect device.*** The coil, mounted in a stationary position, has voltage induced into it by a passing magnet. The Hall effect device is a sensor that gives an output voltage in the presence of a magnetic field. It is relatively inexpensive and is used in low-voltage and low-power applications. The optic sensor uses a light-emitting diode (LED)-photodiode pair. The LED illuminates the photodiode at the proper time. The photodiode, in turn, switches on the transistor or thyristor that controls the coil current. The optical sensors are small and light and can react very quickly to changes in rotor position.

The operation of the BDCM motor with a phototransistor sensor is shown in Figure 3–16. The three phototransistor sensors are placed on the end plate of the motor at 120° intervals. A shutter attached to the rotor always blocks light from two of the sensors, as seen in Figure 3–16a. As seen in Figure 3–16b, the north pole of the rotor magnet is aligned with pole *S1* on the stator. With the shutter in the position shown, light strikes the base of phototransistor *Q1*, turning it on. Transistor *Q1* turns on transistor *Q4*, causing current to flow in stator coil *S1*. The north pole created at the edge of the pole face will attract the south pole of the permanent magnet rotor, pulling it CCW. The rotor will then align itself with the stator pole *S1*. When the shaft rotates, the shutter will revolve, shading *Q1* and *Q3*. Transistor *Q2* will then be exposed to light and turned on, turning on *Q5*. Current then flows in stator *S2* pole winding, creating a north pole on the pole face. The south pole of the rotor will turn CCW to align with the bottom stator pole *S2*. When the rotor turns to this position, the shutter shades *Q2* and phototransistor *Q3* is exposed to light. The switching continues, turning the rotor continuously CCW.

Since the BDCM produces very little starting torque, it is more useful on low torque loads, such as in blowers and fans. Torque can be increased by using a reduction gearing system. A lower gear ratio will increase torque as well as lower speed.

When the BDCM is used in a low-voltage system, it is not very efficient. The switching circuit, usually a transistor or a thyristor, can drop up to 1.5 V across its output terminals, which reduces efficiency. (Transistors and thyristors as switching devices are covered in depth in Chapter 9.) Since the motor has no brushes, the life of the motor is limited by the life of the bearing lubricant.

3–3.3   Characteristics and Applications of BDCMs

Some of the advantages of BDCMs have already been mentioned. BDCMs have a long life and high reliability due to the lack of brushes and mechanical commutator. No maintenance is needed. Since it lacks a mechanical commutator, the motor's radio frequency interference and noise production are low. BDCMs also respond very quickly to applied voltages. Fast response is due to the low mass of the rotor.

Efficiencies of BDCMs are high, often exceeding 75%. Although some wound-field DC motors are this efficient, it is not unusual for wound-field DC motors to have efficiencies of 30%. The majority of the heat in the BDCM is produced in the stator windings, located on the outside of the motor. This construction makes it easier to "heat sink" or conduct heat away from the stator coils. Heat dissipation is much more difficult in the wound-field motor, since the armature must be free to turn on the inside of the motor.

As in every electronic device, advantages are offset by disadvantages. We have already mentioned the low starting torque of these machines. BDCMs are also expensive because of the semiconductor switching and sensing circuitry. The size of the BDCM is approximately the same size as the conventional DC motor.

Generally, BDCMs have top speeds of 30,000 rpm. Some manufacturers claim speeds as high as 100,000 rpm. These speeds are coupled with high reliability and low maintenance. Two particular examples of applications that need these characteristics are the aerospace and biomedical fields. Biomedical applications include cryogenic coolers and artificial heart pump motors. A good aerospace example is the gyroscope motor. High speeds are necessary, and AC power is often not available. Aerospace applications also demand low maintenance. Remember that several thousand hours of operation is all one can reasonably expect from a wound-rotor DC motor. BDCMs are also used as tape drives for video recorders and high reliability tape transport systems. Some stereo record turntables also use this motor. As the popularity of these motors increases, the price should make them more competitive with conventional drives.

3-4 STEPPER MOTORS

Up to this point, we have discussed devices that are basically analog in nature. By analog, we mean that the motor armature turns at a speed proportional to the input voltage. For example, in a permanent-magnet motor, if we increase the input voltage, the armature speed will also increase. The speed of the permanent-magnet motor can also be continuously varied. The speed does not have to be adjusted in steps.

A motor that takes electrical pulses and changes them into mechanical movement is called a ***stepper (or stepping) motor.*** It is one of the few motors that is essentially digital in nature. A device is digital in nature when it can respond to digital information, a series of high and low voltages. This ability makes it an attractive motor for engineers designing computers and computer peripheral equipment. Conventional motors, as we have seen, rotate continuously when power is applied to them. The stepper motor, when pulsed, rotates or steps in fixed angles. The output shaft (or rotor) rotates through a specific number of degrees for each input pulse. For example, with one input pulse, the rotor may move an angle of 7°. After the pulse is applied, the rotor steps, then

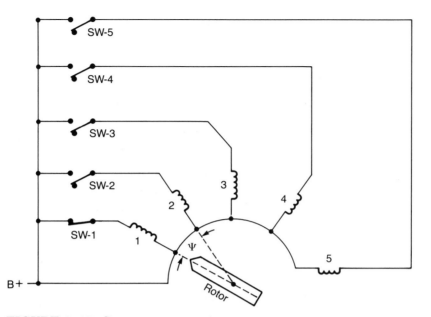

FIGURE 3-17 Stepper motor motion

waits for the next controlling pulse. The angle is repeated precisely with each following pulse. All stepper motors work in this fashion.

The basic operation of a stepper motor is shown in Figure 3–17. The stepper can be compared to a series of electromagnets arranged in a circle. When switch *1* is closed, the permanent magnet will align itself with electromagnet *1*. If that switch is then opened and switch *2* is closed, the permanent magnet will be attracted to electromagnet *2*. If the correct switches are closed in the correct sequence, the permanent magnet will revolve in a complete circle.

3-4.1 Stepper Motor Construction and Operation

There are two basic types of stepper motors: the ***permanent-magnet (PM)*** and the ***variable-reluctance (VR)*** motors.

The Permanent-Magnet (PM) Stepper Motor The PM stepper operates on the reaction between an electromagnetic field and permanent magnet rotor. In its simplest form, the PM unit consists of a two-pole permanent magnet rotor revolving within a four-pole slotted stator. The basic elements of the PM stepper are illustrated in Figure 3–18.

The rotor is actually cylindrical and toothed. The teeth concentrate the magnetic flux. If the flux were evenly distributed, the motor would not have its characteristic and repeatable steps. The bar shape shown in the figure makes it easier to explain its operation. Let us apply DC power to phase *1* in this

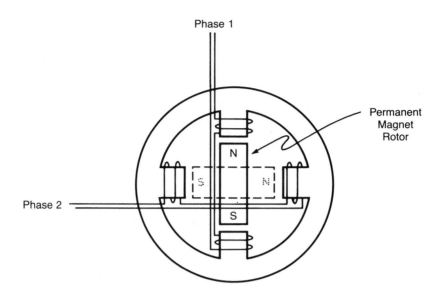

FIGURE 3–18 Permanent-magnet stepper motor

figure so that the coils produce a south pole at the top and a north pole at the bottom. Since unlike poles attract, the rotor will align itself as shown in the figure. Let us then take power from phase *1* and apply it to phase *2* so that the south pole is at the coil on the right, and the north, on the left. The rotor will follow this movement and align itself horizontally as shown in the dotted line. If we were to continue, we would energize the phase *1* coils in the opposite direction. The rotor would move down, in a vertical position, with its north pole down. By changing the excitation sequence to the windings, the motor can be made to operate either CW or CCW.

The stator coils of the PM stepper motor are usually energized with square wave DC pulses. If the pulses are applied in sequence, the rotor can be made to turn like a conventional DC motor. The torque developed by the PM stepper motor is proportional to the current in the stator coils. If more current flows in the stator coils, the electromagnetic field is stronger. This produces more torque. We cannot get an increase in torque if we saturate the stator pole. This proportion works short of *pole saturation* in the stator. One significant difference between stepper motors and conventional DC motors lies in the torque generated by energizing the stator coils. If we keep power applied to the stator coils, the rotor will be held in position by magnetic attraction. This attraction is called *holding torque.* It is the amount of torque needed to break the shaft away from its holding position. Holding torque is measured at the motor's rated voltage and current. The conventional DC motor has no comparable electrical parameter. If we need to hold a conventional motor rotor stationary, we do it mechanically. Another PM stepper motor torque parameter you may encounter is called *residual or detent torque.* It is the rotor's resistance to movement when no current is flowing in the stator windings. This resistance to movement is caused by a CEMF generated in the stator windings by the flux from the permanent magnet moving across the coils. At slow speeds, the resistance is due to residual magnetism.

The rotor of the PM stepper motor must be large in order to produce enough field strength to interact with the stator field. This large rotor construction gives the motor a large inertia. You will recall from basic physics that inertia is an object's resistance to being moved. The large inertia means that the motor will react slowly. This limits the motor's maximum stepping rate.

A new type of permanent-magnet stepper motor has overcome the rotor size and weight problem. The rotor of this permanent-magnet motor is a disk rather than the more typical cylindrical rotor. The rotor is essentially a thin-motor disk (approximately 1 mm) made of a rare-earth magnetic material (Figure 3–19). Because the disk is thin, it can be magnetized with up to 100 individual tiny magnets, evenly spaced around the edge of the disk. Conventional PM steppers are generally limited to a minimum step angle of 30° for a

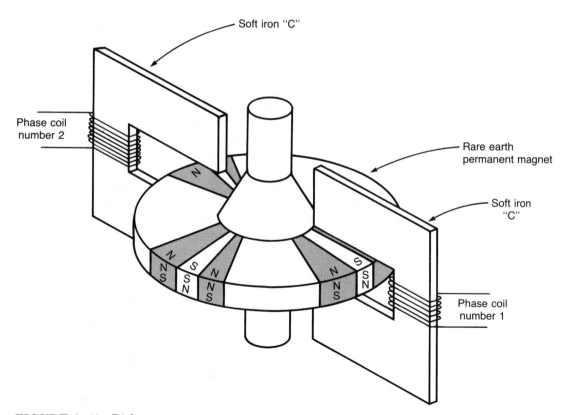

FIGURE 3–19 Disk-type permanent-magnet stepper motor

maximum of 12 steps/revolution. These new thin-disk motors are generally half the size of hybrid motors and weigh 60% less.

The disk is supported on a non-magnetic hub, making up the rotor. The disk magnet is polarized with alternating north and south poles as shown in the figure. A simple C-shaped electromagnetic field pole forms one phase of the motor. A second, identical field pole, offset by half a rotor pole, forms the second phase.

When one of the phases is energized, the rotor will align itself with the electromagnetic field generated. Then, when the first phase is turned off and the second is turned on, the rotor will turn by one-half of a half (or one-quarter) of a rotor pole to align itself with the field from the second phase. To keep the rotor turning in the same direction, the second phase is turned off and the first phase is turned on again. As in other steppers, disk motors are half-stepped by pulsing both coils simultaneously every other step. A photo of a conventional PM stepper is shown in Figure 3–20.

Variable-Reluctance (VR) Stepper Motors Like the PM stepper motor, the VR stepper motor has a stator with several electromagnetic poles. Unlike the PM stepper motor, no permanent magnets are present in the rotor. The VR stepper motor rotor is made of unmagnetized soft steel, with teeth and slots. The teeth are formed in the surface by slots cut parallel to the axis of the cylinder. A diagram showing a typical three-phase VR stepper motor is shown in Figure 3–21. As you can see from the diagram, the stator has fewer teeth than poles. In our example, we have 12 poles and eight teeth. The stator poles are spaced 30° apart, while the rotor poles are spaced at 45° intervals.

If we energize phase *1*, the teeth on the rotor will line up as shown in Figure 3–21a. Note that the teeth under the other unenergized poles are not lined up. Let us now energize the set of poles to the right, which we will call phase *2*. The rotor moves CCW to the new position shown in Figure 3–21b. The rotor moves because the new magnetic path has a higher reluctance than the previous one when the rotor was misaligned. The rotor then moves to the new position to minimize this reluctance. When phase *2* windings were energized, the minimum reluctance point occurred at a different set of poles and teeth.

FIGURE 3–20 Photo conventional PM stepper

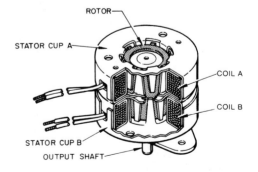

FIGURE 3–21 Variable reluctance stepper
motor

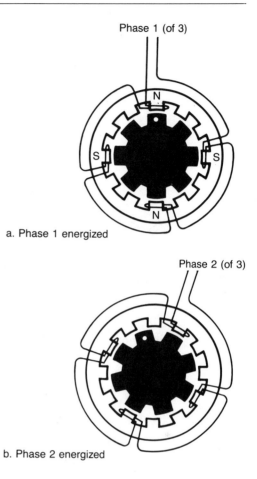

Phase 1 (of 3)

a. Phase 1 energized

Phase 2 (of 3)

b. Phase 2 energized

The stator poles are energized in sequence in a three-phase system. When current is applied to phase *1*, the rotor teeth closest to the four energized (magnetized) stator poles are pulled into alignment. The four remaining rotor teeth align midway between the non-energized stator poles. This is the position of least magnetic reluctance between rotor and stator field.

Energizing phase *2* produces an identical response, the second set of four stator poles magnetically attracting the four nearest rotor teeth, causing the rotor to advance along the path of minimum reluctance into a position of alignment. This action is repeated as the stator's electromagnetic field is shifted in sequence around the rotor. Energizing the poles in a definite sequence gives either a CW or CCW stepping motion.

Since the rotor has little residual magnetism, it has no residual torque. It does, however, have holding torque, just like the PM stepper motor. Torque in the VR stepper motor is proportional to the current squared. Because the rotor

FIGURE 3–22 Cutaway photo of a variable-reluctance stepper motor (courtesy of Warner Electric Brake and Clutch Company)

does not have to be magnetized, it can be made light and small. Its size gives it low inertia, which means that it can respond quickly to applied voltages. It can start, stop, and step faster than the PM stepper motor. A VR stepper is shown in Figure 3–22.

3–4.2 Stepper Motor Parameters

The stepper motor is different from the motors we have discussed so far. Understanding some of the important stepper motor parameters will help us see the differences between the stepper and the other motors we have studied.

In general, stepper motors are chosen for application below 1 kW in power with a maximum speed of approximately 3000 rpm. One of the most important parameters relates to its speed of response. When given a command pulse, the motor will respond within a certain time period. The amount of time it takes for the stepper to respond is called the **step response.** This rating is normally given in milliseconds with the motor unloaded.

Another related parameter is the **stepping rate,** or maximum number of steps per second the motor can make. Typical maximum stepping rates are approximately three hundred steps per second, with a maximum of 800 steps per second. This parameter corresponds to the rpm rating of conventional motors.

We have already mentioned another important parameter—the **step angle.** The step angle is the number of degrees of arc that the rotor moves with a single pulse applied. Normally, only 45° and 90° steps are possible with a two-pole rotor. If the number of poles is increased, the step angles can be made smaller. Most PM steppers have either 12 or 24 poles, giving step angles of

3.75, 7.5 or 15°. Manufacturers can make steppers with other angles by varying the number of rotor and stator poles. The step angle in a variable-reluctance stepper is the difference in angular pitch between stator and rotor teeth. In our VR example, the stator pitch was 30° and the rotor 45°, for a net difference of 15°. The variable-reluctance motor's step angles are small, making finer resolution possible than with the PM type. Related to this parameter is *step angle accuracy.*

Step angle accuracy refers to the amount of error that occurs every time the rotor makes a step. It is usually expressed as a percentage. It is important to note that this parameter refers to the *total* error made by the rotor in a single step movement. This error is not cumulative; i.e., it does not increase as more steps are taken.

The actual speed of the rotor in rpm depends on two of the parameters just discussed, the step angle and the step rate. The rotor turns at an average speed, N, and can be calculated thus,

$$N = \frac{\psi(s/s)}{6}$$

where N is the speed in revolutions per minute, ψ is the step angle in mechanical degrees, and s/s is the number of steps per second. An example may be useful at this point. Let us say we would like to know the speed of a stepper rotor in revolutions per minute. The stepper's stepping rate is 12 steps per second, and it has a 90° stepping angle. The speed is

$$N = \frac{\psi(s/s)}{6}$$

$$= \frac{(90°)(12 \ s/s)}{6} = 180 \ \text{rpm}$$

The nomogram in Figure 3–4 contains important information. We can find the speed in rpm or r/s, step size, and step rate of a stepper motor. A straight edge laid along two points on the nomogram will indicate the speed on the third line. For example, a stepper motor has a step size of 20 steps/revolution and is stepping at 100 steps/second. The rotor speed is approximately 5 r/s or 300 rpm. We can also use this nomogram to find the output power of a stepper motor in HP or W. First, find the motor's speed and torque. Then, lay a straight edge on the two corresponding points on the nomogram. The straight line will intersect the output power line at the appropriate place. For example, let's suppose we were working with a stepper motor with a speed of 100 rpm that produced 8 oz · in at that speed. The motor will produce approximately 65 W at that speed and torque.

The faster we apply pulses to the stepper motor, the faster it steps. With this increase in stepping speed comes an unwanted characteristic called *over-*

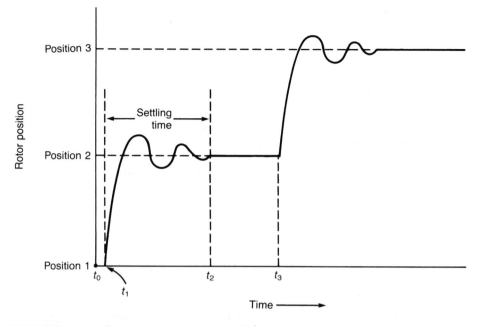

FIGURE 3–23 Stepper motor stepping behavior

shoot. Figure 3–23 illustrates the problem in overshoot. The motor rotor is at rest at time t_0 at position *1*. At time t_1, power is applied to the appropriate stator coils. This action makes the rotor move toward position *2*. The rotor, because it has inertia, does not stop exactly at position *2*. It actually continues on a bit, or overshoots. The magnetic attraction between the two poles makes the rotor come back, but it overshoots again. Typically, the rotor will overshoot several times before it comes to rest at position *2*. This oscillating wave form is called an underdamped response. The time it takes the rotor to come to rest after a pulse has been applied is called the **settling time.** Some applications demand that the rotor come to rest quickly, with a minimum of overshoot. To do this we need to provide **damping.**

3–4.3 Methods of Damping

Resonance is an increase in vibration level and oscillation, caused by the motor running at its natural frequency or a harmonic. Resonance can be a problem with VR steppers and, occasionally, with PM steppers. It can cause the motor to lose step or be unable to follow the step input command. A brief description of the stepper's dynamic movement will be helpful in further defining this condition. As shown in Figure 3–24, when the stepper moves from position *1* to position *2*, the rotor oscillates around its final position. During the time the rotor is oscillating, the kinetic energy of the rotor is dissipated. In PM steppers these oscillations are damped by the rotor's interaction with the sta-

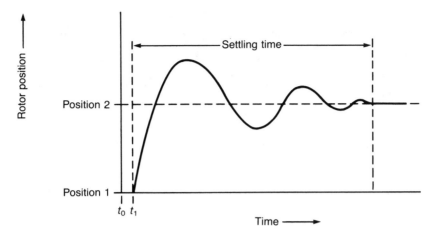

FIGURE 3–24 Stepper motor overshoot

tor magnetic field, eddy currents, and hysteresis loss. However, in VR units with no permanent magnet, eddy currents and hysteresis losses alone do not give much damping effect. If the stator becomes energized while the rotor is in a high undershoot condition, resonance will occur. Under resonant conditions, the rotor will oscillate at random and lose pulse step integrity. This means that the rotor will no longer be following the stator step commands.

Figures 3–25 and 3–26 graphically illustrate this situation in a 15° VR stepper. Rotor pole 2 is in an undershoot position 7.5° away from its final position in line with energized stator winding 2. In this condition, rotor poles 2 and 3 are an equal distance from stator winding 3, 22.5° apart on either side. If

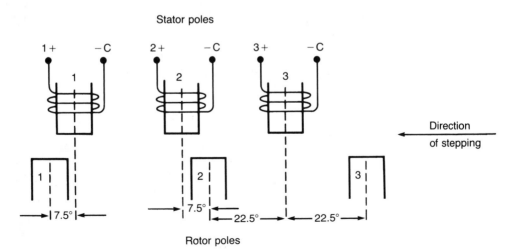

FIGURE 3–25 Stepper motor resonance viewed with respect to stator and rotor poles

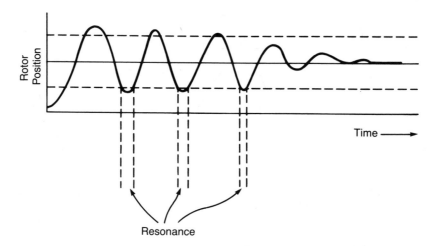

FIGURE 3–26 Resonance in a stepper motor as a function of rotor position

at this precise moment current is switched from stator winding *2* to winding *3*, it is possible for the rotor to step in either direction regardless of the command.

In a 15° VR stepper, resonance can occur at several points, as shown in Figure 3–25. To eliminate resonance in VR steppers, a method of damping can be used. There are many ways to provide damping in a stepper motor. They can be broken down into two classifications: mechanical and electrical. *Slip-clutch damping* and *viscous damping* are two forms of mechanical damping widely used on stepper motors. The other types of damping are electrical.

Slip-Clutch Damping Slip-clutch damping uses a heavy inertia wheel sliding between two collars. As the rotor moves, the inertia wheel resists movement and adds a friction load to the system. This action has the effect of reducing rotor speed, consequently decreasing overshoot and undershoot. The amount of friction is controlled by spring pressure. Wear is reduced by using Teflon discs to separate the steel members.

Many VR steppers are supplied with a rear shaft extension that can be added to this type of damper. Figure 3–27 shows a typical slip-clutch damper

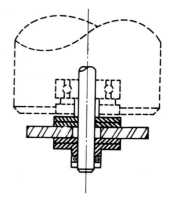

FIGURE 3–27 Slip-clutch damping
(courtesy of IMC Magnetics)

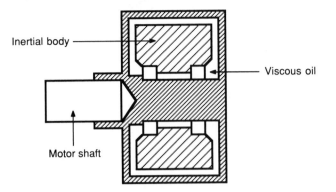

FIGURE 3–28 Viscous damping

installed on the rear shaft of a VR stepper. The disadvantage in using friction to damp resonance is that it will vary as the system wears, slowing down response.

Viscous Damping All objects have *inertia,* a resistance to a change in motion. The viscous damper uses an inertial wheel, surrounded by a viscous oil, inside a cylinder. An inertial body tries to continue rotating at a constant speed. When the rotor starts to oscillate, a difference in speed develops between the rotor and the inertial body. This relative motion between the rotor and the inertial body causes drag forces on both, tending to reduce oscillations. A diagram of a viscous damper is shown in Figure 3–28.

Resistive Damping External resistors across the stator windings (Figure 3–29) allow rated current through one phase while limiting current through the remaining two phases. As a result, a slight reverse torque is applied to the rotor by the two windings not carrying rated current. This prevents the rotor from accelerating as quickly and limits overshoot. Resistive damping increases power consumption an average of 20% for most loads.

Capacitive Damping In place of resistors, capacitors can be used to apply a reverse torque to the rotor as shown in Figure 3–30. At the moment phase *1* is de-energized and field *2* energized, the capacitor on phase *1* slowly discharges. The discharging current applies reverse torque to the rotor. This action is repeated as the remaining phases are shifted. This method offers the advantage to resistive damping of lower power consumption.

Two-Phase Damping In this method of damping, two stator windings are excited simultaneously, causing the rotor to step to a minimum reluctance position, midway between the stator poles, as shown in Figure 3–31. While advancing to the final step position, the two adjacent rotor teeth exert equal and opposing torque, which in combination with the stator's magnetic field, produces twice the damping effect of single-phase excitation.

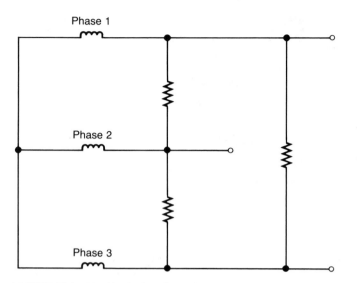

FIGURE 3–29 Resistive damping

The disadvantage of two-phase damping is that power consumption is approximately double that of single-phase excitation, while speed and torque delivery remain essentially unchanged.

Retrotorque Damping A good method of damping the VR stepper is to use a retrotorque controller, which drives the VR stepper in single-phase mode and provides damping by supplying a pulse of power to the stator winding previ-

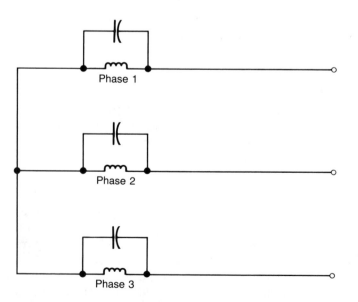

FIGURE 3–30 Capacitive damping

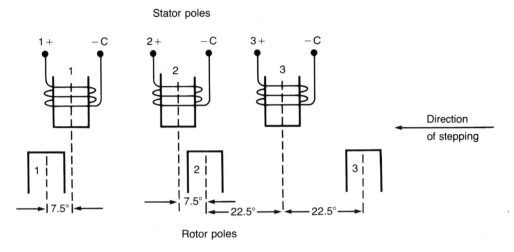

FIGURE 3–31   Two-phase damping

ously (last) energized. Both overshoot and undershoot tendencies are reduced.

A look at Figure 3–32 shows how this type of damping works. We apply a pulse at time t_1. At time t_2, we apply a reverse pulse that produces a reverse torque or retrotorque. This pulse absorbs much of the power that would have gone into the overshoot. A final forward pulse is applied at time t_3. This final pulse moves the rotor to position 2 with a minimum of overshoot. Today, microprocessors control the timing sequences in retrotorque stepper motor damping.

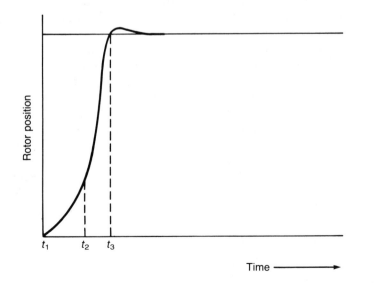

FIGURE 3–32   Retrotorque damping

Retrotorque damping does not increase power consumption or noticeably affect performance. The controller's greater complexity and increased expense may be a factor for consideration.

3–4.4 Stepper Modes of Operation

Up to this point, we have assumed that the stepper motor has only one mode of operation. In this section, we will see that it has a number of operating modes.

The first operating mode is the *rest mode.* When the stator is not energized, PM steppers (due to the interaction between the permanent magnet rotor and the stator) have a resistance to movement called *detent torque.* Sometimes called position memory, this characteristic is valuable when final position must be known in the event of electrical failure in a system. VR steppers do not have this characteristic.

The *stall mode* exists when the stator winding is energized, but the rotor is not moving. Both VR and PM steppers resist rotor movement. Some data sheets call this parameter holding torque or static stall torque.

The third operational mode is called the *bidirectional mode.* Stepper data sheets indicate this mode by the maximum speed at which a given load can be run bidirectionally without losing a step. The maximum load that the motor can drive occurs at a rate of approximately five steps per second. It is listed in some data sheets as the maximum running torque speed. The maximum speed, in steps per second that a stepper will run in bidirectional mode, occurs with no load.

The last operational mode is called *slewing.* In bidirectional mode, the shaft advances, stops momentarily, starts again, stops for a moment, and so on. The direction of rotation can be reversed instantaneously. The motor can be accelerated beyond this range in a unidirectional slewing mode. The rotor pulls into synchronism with the rotating stator field, much like a standard synchronous motor. When operated in the slewing mode, the motor is rotating continuously and is not stepping. The goal in the slewing mode is speed control, not position control. Since the motor, in this mode, is beyond its bidirectional start-stop speed range, it cannot be instantaneously reversed and still maintain pulse and step integrity. Nor can the motor be started in this range. To attain slewing speed, whether from rest or from bidirectional mode, the motor must be carefully accelerated, a condition called *ramping.* In a similar way, to stop or to reverse in slewing mode, the motor must first be carefully decelerated to some speed within its bidirectional capability. When ramping is supplied for acceleration and deceleration, there is no loss of pulse-to-step integrity in slewing mode. Each stepper's modes of operation—detent, stall, start-stop, bidirectional, and slew—are shown in Figure 3–33a and 3–33b.

3–4.5 Excitation Modes

Stepper motors have great flexibility in excitation as well as operation. Changing the excitation mode can change the maximum power output, maximum response, minimum power input, and maximum efficiency.

FIGURE 3–33 Stepper motor modes of operation

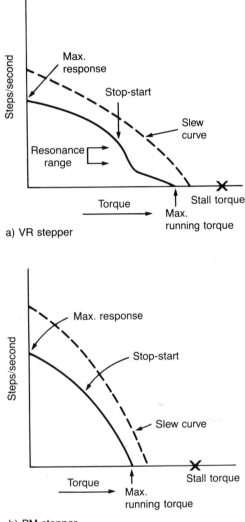

a) VR stepper

b) PM stepper

Depending on the stator winding and performance desired, a stepper motor can be excited in several different modes: two-phase and two-phase modified, three-phase and three-phase modified, and four-phase and four-phase modified. *Phase* refers to a stator winding; *modified* means that two windings are driven simultaneously.

Some stepper motors are constructed with two stator windings that are center-tapped, as shown in Figure 3–34.

When the center taps are connected to the controller, this is a four-phase motor. When the center taps are either eliminated or open-circuited and only the end taps connected, this is a two-phase motor.

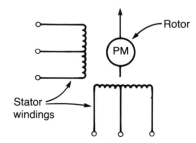

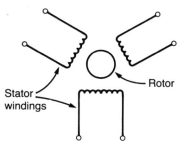

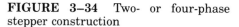

FIGURE 3–34 Two- or four-phase stepper construction

FIGURE 3–35 Three-phase stepper

Many steppers, due to geometry and construction, lend themselves to a convenient "Y" (wye) winding arrangement, shown in Figure 3–35. They are excited in three-phase or three-phase modified mode.

Two-Phase Excitation Mode In two-phase excitation, one entire phase of the motor (end-tap to end-tap) is energized at a given moment in time. Compared to standard four-phase excitation, rotor resistance is doubled. Input current and wattage are, therefore, halved. Heat dissipation is also increased, since more copper of the motor winding is used. Because of the reduced input and greater heat dissipation, the motor's output can be improved by as much as 10% over the standard four-phase mode of excitation. See Table 3–1 for input sequence and rotor position.

Two-Phase Modified Mode In this mode, both windings (i.e., considering one winding to be end-tap to end-tap, ignoring the center taps shown in Figure 3–34) are energized simultaneously. Note that in the first position, 3–1 and 6–4, current flows in one direction in the 3–1 winding in the stator coil. Dur-

TABLE 3–1

Excitation Mode	Energized Winding (Fig. 3–36)	Rotor Position (Fig. 3–36)	Motion Sequence
2-phase commutation of B+ and B–	3–1	f	Index
	6–4	h	CCW
	1–3	b	CCW
	4–6	d	CCW
2-phase modified commutation of B+ and B–	3–1 & 6–4	g	Index
	1–3 & 6–4	a	CCW
	1–3 & 4–6	c	CCW
	3–1 & 4–6	e	CCW

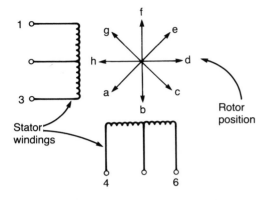

FIGURE 3–36 Two-phase excitation

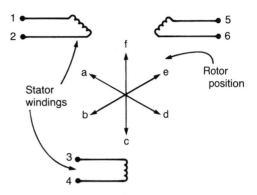
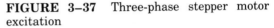

FIGURE 3–37 Three-phase stepper motor excitation

ing the next step, the current flows in the opposite direction in the stator coil, seen in the 1–3 notation. Energy input in this mode is the same as standard four-phase mode. Torque output performance, however, is increased by approximately 40%. The controller is more complex and costly for this mode. Note the need for a polarity reversal switch in Table 3–1. With the rotor in position *g*, the windings 3–1 and 6–4 are energized. In position *a*, 3–1 is reversed to 1–3 indicating that current is flowing in the opposite direction in the winding. A polarity reversal switch will cause current to flow in the opposite direction in the winding.

Three-Phase Excitation The three-phase excitation system is very popular in stepper motor driving circuitry. In this excitation mode, stator fields are excited one phase at a time. The energizing sequence and consequent motion are shown in Table 3–2.

TABLE 3–2

Excitation Mode	Energized Winding (Fig. 3–37)	Rotor Position (Fig. 3–37)	Motion Sequence
3-phase commutation of B+ only	2–1 3–4 5–6	a c e	CCW CCW CCW
3-phase modified commutation of B+ only	2–1 & 3–4 3–4 & 5–6 5–6 & 2–1	b d f	CCW CCW CCW

TABLE 3–3

Excitation Mode	Energized Winding (Fig. 3–38)	Rotor Position (Fig. 3–38)	Motion Sequence
4-phase commutation of B+ only	2–1	f	Index
	5–4	h	CCW
	2–3	b	CCW
	5–6	d	CCW
4-phase modified commutation of B+ only	2–1 & 5–4	g	Index
	2–3 & 5–4	a	CCW
	2–3 & 5–6	c	CCW
	2–1 & 5–6	e	CCW

Three-Phase Modified Mode   In this mode, the two adjoining phases of a three-phase motor are excited simultaneously. The rotor steps to a minimum reluctance position corresponding to the resultant of the two magnetic fields. Since two windings are excited, as shown in Table 3–2, twice as much power is required as in the standard mode. You will recall that the standard mode energizes one phase at a time. The output torque is not significantly increased, but damping characteristics are noticeably improved.

Four-Phase Excitation Mode   In this mode of excitation, each half winding (end-to-center-tap, Figure 3–34) is regarded as a separate phase. The phases are energized one at a time. The sequence of excitation is shown in Table 3–3.

Although the four-phase mode of excitation is less efficient than others, the controller is not very complicated, being a simple four-stage ring counter.

Four-Phase Modified Mode   Phases (halves) of different windings are simultaneously energized in this mode, as shown in Table 3–3. Since two phases are simultaneously excited, twice the input energy of single-phase excitation is required. Torque output is increased by about 40%, and the maximum response rate is also increased, compared to the single-phase excitation.

FIGURE 3–38   Four-phase excitation

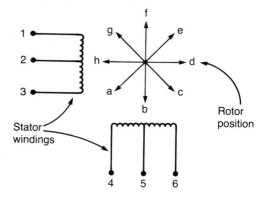

3-4.6 Applications of Stepper Motors

Stepper motors are used in applications where precise positioning is required, especially in combination with microprocessor control. The diagrams in Figure 3–39a, b, and c show three common applications in which stepper motors are used today. Figure 3–39a shows a stepper motor driving a tractor feed mechanism for a computer printer. On command from a computer, the stepper motor will advance the paper by a specified amount. Figure 3–39b shows two stepper motors controlling an X-Y table. The table can be positioned in any combination of X-Y coordinates by pulsing the X and Y stepper. Figure 3–39c shows a stepper motor controlling the head of a computer disk drive. Proper pulsing of the stepper will allow the head to be positioned over any track on the disk drive.

3-4.7 Advantages and Disadvantages of Steppers

Stepper motors offer significant advantages over other types of motors:

1. Feedback is not ordinarily required when a stepper is properly applied, as it is in the BDCM. The stepper is, however, perfectly compatible with either analog or digital feedback, whether for velocity control, position control, or both.

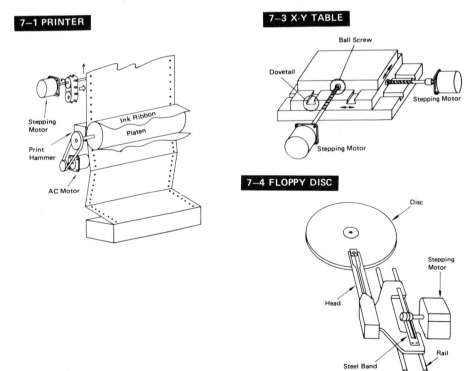

FIGURE 3–39 Various stepper applications

2. Error is non-cumulative as long as pulse-to-step integrity is maintained. A stream of pulses can be counted into a stepper, and its final shaft position can be known within a very small percent of one step. DC servo accuracy, in contrast, is subject to the sensitivity and phase shift of the loop.

3. Maximum torque occurs at low pulse rates. The stepper can, therefore, accelerate its load easily. When the desired position is reached and command pulses cease, the rotor shaft stops. (If there is too much inertia, overshoot can occur.) There is no need for clutches and brakes. Once stopped, there is little tendency to drift. In fact, PM steppers are magnetically detented in the last position. A load can be started in either direction, moved to a position, and it will remain there until commanded again.

4. A wide range of step angles is available off-the-shelf for most manufacturers: 1.8°, 7.5°, 15°, 45°, and 90° angles.

5. Low velocities are possible without gear reduction. For example, a typical stepper driven at 500 pulses per second turns at 150 rpm, a very low velocity.

6. Steppers are truly digital in nature. They do not require a digital-to-analog conversion at the input, as do conventional servos, when used with a computer or digital system.

7. They offer close speed control and can be reversed over a wide range.

8. Their starting current is low.

9. The stepper rotor moment of inertia is usually low.

10. Multiple steppers driven from the same source maintain perfect synchronization.

These advantages are offset by certain disadvantages:

1. The efficiency of a stepper is low. Much of the input energy must be dissipated as heat.

2. Loads must be analyzed carefully for the best stepper performance. Inputs (pulse sources and controllers) must also be matched to the motor and load.

3. Resonance can be a problem with VR motors and, in rare instances, with PM or special PM types when load inertia is exceptionally high. Damping may be required.

CHAPTER SUMMARY

- The field of a PM motor is produced by a permanent magnet.
- Stator magnetic flux in a PM DC motor is constant.

- PM DC motors are classified into the following three groups: 1) conventional, 2) moving coil (MCM) and 3) torque.
- Permanent magnets are made of three different materials: 1) Alnico, 2) ceramic or ferrite and 3) rare earth-cobalt (usually samarium cobalt, more recently neodymium cobalt).
- MCMs differ from conventional PM motors in the construction of the armature.
- MCMs are classified into two basic groups: disc and shell.
- MCMs have high rotation speeds and high accelerations.
- Torque motors are designed to run at low speeds.
- BDCMs use electronic commutation rather than conventional brush-commutator construction.
- BDCMs have low maintenance requirements due to lack of brushes.
- BDCMs use either photodetectors or Hall effect devices to detect the position of the rotor.
- Unlike most motors, which are analog in nature, stepper motors are primarily digital in nature.
- Stepper motors are classified into two basic types: PM and VR.
- Oscillations in stepper motor rotors are reduced by damping methods.

QUESTIONS AND PROBLEMS

1. Define a servo motor.
2. What is the difference between a permanent-magnet DC motor and a wound-field DC motor?
3. We apply 10 V to a PM motor and measure an armature current of 1 A at 1500 rpm. After loading the motor, we slow it to 1000 rpm and measure an armature current of 3 A. What is the factor of merit of this motor? The maximum motor efficiency? The input current at maximum efficiency? The output power at maximum efficiency in watts and horsepower? The speed at maximum efficiency? The output torque at maximum efficiency?
4. How are PM motors classified?
5. A PM DC motor is turning at a speed of 100 rpm and producing a torque of 10 N · m. How much power is the motor producing? (use the nomogram in Figure 3–4)
6. Describe the construction of the two types of moving coil motors. What advantage does the MCM have over conventional DC motors?
7. What is a torque motor?
8. What is electronic commutation? How is it achieved in a BDCM?

9. Name the four parts of a BDCM.

10. Using Figure 3–8, describe how the BDCM works.

11. What kinds of sensors are used in the BDCM? What is their purpose?

12. What is a stepper motor? How is it different from other DC motors?

13. Describe the construction and operation of the two types of stepper motors.

14. Define the following terms relating to a stepper motor.
    a. holding torque
    b. residual torque
    c. step response
    d. stepping rate
    e. step angle
    f. step accuracy
    g. overshoot
    h. settling time
    i. damping

15. A stepper motor has a step size of 100 s/rev and a step rate of 1000 s/second. Using the nomogram in Figure 3–4, find the speed in rpm.

16. With the same speed as in the previous problem, the torque produced is 10,000 g · cm. What is the output power?

4

TRANSFORMERS

At the end of this chapter, you should be able to

- describe the procedure for checking the polarity of a transformer with a voltmeter.
- identify the parts of a hollow and laminated-core transformer.
- calculate turns ratio and primary and secondary voltages and currents.
- describe the losses in a transformer.
- draw the schematic diagram for and explain the operation of the
  - autotransformer
  - multiple-winding transformer
  - saturable-core reactor.
- draw the following three-phase connections using single-phase transformers:
  - delta-delta
  - wye-wye
  - delta-wye
  - wye-delta.
- calculate the efficiency of a transformer.
- describe maintenance procedures on transformers.

4-1 INTRODUCTION

The transformer, like the motor, is a power conversion device. The motor converts an electrical input to torque, a form of mechanical energy necessary to rotate mechanical loads. The transformer, however, does not produce mechanical energy. It produces an AC electrical output voltage with a given AC input voltage. The transformer is a device that converts electrical power from one voltage-current level to electrical power at another voltage-current level. Some loads require voltage and current levels different from those produced by

FIGURE 4–1 The transformer as a power conversion device

a source (like the 120-VAC line) to operate properly. The transformer is widely used to produce these required voltages and currents. It should be stressed at this point that the transformer is a power conversion device, as shown in Figure 4–1. The transformer does not produce a power gain; nor does it change the input frequency. It transforms the available input power to an output suitable to drive an electrical load.

Because it has no moving parts, the transformer is rugged, reliable, and efficient. It requires little preventive and corrective maintenance, an important consideration in industry. Because of its efficiency of 90% or higher, the transformer has contributed to the widespread use of AC as a primary power source. Direct current (DC) was used in the early days of electrical power generation until it was discovered that power was more efficiently transmitted if it was high voltage AC. The transformer today is used to convert the high voltage of transmission to a lower voltage for the end user.

4–2 TRANSFORMER PRINCIPLES

The transformer transfers energy from one circuit to another by electromagnetic induction. As shown in Figure 4–2a, the typical transformer has two windings, the primary and the secondary, that are electrically separate from each other. The input is normally placed across the primary winding, while the load is connected to the secondary winding. Note that the windings, although separated electrically, are connected magnetically through a common core. That is, they are wound on the same core. (The core is normally made of iron or an alloy that contains iron.) If we connect an AC generator to the primary with the instantaneous polarity shown, current will flow in the primary winding. (We will leave the secondary open at this time so that it has little effect on our analysis). As current flows in the primary winding, magnetic flux will be created. The primary winding can then be seen as a simple inductor. The primary offers inductive reactance to the source. Depending on the size of this inductive reactance, current will flow in the secondary, causing flux to be created.

The expanding magnetic flux lines cut the secondary windings producing the voltage shown (V_s). Note that the voltage induced into the secondary is 180° out of phase with the primary. This relationship of voltages is shown in the vector diagrams in Figure 4–2b. Note that the voltage V_p is 180° out of phase with the secondary voltage V_s. Note also that the current produced in the primary winding I_p is small and lags V_p by 90°. Recall that the current through an inductor lags the voltage across it by 90°. The current I_a is small because the circuit is highly inductive. (It is also small because there is no load on the secondary.) This current is called the magnetizing or **excitation current** because it produces the magnetic flux in the core.

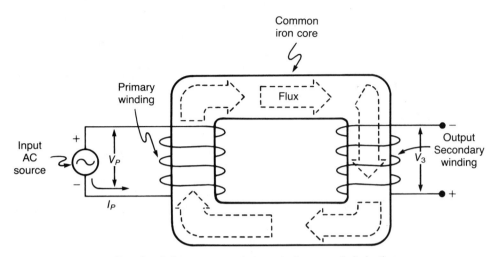

a. Functional diagram - transformer electromagnetic induction

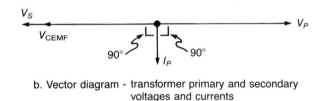

b. Vector diagram - transformer primary and secondary
voltages and currents

FIGURE 4–2 Transformer operation—no load current flowing

The expanding flux in the core does more than create a voltage in the secondary. It also cuts the primary windings, producing a counter EMF, just as in the DC motor. The voltage V_{CEMF} induced in the primary is 180° out of phase with the primary voltage and in phase with the secondary voltage, as can be seen in Figure 4–2b. The values are only approximate and are not drawn to scale.

Because no current is flowing in the secondary of the transformer shown in Figure 4–2, the primary current I_p produces all the flux in the core. The transformer is acting like a simple iron core inductor, providing a lot of inductive reactance to the source but not much resistance. The amount of current flowing in the primary depends on how much inductive reactance the primary voltage sees.

4–2.1 Loaded Transformer

Figure 4–3 shows what happens in the transformer when a load is connected to the secondary. When switch SW-1 is closed, the current I_s will flow in the secondary. The amount of current flow in the secondary will depend on the size of the induced secondary voltage, V_s, and the impedance of the secondary

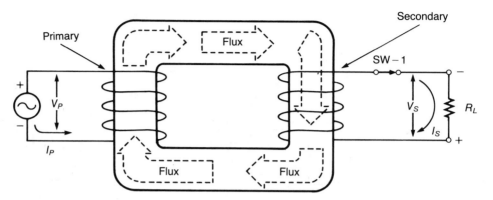

a. Functional diagram - transformer electromagnetic induction

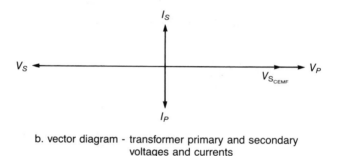

b. vector diagram - transformer primary and secondary
voltages and currents

FIGURE 4–3 Transformer operation—with load current flowing

winding. The current flow in the secondary generates flux that opposes the flux generated by the primary current. This action reduces the total core flux, which decreases the impedance of the primary. The decrease in primary impedance causes more primary current to flow. Any changes in secondary load current will cause corresponding changes in primary current. For example, if the load resistance decreases, the secondary load current will increase, causing more current to flow in the primary circuit. A decrease in secondary impedance causes a corresponding decrease in primary impedance.

4–2.2 Mutual Flux

At any time, the total flux in the core of a transformer is common to both primary and secondary. This common flux is called **_mutual flux._** The inductance that produces this flux is called **_mutual inductance._** The amount of mutual inductance that exists between primary and secondary windings depends on the mutual flux linking the windings. In an iron core transformer, the flux lines produced by the transformer action are confined to the iron core. Almost all of the flux is considered to be mutual flux. If the iron core is removed, however, the flux is no longer confined to a path linking the two windings. The magnetic flux will take the path of minimum reluctance through the air separating the windings. Not all the flux produced will be used to link the windings. The mutual flux between the windings in an air core transformer is considerably less than a comparable iron core transformer. To maximize mutual flux, the secondary winding is frequently wound directly on top of the primary winding. The only separation between primary and secondary windings is the insulation surrounding the wire. The amount of flux linkage between the windings is represented by a factor called the ***coefficient of coupling (k).*** When perfect linkage exists (that is, when all of the flux produced by the primary coil is cut by the secondary and vice versa), $k = 1$. Since it is impossible to have all the flux produced link the two windings, k can never equal 1. Some flux produced is effectively lost to the system. This lost flux is called flux leakage. When using high permeability material for the core and proper winding techniques, coupling coefficients can approach 0.97.

4–2.3 Phase Relationships in Transformers

The secondary voltage of a transformer can either be in phase or 180° out of phase with the primary voltage. The actual phase relationship depends on the orientation of the transformer windings. The phase relationship between primary and secondary is indicated in schematic diagrams by a dot convention, shown in Figure 4–4. In Figure 4–4a, both the primary and secondary are wound in a CW direction from top to bottom, if we take a top view of the windings. When in this configuration, the tops of both the primary and secondary have the same phase relationship. For example, if the top of the primary is going positive, the top of the secondary will be in phase and also going positive.

In Figure 4–4, we see the primary wound in a CW direction and the secondary wound in a CCW opposite direction. Since the expanding flux in the primary cuts the secondary in the opposite direction, the polarities will be opposite. When the top of the primary goes positive, the top of the secondary goes negative. Note the placement of dots in Figure 4–4, indicating the different phase relationships primary-to-secondary.

4–2.4 Voltage and Turns Ratio

The expanding field from the primary winding cuts the secondary winding, inducing a voltage there by electromagnetic induction. The amount of voltage induced in the secondary depends on the ratio of the number of turns in the primary to the number of turns in the secondary. The secondary voltage produced also depends on the amount of voltage applied to the secondary. Referring to the diagram in Figure 4–5, we can see that the primary winding is made of ten turns of wire. The secondary consists of one turn of wire. As the magnetic flux lines expand and collapse, they cut both the primary and secondary windings. Since the length of one turn in the primary is approximately the

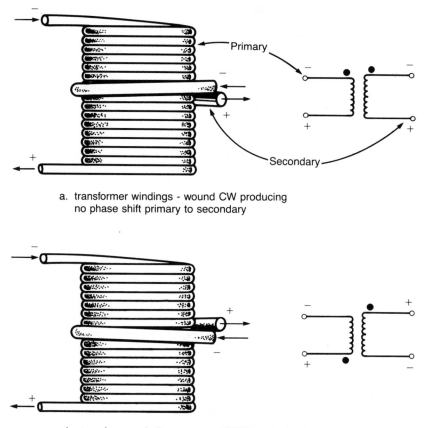

a. transformer windings - wound CW producing
   no phase shift primary to secondary

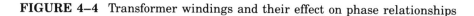

b. transformer windings - wound CCW producing
   180° phase shift

FIGURE 4–4 Transformer windings and their effect on phase relationships

same in the secondary, the EMF induced in the secondary will be the same as the EMF induced in each turn in the primary. Each turn in the primary will then have approximately 1 V per turn or loop of primary winding. Since the same flux lines cut both primary and secondary, each turn of the secondary will also have the same ratio of volts per turn, approximately 1 V per turn. Since the secondary has only one turn, it produces 1 V of potential.

The transformer in Figure 4–5 shows a secondary made of two turns of wire. If we apply the same 10-V potential to the ten turns of the primary, how many volts per turn will we induce in the secondary? Since we have the same number of volts per turn in the primary, we will have that same ratio in the secondary, 1 V per turn. The secondary will produce a total of 2 V. Since the CEMF in the primary is approximately equal to the applied voltage, we can set

FIGURE 4–5 Transformer windings and their effect on turns ratio

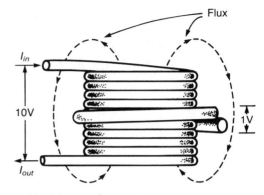

a. 10 : 1 turns ratio

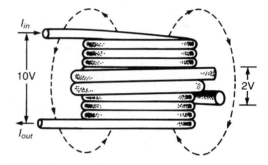

b. 10 : 2 turns ratio

up a ratio to express the value of the voltage induced in the secondary in terms of the primary voltage and the number of turns in each winding. This ratio, a, is expressed by the following equation,

$$a = \frac{V_p}{V_s} = \frac{N_p}{N_s} \qquad \text{(eq. 4–1)}$$

where a is the ***turns ratio***, V_p is the voltage applied to the primary, V_s is the voltage induced into the secondary, N_p is the number of turns in the primary, and N_s is the number of turns in the secondary. Transposing this formula for V_s,

$$V_s = \frac{N_p V_p}{N_s} = \frac{1}{a} V_p \qquad \text{(eq. 4–2)}$$

and

$$V_p = a V_s \qquad \text{(eq. 4–3)}$$

We can use this equation to solve for the induced secondary voltage, given the other three values. For example, let's say we were working with a transformer with 200 turns in the primary, 50 turns in the secondary ($a = 4$), and 120 V AC applied to the primary. What voltage could we expect across the secondary? Using equation 4–2, we see that:

$$V_s = \frac{V_p}{4} = \frac{120 \text{ V}}{4}$$

$$= 30 \text{ V}$$

In this case, we have a ratio of 200:50 primary-to-secondary turns. This works out to a ratio of four to one, or 4:1, as it is sometimes symbolized. Using a for our turns ratio, we can also say that $a = 4$.

The transformer in the example above had fewer turns in the secondary compared to the primary. This means that less voltage is induced in the secondary than is applied to the primary. A transformer in which the voltage induced in the secondary is less than that applied to the primary is called a *step-down transformer.* A transformer with fewer turns in the primary than in the secondary will produce a higher voltage in the secondary than is applied to the primary. This type of transformer is called a *step-up transformer.* If a transformer primary had four times more windings in the secondary than in the primary, the secondary voltage would be four times higher than the primary. This ratio would be called a one-to-four ratio, symbolized by 1:4, or $a = 0.25$.

4–2.5 Current and Turns Ratio

The number of flux lines produced in the core is proportional to the magnetizing force of the primary and secondary. This magnetizing force is measured in ampere-turns. The *ampere-turn* is a measure of EMF and is equal to current times the number of turns of an electromagnet's coil. One ampere-turn is defined as the *magnetomotive force* (**MMF**) or magnetizing force developed by one ampere of current flowing in a coil of one turn. The flux produced in the core surrounds both the primary and the secondary windings. Since the flux is the same for both windings, the ampere-turns for both primary and secondary are the same. We can, therefore, write the following equation:

$$I_p N_p = I_s N_s \qquad \text{(eq. 4–4)}$$

where $I_p N_p$ is the ampere-turns in the primary windings, and $I_s N_s$ is the ampere-turns in the secondary windings. By substituting equation 4–2 into equation 4–4, we can arrive at the following equations:

$$\frac{V_p}{V_s} = \frac{N_p}{N_s} \qquad \text{(eq. 4–5)}$$

and

$$\frac{V_p}{V_s} = \frac{I_s}{I_p} = a$$
(eq. 4–6)

Note that these equations show that current ratio is the inverse of the voltage ratio. This means that a transformer with more turns in the secondary than in the primary would step up the voltage but step down the current. As an example, a transformer has a 6:1 ratio and a primary current of 200 milliamperes (ma). We can find the secondary current using equation 4–6.

$$\frac{V_p}{V_s} = \frac{I_s}{I_p}$$

Rearranging, we get

$$I_s = \frac{V_p I_p}{V_s}$$
(eq. 4–7)

Substituting 6:1 and 200 milliamperes (ma), we get

$$I_s = \frac{V_p I_p}{V_s} = \frac{(6)\,(200\ \text{ma})}{1}$$

$$= 1.2\ \text{amps}$$

We can see that a step-down transformer (6:1) steps up the current in the secondary by the same ratio, 6:1.

4–2.6 Transformer Impedance and Turns Ratio

Let us consider what impedance is "seen" by a source looking into the primary terminals, with a secondary load impedance, Z_s, connected, as shown in Figure 4–6. We know that the impedance, Z, of a device is equal to the voltage across the device divided by the current through it. Thus,

$$Z_p = -\frac{V_p}{I_p}$$
(eq. 4–8)

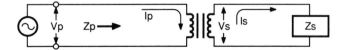

FIGURE 4–6 Impedances in a transformer

This equation holds true for the secondary also:

$$Z_s = \frac{V_s}{I_s}$$

(eq. 4–9)

By substitution, we find Z_p in terms of secondary voltages and currents:

$$Z_p = \frac{a\,V_s}{\dfrac{I_s}{a}} = a^2\,\frac{V_s}{I_s} = a^2\,Z_s$$

(eq. 4–10)

An example may help to clarify this concept. We have a transformer with a secondary current of 1.2 amps, a secondary voltage of 1 V, and a turns ratio of 6:1 ($a = 6$). Using equations 4–9 and 4–10, the secondary impedance will be

$$Z_s = \frac{V_s}{I_s} = \frac{1\text{ V}}{1.2\text{ A}} = 0.833\ \Omega$$

$$Z_p = a^2\,Z_s = (6)^2\,(0.833\ \Omega) = 30\ \Omega$$

It is obvious from this equation that the secondary impedance, caused by the secondary load, determines the impedance of the primary. A shorted secondary can, therefore, reflect a low impedance back into the primary. The primary can then draw dangerously high currents.

4–3 TRANSFORMER CONSTRUCTION

Since we now have an understanding of how a transformer works, we can focus on how actual transformers are constructed. The transformer is one of the simplest power conversion devices. It consists of two coils of wire wound around a core material that is usually either air or iron. In both cases, the parts of the transformer can be summarized as follows:

- The primary winding is connected to the AC source and provides power to it.
- The secondary winding is connected to the load and provides power to it.
- The core provides a path for the magnetic flux, linking the primary and secondary.

4–3.1 Transformer Windings

As stated in the preceding description, the primary winding of the transformer is connected to the AC source. Figure 4–7 shows an exploded view of the shell-type transformer. It gets its name from the cardboard form used as a support

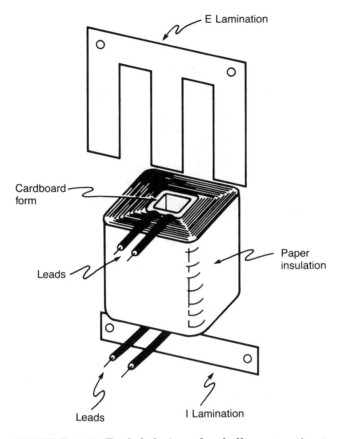

FIGURE 4–7 Exploded view of a shell construction transformer

for the primary and secondary windings. Note how the laminations fit around the windings. The construction of the primary and secondary windings is shown in the cutaway view of this same transformer in Figure 4–8. The primary windings are the smaller gauge wires at the center of the transformer. These primary wires are usually coated with varnish to electrically insulate the wires from each other. In high voltage transformers, paper insulation is also added between layers of windings to give additional insulation. After the primary is completely wound, the manufacturer covers it with a layer of insulating material, usually paper. The secondary winding, shown as the heavier wires in Figure 4–8, is then wound on top of the primary. The manufacturer coats these secondary windings with varnish to keep them from short circuiting other windings. When the secondary winding is complete, it too is covered with paper insulation.

Figure 4–8 shows the four leads extending from the windings. Two of these leads are for the primary, and two are for the secondary. These leads can

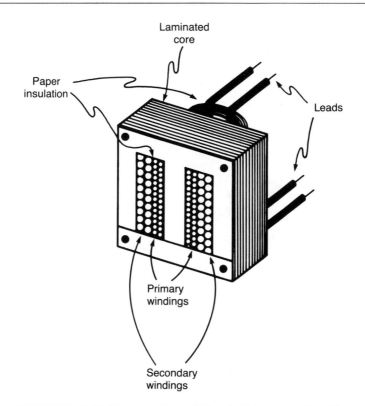

FIGURE 4–8 Cross-section of the shell-type construction transformer

be connected directly to the source and load, or they can be brought out to terminals provided on the transformer enclosure.

4–3.2 Transformer Cores

Transformer cores are made of different substances, depending on the application in which the transformer will be used. The final application will determine such characteristics as frequency, voltage, and current demands. Generally, for frequencies above 20 kHz, air-core transformers are preferred. Iron-core transformers are used for frequencies below 20 kHz. Steel is often used as a core because it conducts and dissipates heat quickly. When steel is used as a core material, the core is sliced into thin plates, much like sliced cheese. These laminations help reduce **eddy current** and **hysteresis** losses in transformers.

Eddy currents are generated in the transformer core as the alternating flux lines cut the iron core material, producing current flow. These currents consume power and produce no useful work. The laminations and the insulating varnish between the laminations cut down on these losses by reducing the current flow. Hysteresis losses are caused by the friction developed between magnetic particles as they rotate during each cycle of magnetization. These losses are minimized by using a special grade of heat-treated steel that is less resistant to magnetization.

An example of a steel-laminated core is shown in Figure 4–9. Approximately 50 laminations make up one inch of a steel-laminated core. Bear in mind that an efficient transformer is one that can offer the best path for the most flux lines with the least loss in magnetic and electrical energy.

4–3.3 Types of Transformers

Three types of cores are used in transformers: the hollow-core, the shell-core, and the H- or cross-type core. The hollow-core, shown in Figure 4–10, is so named because the center of the core is hollow. Note that the primary and secondary windings are wrapped around the sides of this core. In the hollow-core transformer, the copper windings surround the laminated iron core. The shell-core, which is more efficient and more popular, is illustrated in Figure 4–9. Each laminated layer is made up of an E- and an I-shaped piece. These pieces are connected to form the total laminated section. As in the hollow-core structure, the shell-core laminations are insulated from each other, usually by varnish. In the shell-type transformer, the core surrounds the copper windings.

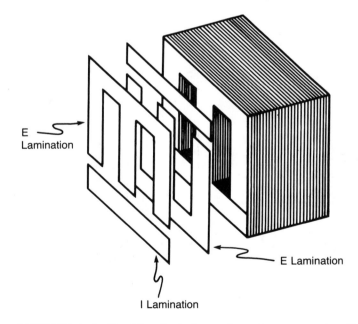

E Lamination

E Lamination

I Lamination

FIGURE 4–9 Steel-laminated core transformer construction

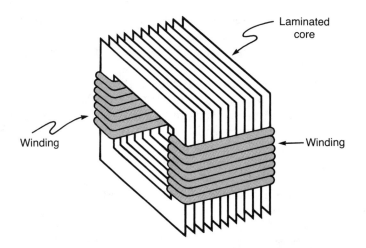

FIGURE 4–10 Hollow-core transformer construction

The third type of core, called the H- or cross-type core (Figure 4–11), is a modification of the shell-type. Two shell-type cores are placed at right angles to each other. Because of its compact design, the H-type core is easily coded. H-type cores are most often found in large power transformers that are filled with oil for more efficient cooling.

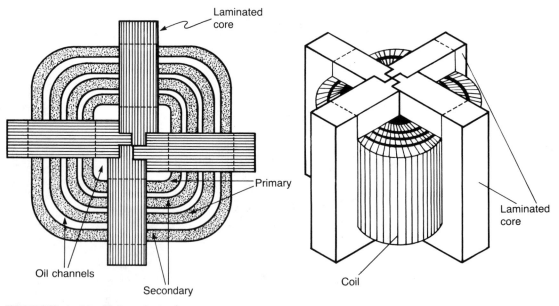

FIGURE 4–11 H-type transformer

If it is important in circuit operation, polarity is usually shown by the dot convention. In Figure 4–12a, note that both dots are at the top of the transformer's windings. The dots so placed indicate an in-phase relationship between primary and secondary voltage. This occurs, for example, when the voltage in the primary and secondary reach a peak at the same instant. Primary and secondary voltages are then said to be in phase. In Figure 4–12b, the dots are on opposite ends of the secondary. This indicates that the primary and secondary are 180° out of phase.

Manufacturers also use another method to indicate polarity in transformers. In Figure 4–13a, the primary winding is labeled at either end with X-1 and X-2. The secondary, on the other hand, is labeled H-1 and H-2. Since most transformers are either step-up or step-down transformers, one winding can be called the high-voltage winding and the other the low-voltage winding. The low-voltage windings are labeled *X*, while the high voltage windings are labeled *H*. Polarity is indicated by the orientation of the numbers associated with the letters. In Figure 4–13a, the top of the primary and the top of the secondary are in phase. Note that the same numbers are adjacent to each other. Figure 4–3b shows the opposite, where the top of the primary and secondary are 180° out of phase. The same numbers are opposite each other in this case. These standards are part of those set by the ***American National Standards Institute (ANSI).***

Determining Transformer Polarity Sometimes we need to know the polarity of a particular unmarked transformer. Let's assume we have a transformer that we know is a 10:1 step-down transformer, but we are unsure of its polarity. We can determine the high voltage side by placing a small AC potential across one of the windings. If the voltage is stepped down across the other set of

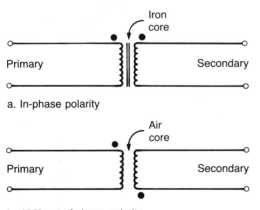

FIGURE 4–12 Transformer polarity dot convention

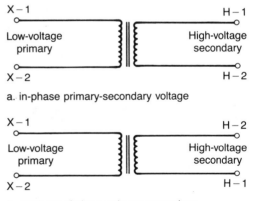

FIGURE 4–13 Transformer polarity—letter convention

FIGURE 4–14 Determining transformer polarity with a voltmeter

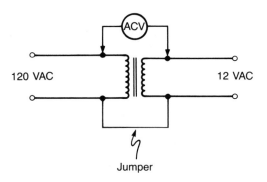

120 VAC   12 VAC

Jumper

windings, we know that the terminals to which we are applying the voltage
must be connected to the high side of the transformer. Next, we can apply
120 VAC to the transformer primary (high side). We expect to see approximately 12 VAC across the secondary. Now we will introduce a jumper from the
bottom of the primary to the bottom of the secondary winding, as in Figure
4–14. If we place an AC voltmeter across the tops of the secondary windings,
we will see a voltage that will be the sum or the difference of the primary and
secondary voltages. If the voltages are summed, we read 132 VAC on our
meter. A summation indicates that the transformer windings are 180° out of
phase. We can see this clearly if we look at instantaneous voltages for a moment.

Figure 4–15 shows us what we expect to see if we look at the transformer
primary at the instant when the voltage across it is 120 V with a positive
potential at the top of the primary. We should expect to see a negative to
positive 12 V at the top of the secondary with respect to the bottom at this
instant of time, if a phase shift were occurring. Redrawing this circuit, we see
that the potentials are indeed additive. At this instant, we would expect to see
132 V from the top of the circuit with respect to the bottom. If the transformer
primary and secondary were in phase, the voltages would subtract. We would,
therefore, see 108 VAC across the tops of the windings, instead of 132.

4–4 TRANSFORMER CONNECTIONS

Transformers are classified according to their use in industry as instrument,
power, and distribution transformers. Power transformers are the large transformers used by power companies for power transmission. Power transformers
step up the AC voltage for transmission down power lines. Power transformers
also step down the high transmission voltage for distribution at power substa-
tions. Figure 4–16 shows a three-phase power transformer. The accepted cutoff
for power transformers is approximately 500 kVA. A transformer rated at
500 kVA or over is a power transformer.

FIGURE 4–15 Equivalent circuit when measuring transformer polarity with a voltmeter

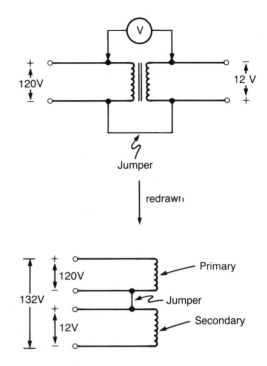

Distribution transformers are used to step down the AC voltage into the
final form for end users. For industrial plants, the final voltage is usually 480,
600, or 240 VAC.

Finally, transformers are classified as instrument transformers. Instrument transformers can be subclassified as current or potential (voltage) transformers. Potential transformers range up to approximately 500 VAC. Potential

FIGURE 4–16 Power transformer

transformers change the high distribution voltage to a lower voltage, such as 120 VAC. The lower voltage secondary is often connected to an instrument used in measuring voltages. Potential transformers are normally single-phase devices. If they are to be used in polyphase circuits, one potential transformer is used for each phase. Potential or voltage transformers are used in combination with voltmeters, wattmeters, power factor meters, and some electromagnetic relays.

Current transformers are rated up to approximately 200 kVA. As their name suggests, they are usually connected in series with the line, between a source and a load. Because they are a series element, the current transformer must carry full load current. As the name *instrument transformer* implies, the current transformer is used mostly in measurement or sensing instruments. The current transformer usually steps down line currents for use in lower current instruments. Current transformers are used in ammeters, wattmeters, and circuits which sense motor overloads. Figure 4–17 shows a current transformer used to measure circuit current.

The insulation between the primary and secondary of a current transformer must be sufficient to withstand full-circuit voltage. The current transformer differs from other types of transformers in that the primary current is determined by the load on the circuit to which the primary is connected. In other types of transformers, the primary current depends on the secondary impedance and the turns ratio. If the secondary in the current transformer

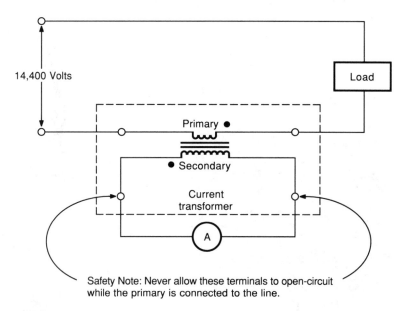

FIGURE 4–17 Current transformer connection in a 14,400-V circuit measuring current

becomes open-circuited, a dangerously high voltage will exist across the secondary. The current transformer produces the high secondary voltage because it is a step-up transformer. SAFETY NOTE: THE SECONDARY OF A CURRENT TRANSFORMER SHOULD NOT BE OPEN-CIRCUITED UNDER ANY CIRCUMSTANCES. The potential across the secondary could be lethal!

Transformers can be classified into two groups: single-phase and three-phase. A single-phase transformer converts a single-phase voltage from one level to another. A three-phase transformer has three separate windings for the primary and for the secondary, all usually wound on the same core.

4–4.1 Single-phase Connections

The construction of a standard single-phase transformer is shown in Figure 4–18a,b. Most distribution transformers have connections similar to this one. Note that the secondary is divided into two sections. In Figure 4–18a, the secondaries are connected in series. In this case, the voltages of the two windings are series aiding and add. In Figure 4–18b, the windings are connected in parallel, with currents, not voltages, adding. In this example, let's say that each of the secondary windings is rated at 120 VAC, with 100 amps of current flowing. In the series-connected windings, the output voltage will be 240 VAC at a maximum of 100 amps, for a power rating of 24 kVA. If, by connecting the secondaries in parallel, we double the current to 200 amps, the voltage remains at 120, since voltage is common in a parallel circuit. Note that the kVA rating is the same, 24 kVA, providing we have 100% efficiency.

In both of these connections, the polarities must be correctly observed. What would happen if we reconnected the transformer into the connection

FIGURE 4–18 Transformer series and parallel connections

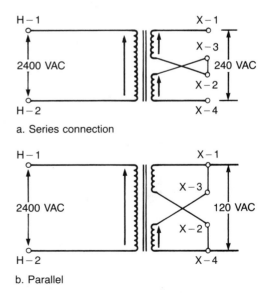

a. Series connection

b. Parallel

shown in Figure 4–19a? If the voltages were exactly equal in both windings, we would see zero volts across the secondary. Since both windings are unlikely to have exactly the same characteristics, we would probably see a small total potential across the two secondaries. The reason for this small potential can be found by examining the circuit in Figure 4–19b, which is a redrawing of the circuit in Figure 4–19a. Note that the potential developed across X-1 and X-2 exactly opposes the potential developed across X-3 and X-4. The total instantaneous voltage across the entire circuit will be near zero.

In the parallel connection of Figure 4–18b, we must connect the coils so that their voltages are in opposition, with the currents adding. The correct configuration is indicated in Figure 4–18b. If we trace through the secondary windings from *X–1*, we can see that the windings are opposing each other. In other words, the instantaneous polarities have values exactly opposite to each other. No current will circulate since there is no load. If we were to reverse the direction of either winding, as in Figure 4–20, short circuit current would flow. The short circuit in the secondary would be reflected back into the primary, which would damage the transformer and possibly the source. When connecting a transformer in parallel, make sure that ratios and impedances are approximately equal.

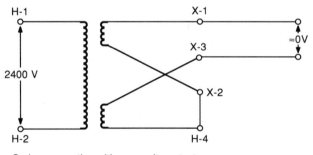

a. Series connection with zero volts output

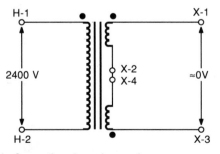

b. Connection shown in a. redrawn

FIGURE 4–19 Transformer secondaries connected in series

FIGURE 4–20 Improper transformer connection—short circuit condition

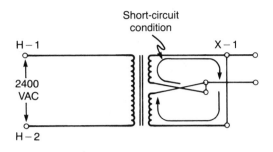

4–5 SPECIAL TRANSFORMERS

4–5.1 Multiple-Winding Transformers

A transformer can have more than two windings on its core. The single-phase transformer has both primary and secondary windings. Multiple-winding transformers are used when power is required at more than one voltage level. A computer, for example, may need a +5V supply for *transistor-transistor logic (TTL)* as well as +12 for the motor in a disk drive. Figure 4–21 shows a multiple-winding transformer. Note that it has one primary and three secondaries. In single-phase transformers, the primary is often connected to the 115-VAC line. The secondaries have voltages that depend on each one's primary-to-secondary turns ratio. Each secondary behaves as if it were an independent transformer with the same primary in respect to voltage output. Power in the primary circuit is equal to the power delivered to each secondary circuit plus the transformer's losses. For example, let's assume that we are delivering 100 W to the primary. If the secondaries were consuming 60 W, 20 W, and 15 W, respectively, the transformer losses would be 5 W or 5%.

The true power in a circuit is equal to the current times the voltage times the power factor. When dealing with transformers with a 110-VAC primary, the true power can be approximated if we know or can calculate the primary current and if we can assume, as is usual for a power transformer, a power factor of 0.9. If we multiply the primary current by 100, the result is the ap-

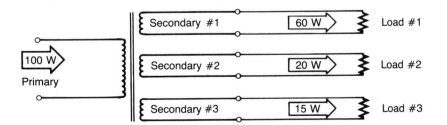

FIGURE 4–21 Multiple-winding transformer

proximate true power, since 110 times 0.9 is approximately 100. For example, if primary current is 2 A, the approximate true power is 2 kW. One of the secondary windings is center-tapped in some multiple-winding transformers to provide two equal voltages 180° out of phase.

4–5.2 Constant-Current Transformers

Constant-current transformers, as their name implies, provide a constant current to circuits whose circuit resistances can change. The primary of a constant-current transformer is stationary and connected to an AC supply. The secondary is movable, responding to changes in load. The secondary coil moves until the current is brought back to its original value.

4–5.3 Autotransformers

In the normal transformer, primary and secondary are electrically insulated and isolated from each other. The autotransformer is a special type of transformer in which the same winding doubles as both primary and secondary. To use both primary and secondary, the winding must be tapped as shown in Figure 4–22. The voltage induced in the top part of the secondary is added to the voltage in the primary, since they are in series. The secondary voltage is higher than the voltage in the primary. This transformer is a step-up transformer. Examination of the current in an autotransformer shows that it steps down current in the secondary (Figure 4–23). Current I_1 splits to form I_2 and I_3.

Note that currents I_2 and I_3 oppose each other in the secondary. This opposition creates a net current flow in the secondary that is less than that in the primary. You will recall that normal power transformers are designed for a specific current load. If the load on the transformer is decreased from that design level, high voltages may be induced, which could result in arcing and burning the transformer. The autotransformer's advantage lies in its ability to allow variations in the load with no arcing and little change in the output voltage. This advantage comes from the out-of-phase relationship between primary and secondary currents.

The same autotransformer shown in Figure 4–23 can also be used as a step-down transformer by connecting the primary voltage to terminals *3* and *4*. A smaller voltage will be found at the secondary terminals *1* and *2*. As shown

FIGURE 4–22 Autotransformer schematic diagram

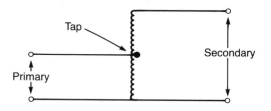

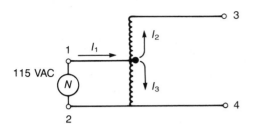

FIGURE 4–23 Current division in an auto-transformer

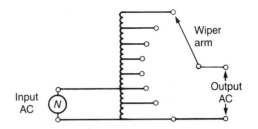

FIGURE 4–24 Variable autotransformer—schematic diagram

in Figure 4–24, autotransformers can have many transformer taps. These taps are brought out to a sliding contact so that a large number of voltages can be chosen between zero and maximum. This type of autotransformer is called a variable autotransformer (Figure 4–25). It is used where line voltages need to be manually adjusted.

Autotransformers are cheaper to manufacture than a comparable two-winding transformer and have greater efficiency and better voltage regulation. They are used in motor starters where applied voltage must be reduced at starting to prevent damage to the motor. The autotransformer is not often used in voltage distribution systems due to the lack of electrical isolation between primary and secondary. It is used in transmission systems when the values of voltages are close together. The autotransformer is also used to economically step up or step down voltages on high-voltage lines.

FIGURE 4–25 Variable autotransformer

4–5.4 Saturable-Core Reactor

The *saturable-core reactor* is a special type of inductor used to control the amount of AC power delivered to a load. Both AC and DC are supplied to this device, which controls the AC by varying the DC voltage. Although it is not strictly a transformer, it works in a similar fashion by changing the flux within its core.

A generalized saturable-core reactor is shown in Figure 4–26. You will notice three windings, labeled *L1*, *L2*, and *L3*, each wound on a different part of the core. Variable DC power is applied to the central winding, called the control winding. The windings *L1* and *L3* are called the load windings. Both of these windings are connected in series with the load and have the same number of turns in each load winding. Note that we have a complete series circuit including the source, winding *L1*, the load, and winding *L3*. Since current is common in a series circuit, the same AC current flows in the load windings and the load.

Let's assume that no current is flowing from the DC source and the core is in an unsaturated condition. When we apply AC from the AC source, the in-

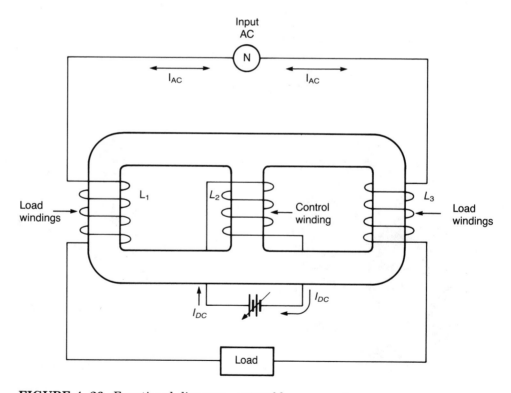

FIGURE 4–26 Functional diagram—saturable-core reactor

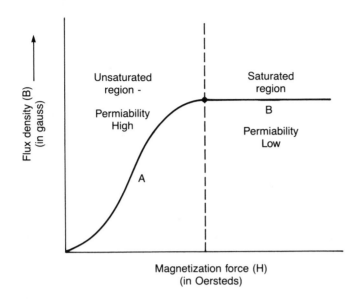

FIGURE 4–27 Graph plotting flux density in a saturable-core reactor versus magnetization force

ductive reactance of the load windings is large due to the large inductance of *L1* and *L3*. The large inductive reactance of these two windings effectively acts as a voltage divider, dropping almost all of the voltage provided by the source. Little voltage is then present at the load.

When we apply DC to control winding *L2*, flux is generated in the core. If we increase the DC to a high enough level, the core will saturate. Core saturation means an increase in DC will not produce an appreciable increase in flux. At this point, with the core saturated, if we apply AC to the system, the windings *L1* and *L3* have a low reactance due to the saturated core. These windings will then drop little voltage, the large part of the voltage being delivered to the load with the core saturated.

We can use the chart in Figure 4–27 to examine this phenomenon in more depth. You will recall that the permeability of a core changes with changes in the saturation of the core. The region labeled *A* in the figure shows the area of the curve where the core is in its unsaturated condition and where the core permeability is high and flux density low. The inductive reactance of the coil in this case is high. In the case of the saturable core reactor, most of the AC source voltage is dropped across the saturable core reactor. In the saturated region of the curve, labeled *B*, the core permeability is low and the inductive reactance high. Any AC power applied will be felt across the load, rather than across the saturable core reactor.

4–6 THREE-PHASE TRANSFORMERS

Three-phase transformers are common in industrial power distribution systems. In general, three-phase power distribution systems use either single-phase transformers connected in a three-phase arrangement or an actual three-phase transformer. When using single-phase transformers, ensure that the devices have the same transformation ratio and impedances. If polarity is not indicated on the transformer, check it with a voltmeter. (Caution: check transformers carefully; dangerous voltages can be present.)

Three-phase power can be generated from a delta-connected transformer or a wye-connected transformer. Single-phase transformers used in three-phase systems follow these same connections. In single-phase systems, there are seven standard ways of connecting single-phase transformers in a three-phase bank: wye-wye (also called star-star), delta-delta, wye-delta, delta-wye, open delta, double T, and Scott.

4–6.1 Wye-Wye Banks

In the wye-wye transformer bank, both the primary and secondary are connected in a wye or star configuration. Each of the single-phase transformer's primary leads is connected to one phase of a three-phase source. The neutral lead of the source is connected to the junction of the high-voltage H-2 leads, which are connected together. The wye-wye configuration is illustrated in Figure 4–28. Note that the neutral in the secondary is formed by connecting one lead (X-2) of each of the primaries together. If we measure the voltage between any secondary X-1 lead and the secondary neutral, we will find that the voltage is equal to 0.5774 times the voltage between any two X-1 leads. As an example, we will measure 208 V between two X-1 leads (line-to-line). What is the line-to-neutral secondary voltage?

$$V1 - n = V1 - 1 \ (0.5774) = 208 \text{ V} \ (0.5774) = 120 \text{ V} \qquad \textbf{(eq. 4–11)}$$

By studying this configuration, we see that there is no phase shift between primary and secondary voltage. When connecting transformers in this configu-

FIGURE 4–28 Three single-phase transformers connected wye-wye, three-phase

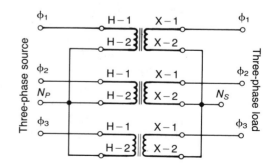

ration, you must observe proper polarities. The best way to do this is to connect the transformers and then measure between each phase without adding the load. Each line-to-line voltage should be 1.73 times the line-to-neutral voltage.

4–6.2 Delta-Delta Banks

As its name implies, the delta-delta (shown in Figure 4–29) transformer bank uses single-phase transformers with both primaries and secondaries connected in delta configuration. In this connection H-1 primary leads are wired to the adjacent H-2 primary lead, as are the X-1 and X-2 leads in the secondary. If we measure between any two leads in the primary or in the secondary, we can expect to measure equal voltages.

As in the wye-wye transformer, the proper phase relationships must be observed. In a properly connected delta-delta bank, little or no current should flow in the secondary windings (without the load), since the secondary voltages will be equal and 120° apart. If we were to sum these voltages, we should get zero volts. In an improperly connected transformer bank, excessive currents will flow in the loop, possibly damaging the transformers. In a case where the transformer is unmarked or you are unsure of the connection, measure as shown in Figure 4–30. Before connecting the bank, separate any two leads, placing a voltmeter between them. In a properly connected bank, the voltage should read zero volts or close to it without the load. In this case it is safe to close the connection, since no short circuit will occur. If the potential between the points is high, closing the delta will cause a short circuit condition. This can be prevented by switching the transformer connections until no potential is read on the voltmeter.

4–6.3 Wye-Delta Banks

Three single-phase transformers can be connected in wye-delta, where the primary is connected in wye and the secondary in delta. This transformer is often used to step down voltages, as in the end of a transmission line. Recall that

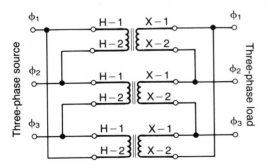

FIGURE 4–29 Three single-phase transformers connected delta-delta three-phase

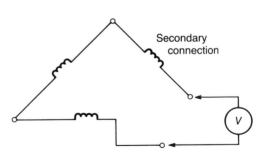

FIGURE 4–30 Checking a transformer for proper delta connections using a voltmeter

FIGURE 4–31 Three single-phase transformers connected wye-delta three-phase

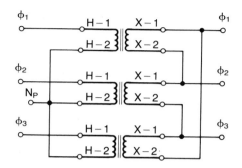

voltage transmissions are normally high and need to be stepped down for use by the load. The high voltage transmission line is connected in the wye configuration. This takes advantage of the fact that the voltage across each coil of the wye is about half of the line voltage, reduced by a factor of 1.73. A diagram of the wye-delta connection is shown in Figure 4–31.

4–6.4 Delta-Wye Banks

The wye-delta bank is used to step down high voltages to lower voltages. The opposite connection, the delta-wye, is used to step up applications, often linking the power generator to the transmission line. The generator outputs are connected in delta with the transmission line in wye. In this case, the line-to-line voltage is the secondary voltage multiplied by 1.73. The same advantage occurs here; the insulation necessary for the high voltage line is only about half the secondary voltage. A diagram of the delta-wye bank is shown in Figure 4–32.

The open delta, sometimes called the *V-V* connection, is used in emergency situations where one coil/phase is damaged. In the normal delta-delta connection, any one-phase voltage is equal to and opposes the sum of the other two voltages. If one of the three-phase transformers is damaged and removed, the output will still be three-phase at the secondary. The open delta configuration, however, can provide only approximately 58% of the capacity of the delta-delta configuration. Power companies sometimes use the open delta configuration on lightly-loaded three-phase service. They then have the option of adding another transformer (providing extra power) if the demand justifies it.

FIGURE 4–32 Three single-phase transformers connected delta-wye three-phase

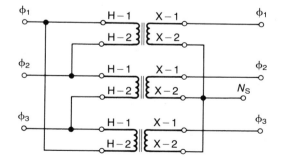

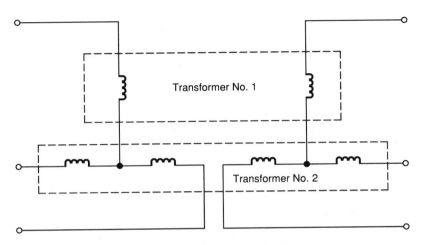

FIGURE 4–33 Double-T transformer connection

The open delta configuration permits three-phase power transformation with only two transformers. Another way to do this is with the T-T or double T configuration shown in Figure 4–33. Note that transformer #2 has the full-line voltage across its center-tapped primary. This transformer is called the main transformer. Connected to the center tap of the main transformer is a transformer known as the teaser. The double T configuration has no advantage over the open delta, which performs essentially the same transformation. The double T requires expensive transformers and special connections. The open delta requires neither. A variation of the double T is sometimes used to create two-phase power from a three-phase source. This connection, shown in Figure 4–34, is known as the T or Scott configuration.

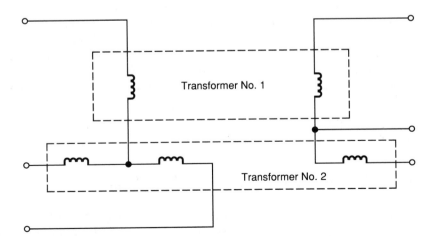

FIGURE 4–34 Scott transformer connection

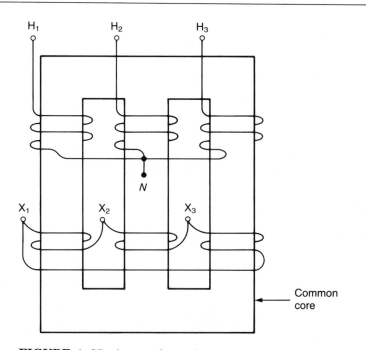

FIGURE 4–35 A true three-phase transformer connected wye-delta

4–6.5 True Three-Phase Transformers

We have seen that three-phase transformer banks can be constructed out of three single-phase transformers. We can also use a true three-phase transformer, which uses one winding from each phase wound on a single core, as in Figure 4–35. This particular transformer is connected in wye-delta; that is, the primary is connected in wye and the secondary in delta. In the transformer shown in Figure 4–35, the high-voltage primary leads connect together at the neutral point N. Note the similarity between this and the example of the wye-delta single-phase 4 connection described previously. Three junctions and three leads from each junction appear in the secondary of Figure 4–35. This transformer secondary is connected in the same way as the one shown in the wye-delta connection.

Three-phase transformers can also be found in delta-wye, wye-wye, or delta-delta.

4–7 TRANSFORMER LOSSES AND EFFICIENCY

The transformer's turns ratio affects the output current as well as the output voltage. If voltage is doubled in the secondary, secondary current is reduced by half. The reverse is also true. If the voltage in the secondary is halved, the current is doubled. In the ideal transformer, all the power delivered to the primary by the source is also delivered to the load by the secondary. In the practical transformer, the power delivered to the load is actually somewhat less than the input from the source. The reason for this is the losses in the transformer. No device is 100% efficient, and the transformer is no exception to this rule.

4–7.1 Transformer Losses

The transformer, like other power conversion devices, has some losses. The small power transformers used in electronics equipment have efficiencies ranging from 80 to 90%. Commercial power transformers, such as those used by local utilities, can have efficiencies that exceed 98%. The actual power delivered to the load will be the power delivered by the source minus the transformer losses. The actual power losses in a transformer come from several sources.

One loss comes from the DC resistance in the windings of the primary and secondary. This loss is called *copper loss* or, sometimes, $I^2 R$ loss. Copper loss varies with the square of the current in the windings. If the transformer has 1200 turns of number 23 copper wire with a length of 1,320 ft, the resistance of the primary winding is 26.9 Ω. If the current in the primary is 0.5 amps, the copper loss in the primary is

$$= I^2 R = (0.5)^2 (26.9) = 6.725 \text{ W} \qquad \text{(eq. 4–12)}$$

In this same transformer, let's say the secondary has 120 turns of number 13 copper wire, with a length of 132 ft. The secondary resistance would be 0.269 Ω. Since we have a 10:1 step down transformer, the current in the secondary is 5 amps. The copper loss in the secondary is

$$= I^2 R = (5)^2 (0.269) = 6.725 \text{ W}$$

The total amount of the copper loss is the sum of primary and secondary losses:

$$= (6.725 \text{ W}) (2) = 13.45 \text{ W}$$

This 13.45 W of power is lost to the system as heat dissipation and, therefore, does no useful work. Copper losses can be minimized by using the correct diameter of wire and core size. Large diameter wire is normally used for large-current windings and smaller wire for low-current applications.

Another loss is called **core loss,** which can be broken down into two subclassifications, hysteresis loss and eddy-current loss. We mentioned earlier that the core of the transformer is constructed of an iron-based material, since this type of material is a good conductor of magnetic lines of force. When current flows in the primary windings, a magnetic field is created in the core. This magnetic field cuts the core material and induces random currents in the

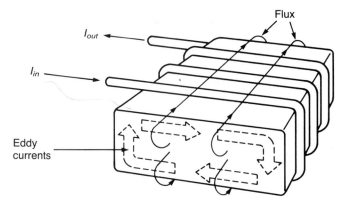

FIGURE 4–36 The effects of eddy currents on transformer operation

core material. These random currents dissipate power in the form of heat. The amount of the eddy current loss depends on how fast the flux changes and on the amount of flux. The eddy current loss can be reduced by making the core out of laminated sections instead of a solid piece of metal.

The effects of eddy currents are shown in Figure 4–36. Note that when current flows through the windings, flux is produced in the direction shown. The flux created, however, induces current flow of its own in the form of eddy currents. Figure 4–37 shows what happens when we laminate the core. Note that each eddy current is limited to a path with a very small cross-section.

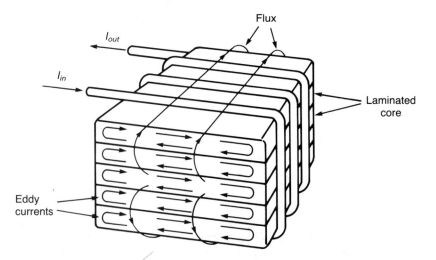

FIGURE 4–37 Reducing eddy currents by laminating the core

Recall from your studies in basic electricity that the resistance to current flow in a conductor is increased if the cross-sectional area is decreased. The eddy currents are limited to a very small value in this type of core. Although the laminations reduce the eddy currents, they do not increase the reluctance of the magnetic circuit since the laminations are parallel to the direction of the flux.

The second type of core loss is called hysteresis loss. Recall that when an iron-based material is magnetized, the domains within the core tend to align themselves with the field. When the direction is reversed, the alignment of the domains also reverses. At 60 Hz, the domains must reverse their orientation 120 times every second. The energy expended to realign the domains is dissipated as heat within the core. This loss, called hysteresis loss, can be thought of as a kind of molecular friction caused by the realignment of the domains. It can be reduced by proper choice of core materials.

4–7.2 Transformer Efficiency

The efficiency of a transformer is the ratio of the input power from the source to the output power delivered to the load. It is also equal to the ratio of the output to the output plus the losses. We could then say:

$$\text{efficiency} = \frac{\text{output power}}{\text{output power} + \text{copper loss} + \text{core loss}} \times 100 \qquad \textbf{(eq. 4–13)}$$

or

$$\text{efficiency (in \%)} = \frac{\text{output power}}{\text{input power}} \times 100 \qquad \textbf{(eq. 4–14)}$$

As an example, we will use a 10:1 step down transformer with 60 W of power applied to the primary from the source. A measurement indicates that we have 57 W at the load. What is the efficiency of the transformer? Using equation 4–14, we can find the efficiency:

$$\text{efficiency (in \%)} = \frac{\text{output power}}{\text{input power}} \times 100$$

$$= \frac{57}{60} \times 100 = 95\%$$

If we measure the current and resistance in both primary and secondary and multiply them together we can find the power loss due to copper loss. Let's suppose the total power loss due to copper loss in primary and secondary is 2 W. What is the core loss? Using equation 4–13, we can solve for the core loss:

$$\text{efficiency} = \frac{\text{output power}}{\text{output power} + \text{copper loss} + \text{core loss}} \times 100$$

$$\text{core loss (in W)} = \frac{\text{output}}{\text{efficiency}} - \text{output} - \text{copper loss} \qquad \textbf{(eq. 4–15)}$$

$$= \frac{57 \text{ W}}{0.95} - 57 \text{ W} - 2 \text{ W} = 1 \text{ W}$$

The core loss is 1 W. This loss, unlike the copper loss, stays relatively constant regardless of load current.

4–8 TRANSFORMER RATINGS

Transformers can be rated in terms of power, current, voltage, and frequency. When choosing or replacing a transformer, the technician should not exceed the maximum ratings. If the maximum voltage rating is exceeded, the windings can arc, destroying the insulation and short-circuiting the windings. Likewise if a load resistance is connected to the secondary, causing too much current to be drawn in the secondary, the transformer can overheat, damaging the insulation.

4–8.1 Power Ratings

Most transformers are rated by the amount of power they can safely carry continuously without exceeding a temperature rise of 80°C. This power is also specified at a given secondary voltage, a given applied frequency, and an ambient temperature of 40°C. The power rating is normally given in either watts or volt-amperes. You will recall that, in AC circuits, the power is calculated by using the formula

$$P = V I \cos \theta \qquad \textbf{(eq. 4–16)}$$

where P is the power dissipated in watts, V the voltage, I the current, and $\cos \theta$ the power factor. If the transformer has a resistive load, the voltage and current are in phase and the power factor is 1. The phase angle is zero degrees, which makes the cosine of that angle 1. In this case, the power is equal to the product of the voltage and current and is expressed in watts. As is often the case, the transformer's load is not purely resistive; it is reactive. This means the current and voltage are out of phase. In transformers where the load is capacitive or inductive, the power factor is less than one. The phase angle is, therefore, between zero degrees and 90 degrees. Since the true power is now *less* than the apparent power, the power in transformers is usually expressed in volt-amperes. Large transformers are rated in kVA or kilovolt-amperes. The kVA rating of a winding is equal to the rated full-load current of the winding, multiplied by the rated voltage at the winding, and divided by 1000.

If transformer power ratings were given in watts, a wattmeter in the system shown in Figure 4–38 would indicate that no real power was being consumed by the load. If the load was purely resistive, we would conclude that no current is flowing in the secondary. If we look closely at this circuit, we can see that secondary current is flowing, but it is out of phase with secondary voltage. If the load were purely inductive, the power meter would indicate that no power was being consumed. If we connected an ammeter to the secondary or to the primary, we would see current flowing. The wattmeter would then give us an incorrect impression about what is happening in the circuit. We can now see the rationale for giving the transformer power ratings in apparent power rather than in true power.

Using the kVA rating of a transformer, an ammeter, and a voltmeter, we can determine if the transformer is overloaded. Working with a 50-VA transformer, we measure a secondary voltage of 12 V and an output secondary current of 3 amps. Our power equation tells us that the voltage in primary or secondary times the current in primary or secondary will give us the apparent power demanded from the transformer. Thus:

$$P \text{ (apparent power)} = I_s V_s$$
$$= (3 \ A)(12 \ V) = 36 \ VA$$

(eq. 4–17)

We can see from this example that our 36-VA demand is not overloading the transformer. Since transformers are efficient, we could have used the primary current and voltage to calculate the load on the transformer. To do this, we can approximate:

$$V_p I_p = V_s I_s$$

(eq. 4–18)

In this equation, we are saying that the power in the primary and the power in the secondary are equal, neglecting losses.

In an earlier discussion, we talked about power factor in AC circuits. Since the power factor is equal to the cosine of the phase angle, the equation 4–16 can be rewritten:

$$P \text{ (true power)} = V I \cos \theta$$

or

$$P \text{ (true power)} = V I \text{ power factor}$$

(eq. 4–19)

FIGURE 4–38 Transformer secondary draws current but the load consumes no power

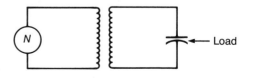

The maximum power a transformer can handle is determined by its ability to dissipate heat. If heat can be removed from the transformer, the capacity to handle power will be increased. Heat is commonly removed from a transformer by immersing it in oil or by using cooling fins. It is important to handle the oil in transformers very carefully. Until a few years ago, polychlorinated biphenyls (PCBs) were used widely as transformer oil. PCBs have been recognized as carcinogens, or cancer-causing substances. If improperly disposed of, they can be an environmental hazard. Always dispose of transformer oil carefully in approved containers.

4–8.2 Frequency

Transformers can be operated at a higher frequency than that for which they were designed. Power losses in a transformer increase, however, when applied frequency increases. If the frequency applied to the transformer increases, the inductive reactance in the primary increases. Increases in the primary inductive reactance will cause more voltage to be dropped across the primary and less across the secondary. The transformer should not be damaged, but the voltage and power to the load will be decreased. Transformers should not be operated much below their rated frequency. When operated more than 10% below rated frequency, most transformers will overheat, since the current in a transformer will increase when frequency is lowered. When frequency is lowered, the reactance of the windings is decreased, causing increased current flow. The increased heat can damage the insulation on the windings.

4–8.3 Current and Voltage

The maximum voltage that transformer windings can sustain depends on the type and size of insulation on the windings. For example, greater insulation thickness on the windings increases the point at which the windings will arc over from too much voltage. If too much voltage is applied to the windings, electrical arcs will occur, weakening the insulation material and possibly destroying it.

The maximum current that transformer windings can sustain depends on the diameter of the wire used in the windings. Larger diameter wire in the windings will allow the windings to carry more current. If the windings are forced to carry an excessive current, the windings can overheat, possibly damaging the insulation. In larger transformers using cooling oil, the cooling oil will be damaged by excessive heat.

4–9 MAINTAINING TRANSFORMERS

Since it does not have any moving parts, the transformer has lower maintenance requirements than other types of electrical devices. It does not require lubrication. Larger transformers, which have lug-type connections, need periodic inspection for the tightness of connections. In addition to tightening connectors, transformers need to be cleaned periodically with approved solvents. Special attention should be given to the bushings and insulators. If the transformer is water-cooled, the cooling coils should be checked and cleaned regularly. Large power transformers are usually cooled with oil. Samples of oil should be taken and tested regularly. If the oil is contaminated, it should be cleaned and dried or replaced as required.

One of the most important maintenance requirements for the transformer is a regular inspection schedule. Large distribution transformers should be disconnected from service and all parts checked. All electrical connections should be given a thorough check for tightness and insulation breakdown or other signs of deterioration. If the transformer is water-cooled, the system should be checked for leaks and regularly flushed with cleaning solvent to clean the coils. One of the most common contaminants is water. If water gets into the oil, the oil loses its dielectric or insulating strength. Remember that the dielectric strength of an insulating material is the highest applied voltage without breaking down the material. A fraction of a percent of water in a transformer's oil can seriously reduce the insulating properties of the oil. For example, if a transformer contains 100 liters of oil, as little as seven grams of water will ruin the dielectric properties of the oil. The technician should regularly check transformer oil for the presence of water. Oil purity should also be checked after transformers have been operated at temperatures, currents, or voltages higher than recommended. High oil temperatures promote the formation of sludge, which is oil that has turned thick and gummy. Sludge can damage a transformer by reducing the ability of the oil to cool the transformer.

The technician should make regular tests of the insulation-resistance of the transformers. The insulation-resistance test is a test of the quality of the insulating material in a transformer. The testing device is called a megohmmeter or megger. Insulation that has absorbed water will have a lower ability to insulate, a condition that can easily be tested by a megger. An example of a megohmmeter is shown in Figure 4–39. All windings except the one under test should be grounded. The megger should be connected between the winding under test and the transformer case. If the resistance is lower than recommended, the transformer insulation can break down under actual applied voltage conditions. If breakdown occurs, the windings can arc and burn, causing a possible fire hazard.

If you are unsure of the insulation resistance of the particular power transformer you are testing, use the formula below:

$$R = \frac{kV \times 30}{\dfrac{kVA}{f}} \qquad\qquad \text{(eq. 4–20)}$$

where R is the resistance in megohms, kV is the voltage rating of the winding

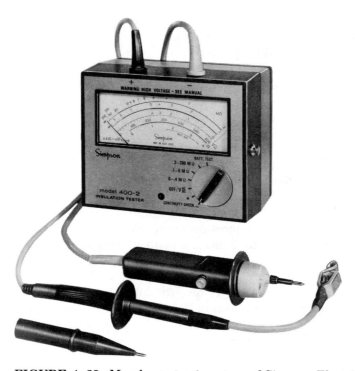

FIGURE 4–39 Megohmmeter (courtesy of Simpson Electric Company)

in kilovolts (kV), kVA is the kilovolt-ampere rating of the winding, and f is the frequency of allied AC in hertz (Hz).

As an example, if we are using a 500-kVA transformer with a winding designed to bear 5000 V at 60 Hz, what is its minimum insulation resistance?

$$R = \frac{5\text{kV} \times 30}{\dfrac{500\text{kVA}}{60\text{Hz}}} = \frac{150}{2.89}$$

$R = 52$ M ohms

The preceding formula is for a measurement taken when the transformer windings are in air at 85°C. If the windings are in oil, the temperature is 40°C. The insulation resistance, however, varies inversely with temperature. As temperature increases, the insulation resistance decreases and vice-versa. The insulation resistance doubles or halves for each 10° temperature change. If the insulation resistance is 100 M ohms at 85°, at 95° the resistance would be 50 M ohms. At 75° the insulation resistance would be 200 M ohms.

The life of a transformer is determined by the state of the insulation. A transformer can no longer be used when the insulation becomes so brittle that the transformer winding short circuits. Increased temperature speeds the aging of a transformer. In general, the higher the temperature, the faster the insulation will deteriorate. A good rule of thumb is that the life of the transformer will halve for each 8°C rise in temperature. For example, if a transformer is to be operated at 72°C, it has a life expectancy of 20 years; the life expectancy would be 10 years if operated at 80°.

It is difficult to predict the useful life of a transformer. Experience has shown that if a typical transformer is operated at 55°C, with ambient temperature seldom above 30°C and no overloads, the life expectancy will be between 20 and 40 years.

CHAPTER SUMMARY

- The three parts of a transformer are: 1) primary, 2) secondary, and 3) core.
- Polarity in transformers is indicated in two ways: 1) by the dot convention, and 2) by the numerical convention.
- Polarity in transformers is important when connecting secondaries in series or in parallel.
- Transformers are classified either as power, instrument, or distribution transformers.
- Current transformers are special instrument transformers used to measure current, voltage, or power in a circuit. NEVER OPEN THE SECONDARY OF A CURRENT TRANSFORMER CONNECTED TO A CIRCUIT. DANGEROUS, POSSIBLY LETHAL VOLTAGES, CAN RESULT.
- The current drawn by the primary of a transformer without a load on the secondary is called the magnetizing current.
- The primary current drawn by a transformer with a loaded secondary depends on the impedance of the secondary.
- The turns ratio of a transformer is the number of turns in the primary divided by the number of turns in the secondary.
- The voltage developed in the secondary of a transformer is proportional to the voltage applied to the primary and the turns ratio.
- A step-up voltage transformer steps down current in the secondary.
- Three types of cores are most often found in transformers today: 1) hollow-core, 2) shell, and 3) H-type.
- Both primaries and secondaries can have multiple windings.
- Autotransformers have one winding that doubles as a primary and secondary.

- A saturable core reactor is a device that can control the AC power delivered to a load.
- Single-phase transformers can be connected in the following ways to achieve three-phase outputs: 1) wye-wye, 2) delta-delta, 3) wye-delta, and 4) delta-wye.
- The open delta permits three-phase power transformation with only two transformers.
- Transformer losses are both AC and DC.
- The loss from the current flowing in the DC resistance of the windings is called copper loss.
- Core (AC) losses come from eddy currents and hysteresis losses.
- Transformers are rated in terms of power, current voltage, and frequency.

QUESTIONS AND PROBLEMS

1. Describe the power conversion of the transformer.

2. Identify the parts of a transformer in the hollow-core and laminated-core construction.

3. Given a transformer with a single primary and secondary, describe how you would check the polarity of this transformer with an AC voltmeter. After checking the polarity, label the transformer with the proper *ASA*, *X*, and *Y* labels.

4. A transformer has 100 turns in the primary and 400 turns in the secondary. If 120 V is applied to the primary and 1 amp of primary current is measured, calculate the following, assuming 100% efficiency:
   a. secondary current
   b. secondary voltage
   c. turns ratio.

5. Describe the operation of the saturable-core reactor and the auto-transformer.

6. Draw the schematic diagrams of the four configurations of single-phase transformers connected in three-phase arrangements.

7. A transformer has the following construction: primary turns = 440, 24 turns of wire in the secondary. If 120 V is placed on the primary and 0.91 amps of current flow, what is
   a. the secondary voltage
   b. the secondary current
   c. the primary volts/turn
   d. the secondary volts/turn

e. the primary resistance
f. the secondary resistance.

8. A 240/120V step-down transformer has a 1500 kVA rating, a core loss of 10 W, a primary resistance of 0.2 Ω, and a secondary resistance of 0.05 Ω. Calculate the efficiency of this transformer.

9. List the types of transformer losses and discuss how they can be reduced.

10. Define the following terms:
a. leakage flux
b. mutual flux
c. turns ratio.

5

AC INDUCTION MOTORS

At the end of this chapter, you should be able to

- describe the principle of the rotating field.
- identify the parts of the single-phase AC motor and the three-phase AC motor.
- explain how the induction motor creates torque.
- draw and explain the speed-torque curve of an induction motor.
- list and describe the different NEMA classifications for induction motors.
- calculate the following for an induction motor:
  - efficiency
  - rated kVA
  - power factor
  - synchronous speed
  - slip.
- explain why the single-phase motor needs to have a special starting method.
- draw the schematic diagram for a split-phase and capacitor-start single-phase induction motor.
- explain how the capacitor-start and split-phase motors produce starting torque

5–1 INTRODUCTION

Because of the easy availability of AC power in industry, the AC motor is the most popular motor in industry. Since the AC motor is generally cheaper than a DC motor of comparable power, it is also popular because of the cost factor.

The most popular type of AC motor does not use brushes and a commutator like the wound-field DC motor, which eliminates a major source of maintenance problems.

The AC motor plays a vital role in industry. Huge AC motors drive pumps and cranes in production plants and grind rock in cement plants. At the same time, small AC motors are used in clocks, fans, and office machinery. Industry uses AC motors in different sizes, shapes, and ratings to perform many different jobs. The AC motors are designed to use either three-phase or single-phase power.

AC motors can be classified in several different ways. One method of classification is by power rating as fractional or integral horsepower machines. As the name suggests, a *fractional HP* machine has a power rating under 1 HP. Fractional HP motors use three-phase and single-phase AC power sources. Single-phase voltages are typically either 120 VAC or 240 VAC. The three-phase voltages found in industry are 208, 240, 480 and 600 VAC. The *integral HP* machine has a rating above 1 HP. *Large-apparatus* AC motors are usually found in sizes from 200 HP to 100,000 HP. Operating voltages are high, starting at 480 VAC and continuing to 2300, 4000, 6900 and 13,200 VAC.

Another method of classifying AC motors is by their physical construction. This chapter and the next chapter will discuss the AC induction motor, the AC synchronous motor, and the universal motor. Under synchronous and induction motors, both single-phase and three-phase motors will be covered.

Single-phase induction motors are used in fractional HP sizes. Larger three-phase induction motors are used in integral HP sizes, although some fractional HP three-phase induction motors are used in industry. Synchronous motors are used in large sizes, in mills, fans, and other constant speed systems. Every type of motor has a particular area where its characteristics are most useful. Because of the induction motor's simplicity, reliability, and low cost, it is used in many industrial areas. There are no hard and fast rules to determine which type of motor/drive system is best for a particular application. Those involved in choosing a motor/drive system must evaluate each application individually.

5-2 THE ROTATING FIELD PRINCIPLE

Both AC and DC motors generate torque by the interaction of rotor and stator fields. In the wound-field DC motor discussed in Chapter 3, current flows in one direction through the rotor windings. The interaction of the rotor and stator fields generates the torque to turn the motor. This principle of torque generated by interacting fields is applied differently in the synchronous and induction AC motors.

The basic principle of the rotating field is shown in Figure 5–1. If we had a permanent-magnet rotor, we could rotate permanent magnets (or electromagnets) around the rotor. The attraction of the rotating permanent magnets

FIGURE 5-1 Two rotating fields

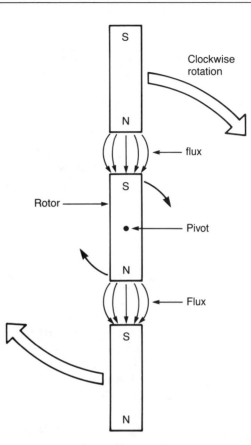

for the rotor field would pull the rotor in a circular fashion. You will immediately see the flaw in this design, the fact that the motor now has two rotating parts, when it can have only one. A motor normally has one rotating part called the rotor and a stationary part called the stator. Even if the stator cannot be made to rotate physically, it can be made to rotate magnetically. This principle is called the principle of the rotating field.

The principle of the rotating field lies at the heart of the operation of induction and synchronous AC motors. Both the synchronous and induction motors rely on rotating magnetic fields in their stators to cause the rotors to turn. The idea is very simple. The magnetic field on the stator is made to rotate magnetically (not physically). Another magnetic field in the rotor can be made to chase the stator field by magnetic attraction and repulsion. The rotor, being mechanically free to turn, follows the field in the stator.

To get a better idea of the rotating field principle, let us turn to Figure 5-2. We will use a simplified example of a two-phase AC machine, seldom seen in industry, for purposes of illustration. The principle of the rotating field is the same for both two-phase and three-phase machines. To establish a rotating

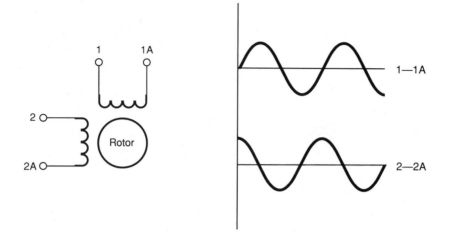

FIGURE 5–2 Two-pole, two-phase AC motor (poles shown without windings)

field in a motor stator, the number of pole pairs must be the same as or a multiple of the number of phases in the applied voltage. The stator poles must also be physically displaced from each other by an angle equal to the phase angle between the individual phases of the applied voltage. In this example, the stator has two windings, placed at right angles to each other, as shown in Figure 5–2.

If the voltages applied to phases AA′ and BB′ are 90° out of phase, the currents that flow in those windings will be 90° out of phase. Since the magnetic fields created in the coils are in phase with the currents, the magnetic fields are also 90° out of phase with each other. These two out-of-phase magnetic fields add together to form a single total field. This total field rotates one complete revolution for every AC cycle.

Let us analyze the rotating magnetic field in more depth, referring to Figure 5–3. Sections 5–3a, b, c, and d are simplified sectional views of the two-phase, two-pole stator winding. Figure 5–3e shows the sine waves of current applied to windings AA′ and BB′ of the stator, shown in Figure 5–2. The currents have a phase difference of 90°. In this example, when we speak of current as being negative, we mean that the current is flowing in one direction. When we say current is positive, we mean that it is flowing in the opposite direction. At time t_0, the current in the AA′ winding, I_0, is zero and the current in the BB′ winding, I_0 is maximum negative. Using the left-hand rule for electromagnets, we can see that a north-south field will be created in the stator in the direction shown.

At time t_1, I_B has decreased but is still negative. The current I_A is now positive and equal in amplitude to I_B. The current in winding BB′ has decreased, but the field caused by this current is in the same direction. Note the

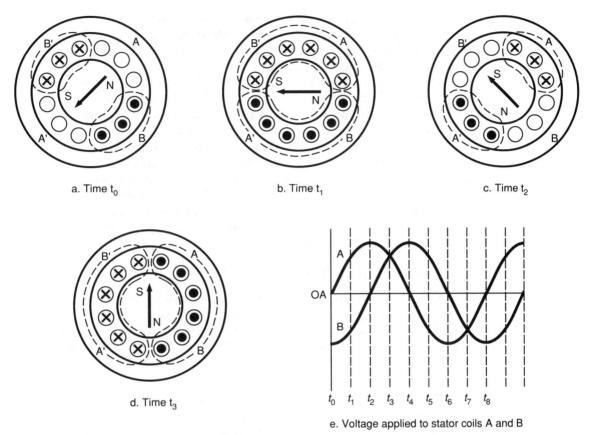

a. Time t_0 b. Time t_1 c. Time t_2

d. Time t_3

e. Voltage applied to stator coils A and B

FIGURE 5–3 Two-phase rotating field

direction of current flow shown in Figure 5–3b, which corresponds to time t_1. The flux due to AA' is at right angles to BB'. The resulting field has the direction shown. Note that the north pole of the resultant field has moved 45° CW. At time t_2, I_A is maximum positive and I_B is zero. With no current flowing in winding BB' and maximum current flowing in AA', the resultant field is as indicated in Figure 5–3c. This field, caused only by the flux from winding AA', has a direction that is 45° in a CW direction to the previous position. Each time the phase of the current advances 45° (electrical), the resultant magnetic field rotates 45° (mechanical).

At time t_3, both currents are positive and equal in amplitude. The direction of the current in winding BB' is the reverse of that in Figure 5–3a and b. The magnetic fields caused by the currents in the two windings combine to form a vertical resultant field, as shown in Figure 5–3d. If we analyzed the position of the resultant field for times t_4 through t_8, we would see that the field continues to move in a CW direction. This rotation continues until the resultant field is back in the position in which it started in Figure 5–3a. In one complete cycle of AC, the resultant field has rotated one complete revolution. The rotating field for a two-phase, two-pole stator makes one complete rotation for each cycle. Its speed in revolutions per second is equal to the supply frequency in cycles per second or Hertz.

A four-pole stator, similar to the one just discussed, can be analyzed in a similar manner. Such an analysis will show that the rotating field advances only 22.5° for every 45° advance in the phase of the current. The rotor speed of a two-phase, four-pole motor will, therefore, be one-half that of the two-pole motor. From the illustrations and discussions on the rotating field, we can see that the angular space between windings in mechanical degrees must be the same as the electrical degrees between the currents in the windings. With two-phase current, the spacing will be 90°. With three-phase current, the spacing will be 60°. Our illustration produced a two-pole rotating field. To produce a four-pole rotating field, the angular space in mechanical degrees must be one-half the number of electrical degrees.

5–2.1 Three–phase Motor Example

The simplified diagram in Figure 5–4 shows sectional views and current windings of a three-phase stator. Note the winding currents shown in Figure 5–4 at time t_0. At this time, the current in phase B, I_B is maximum positive. Let's assume that I_B is 10 amps at maximum. At time t_0, current is flowing into the

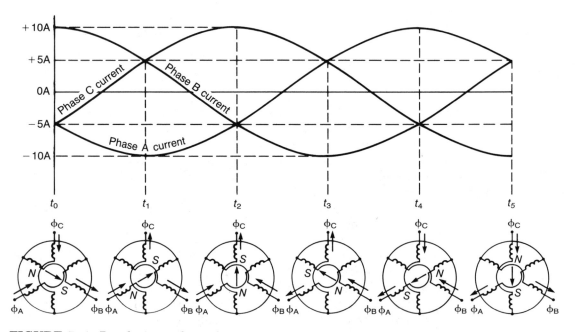

FIGURE 5–4 Developing a three-phase, rotating field

A and C terminals at half value ($-5A$ each in this case). The resulting field at time t_0 is established in the direction shown by the arrow in Figure 5–4. The major portion of this field is produced by the current flowing in phase B, which is at full strength (10A) at this time. The total field is aided by the adjacent A and C fields, which add to it. The field is a two-pole field extending across the space that would normally contain the rotor. This total field is, as shown in the diagram, pointing down and to the left.

At time t_1, the current in phase B is reduced to 5A, half its former value of 10A. The current in phase C has reversed its flow. The current I_C is 5A in the opposite direction. Finally, the current in phase A has increased to 10A without changing direction. The resultant field at time t_1 is now established in a downward position and to the right, as shown in Figure 5–4. The major part of the total field is provided by phase A, which is at full strength at this time. It is aided by the fields created by the lower currents in the B and C windings.

At time t_2, the current in phase C is 10A. The resultant field points to the left. The resultant field continues to rotate in a CCW fashion as time increases. A full rotation of the two-pole field occurs during a full 360° cycle of AC voltage. The direction of rotation is CCW. Interchanging any two of the three line leads to the motor will change the direction of the resultant field rotation.

In Figure 5–4, note that the current sine waves have gone 300° through the six positions shown. The field has also rotated 300° (electrically). If the supplied current were 60 Hz (usual in the USA), the resultant field would be rotating at 60 revolutions per second. To convert this to revolutions per minute, we would multiply the 60 revolutions per second by 60, since there are 60 seconds in a minute. The resultant field would then be rotating at 3600 rpm. If we were to add another set of field windings for each of the three phases of AC applied, we would have a resultant field with four poles instead of two. A four-pole field will rotate half as fast as a two-pole field. As shown in equation 5–1, the speed of the rotating field varies directly as the line voltage frequency and inversely as the number of poles.

$$N = \frac{120\,f}{p} \qquad\qquad \textbf{(eq. 5–1)}$$

where N is the speed of the stator field in rpm (called the synchronous speed), f is the line frequency in Hz, and p is the number of magnetic poles produced by the three-phase winding.

We can use this equation to solve for the synchronous speed of a stator, given the line frequency and the number of poles. For example, what is the synchronous speed of a stator with four poles and an applied frequency of 60 Hz? Using equation 5–1,

$$N = \frac{120\,f}{p} = \frac{120\,(60)}{4} = 1800 \text{ rpm}$$

5–3 INDUCTION MOTORS

The driving torque of both AC and DC motors comes from the same source, the interaction of magnetic fields in the rotor and stator. In the DC motor, the one magnetic field is stationary (the stator field), and the armature, or rotor, with its current-carrying conductors, rotates. The brush-commutator system supplies current from a DC supply to the armature conductors.

The ***induction motor*** is the most commonly used motor in industry. It is simple, rugged, and inexpensive. The induction motor has a rotor that is not connected to an external voltage source. As its name implies, the induction motor rotor gets its power by electromagnetic induction. AC voltages are induced in the rotor by the rotating magnetic field in the stator. The induction in this motor is similar to the type of induction in a transformer. The stator can be considered the primary, while the rotor acts as a rotating secondary.

The induction rotor (shown in Figure 5–5) has a laminated cylinder with slots embedded in its surface. In the most common type of induction rotor, called the *squirrel cage,* bars are embedded in the slots. The rotor bars are composed of aluminum, copper, or another suitable alloy. The rotor bars connect at each end by a metal ring (called the end ring) made of copper or brass. Electrically, this construction (shown in Figure 5–6) gives the squirrel-cage rotor the appearance of an apparatus found in a hamster's cage. The rotor bars carry large amounts of current at low voltages. Although these bars are embedded in an iron cylinder, it is not necessary to insulate the bars from the core. The currents follow the path of least resistance (the bars and end rings) and are confined to the cage.

FIGURE 5–5 Squirrel-cage induction motor rotor

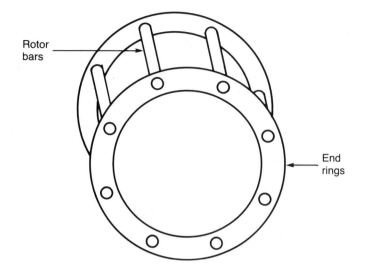

FIGURE 5–6 Basic squirrel-cage rotor construction

5–3.1 Torque Production

Regardless of the type of rotor used, the basic principle of torque production is the same. The rotating magnetic field produced in the stator induces a magnetic field in the rotor. Fields are produced in the rotor bars when the moving stator field cuts the bars, inducing a voltage. Rotor currents flow in the bars because the end rings provide a complete path for current flow. The two fields, rotor and stator, interact and cause torque production. The resulting torque turns the rotor in the direction of the rotating field. To get the maximum interaction between the fields, the air gap between the rotor and stator is small.

To get a better idea of how the rotor produces torque, let us examine Figure 5–7. This simplified cross-sectional diagram shows one pole of the stator sweeping across a rotor bar in a CCW direction. For purposes of illustration, let us state that the rotor bar does not move at first. If we apply the left-hand rule for generators, we will see that the current produced in the rotor bar will be flowing out of the page, symbolized by the dot. Be careful at this point when using this hand rule. The *motion* refers to the relative motion between field and conductor with respect to the field. Since the field is moving and the conductor is stationary, the relative motion is CW with respect to the field. In other words, the same effect would be produced if we held the field still and moved the conductor CW.

A CW field is produced in the rotor bar shown in Figure 5–7. The lines of force will reinforce each other on the right side of the rotor bar. On the left side of the rotor bar, however, the magnetic lines of force will cancel each other out. The resulting distribution of flux will force the rotor CW, producing torque.

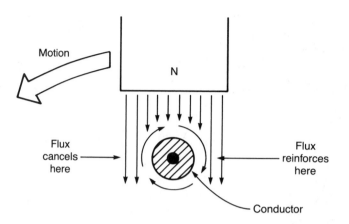

FIGURE 5–7 Torque generation in an induction motor

This torque is proportional to the product of the rotor current, the strength of the stator field, and the rotor power factor. We can confirm the motion of the rotor by applying the right-hand rule for motors. The direction of the flux is down, the direction of the current is out of the page; therefore, the motion of the rotor is CCW.

The induction motor is like a transformer with a rotating secondary. At starting, the frequency of the current in the rotor is the same as that of the stator (primary). The inductive reactance of the rotor bars is large, when compared to the rotor resistance. This means that the current in the rotor lags the voltage across the bars by almost 90°. The power factor of the rotor is said to be low and lagging. The rotor increases in speed in the same direction as the rotating field. As the rotor speed increases, the rate at which the rotating field cuts the rotor bars decreases. Both the voltage induced into the rotor and the frequency of the rotor currents are reduced. If the rotor speed increased to synchronous speed, no voltage would be induced into the rotor bars. No current would be produced, and, therefore, no torque would be generated by the rotor. For this reason, it is impossible for the rotor of an induction motor to turn at the same speed as the rotating magnetic field. If the speeds were the same, there would be no relative motion between the rotor and stator fields. Without relative motion between rotor conductors and stator field, no induced rotor voltage will result. For relative motion to exist between the two, the rotor must rotate slower than the stator field.

5–3.2 Slip

The difference between the stator speed and the rotor speed is called *slip.* The smaller the slip, the closer the rotor speed approaches the stator field speed. Slip is expressed mathematically by the following equation:

$$s = \frac{N_s - N_r}{N_s} \qquad\qquad \textbf{(eq. 5-2)}$$

where s is the slip, N_s is the synchronous speed of the stator in rpm, and N_r is the speed of the rotor in rpm. As an example, what is the slip of an induction motor with a stator speed of 1800 rpm and a rotor speed of 1780 rpm? Using equation 5–2, we find

$$s = \frac{N_s - N_r}{N_s} = \frac{1800 - 1780}{1800}$$

$$= 0.011 \text{ or } 1.1\%$$

We can see from equation 5–2 that the slip at starting is *1*, or 100%. The slip factor can be used to find several useful values. We referred to the fact that the frequency of the currents induced in the rotor varied with the speed of the rotor. The higher the rotor speed, the fewer lines of force are cut by the rotor conductor per unit of time. The rotor current frequency decreases as rotor speed increases. The rotor current frequency can be calculated by multiplying the applied stator frequency by the slip:

$$f_r = s\, f_s$$

where f_r is the rotor current frequency, s is the slip, and f_s is the applied stator frequency (usually 60 Hz). In our previous example, slip was 0.011. If the applied frequency is 60 Hz, the rotor frequency can be calculated thus:

$$f_r = s\, f_s = (0.011)\,(60 \text{ Hz})$$

$$= 0.66 \text{ Hz}$$

5–3.3 Rotor Reactance

You will recall from courses in basic electronics the calculation of inductive reactance:

$$X_L = 2\,\pi f L$$

where f is the applied frequency in Hz and L is the inductance in henrys. Rotor reactance can also be calculated, if the reactance is known. The inductive reactance of the rotor can be calculated from the following equation:

$$X_L = 2\,\pi f_s L\, s \qquad\qquad \textbf{(eq. 5-3)}$$

where X_L is the rotor inductive reactance in Ω, f_s is the applied stator frequency in Hz, L is the inductance in henrys, and s is the slip.

Several observations can be drawn from these equations. We can see that the frequency of the rotor current varies directly with the slip. As slip increases, so does the frequency of the current induced in the rotor by the rotating stator field. Thus, when slip and rotor frequency are close to zero, the rotor reactance and angle of phase lag are very small. When power is first applied to the stator and the rotor is at a standstill, the difference between rotor and stator speed is maximum. The slip is 1. The rotor reactance is, therefore, maximum at starting.

5–3.4 Rotor Torque

The induction motor normally operates between these two extremes of slip. The motor speed under normal conditions, however, is rarely more than 10% below synchronous speed. When slip is 1 (at starting), the rotor reactance is so high that the torque produced by the rotor is low due to the rotor's low power factor. Recall that the power factor is the ratio of real and apparent power. When slip is near zero, which usually happens in an unloaded induction motor, the rotor torque is also low. Rotor torque is low when slip is low because of low rotor current. In an AC induction motor, the torque equation is

$$T = K \phi I_R \cos \theta_R \qquad \text{(eq. 5–4)}$$

where T is the torque produced by the rotor, K is the torque constant, ϕ is the strength of the rotating field, I_R is the rotor current, and $\cos \theta_R$ is the power factor of the rotor current.

Equation 5–4 implies that the product of the rotor current and the rotor power factor for a given strength of magnetic field is maximum when the phase angle between rotor current and voltage is 45° lagging. When rotor current lags rotor voltage by 45°, the reactance of the rotor equals the resistance of the rotor. The rotor power factor is, therefore, 0.707 or 70.7%. Beyond this point, sometimes called the pull-out point, the motor will stall if the load on the rotor is increased.

Another implication of equation 5–4 relates to the torque produced by the induction motor rotor. The flux in the stator is created by the stator current. Stator currents are, in turn, caused by the applied stator voltage. Since the torque produced by the rotor is the result of the interaction of the rotor and stator fields, the torque produced will be directly proportional to the applied stator voltage. As stator voltage increases, so does stator current and stator flux. Increasing flux strength produces more rotor torque. The rotor torque is actually proportional to the square of the stator voltage.

The full-load torque of a motor is defined as the turning force generated by the rotor when the motor develops full-rated load at rated speed. The maximum torque of an induction motor is usually twice the full-load torque. Induction motors can carry loads greater than the full-load torque for a short time. A load in excess of the full-load is called an overload. Momentary overloads can be tolerated for a short time. Overloads that exist for a long time can overheat the motor and damage it.

The starting torque of the average SCIM is from 1.25 to 1.5 times the normal rated torque when started at full voltage. The average SCIM takes approximately five to six times the full-load current at starting. In large SCIMs, starting at full voltage causes starting current to be excessive. The large SCIM usually starts at a reduced voltage. You should be aware that the starting torque is reduced when starting with a reduced stator voltage. As stated previously, the starting torque varies as the square of the voltage. For example, a motor is started at 50% of the full-rated voltage. The motor will start with $0.5 \times 0.5 = 0.25$ or 25% of the starting torque developed at full-stator voltage. The relationship between torque and voltage can be expressed in the form of an equation:

$$T_A = \frac{(V_A)^2}{(V_S)^2}(T_S) \qquad \textbf{(eq. 5–5)}$$

where T_A is the actual starting torque, $V_A{}^2$ is the actual applied stator voltage, $V_S{}^2$ is the full-stator voltage, and T_S is the starting torque developed at rated voltage. For example, a 220-VAC SCIM produces 100 lb · ft of torque at starting. If the voltage is reduced at starting to 132 VAC, what is the starting torque? Equation 5–5 can be used to solve for the new starting torque:

$$T_A = \frac{(V_A)^2}{(V_S)^2}(T_S) = \frac{(132)^2}{(220)^2}(100 \text{ lb} \cdot \text{ft})$$

$$= (0.36)(100 \text{ lb} \cdot \text{ft}) = 36 \text{ lb} \cdot \text{ft}$$

5–3.5 Operating Characteristics of the SCIM

The SCIM can be compared to a transformer with a rotating secondary. In a no-load condition, the rotor cuts the turns of the stator winding. This action generates a CEMF in the stator winding that limits line current to a small value. This no-load value is called the exciting current. The exciting current maintains the rotating field. Because the circuit is highly inductive, the power factor of the motor with no load is poor. The power factor for an unloaded SCIM can be up to 30% lagging. When unloaded, the rotor runs close to synchronous speed, and rotor current is very small. The interaction between the rotor flux on the rotating field is also small.

When a load is placed on the rotor shaft, the rotor slows down slightly. The rotating field, however, continues to rotate at synchronous speed. When loaded, the SCIM slip increases, which causes rotor current to increase. The motor torque increases when the speed decreases and the power output increases. The increased rotor flux, caused by the increased rotor current, opposes the stator flux and lowers it slightly. The primary CEMF, therefore, decreases slightly and stator current increases. This behavior is similar to what happens in a transformer. Because of the low impedance of the stator windings, a small reduction in speed and CEMF can produce large increases in motor current, torque, and power output. We can, therefore, conclude that the SCIM has variable-torque, constant speed characteristics.

If the motor is stalled, the resulting increase in rotor current lowers the stator CEMF and causes excessive stator current. This excessive current can damage the stator windings. When a motor is deliberately stalled or operated in a locked-rotor condition (where the rotor is held stationary), the voltage applied to the stator should not exceed 50% of the rated voltage. For example, if we were operating a 240-VAC motor under a locked-rotor condition, the maximum voltage we should apply to the stator is 120 VAC.

When an SCIM is operated at full load, the stator voltage and current are more in phase than when operated in a no-load condition. The power factor of a motor operated at full load is, therefore, better than one that is operated at no load.

Figure 5–8 shows typical torque and current curves for a three-phase SCIM. Rotor reactance increases with slip and increases its effects on rotor current and power factor as the motor load is increased. The pull-out point on the torque curve occurs at a slip of approximately 0.25 or 25%. The maximum torque at the pull-out point is approximately 3.5 times the full-load torque. At

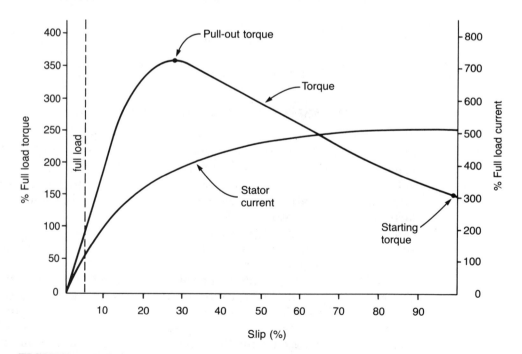

FIGURE 5–8 Induction motor torque-current curves

this point, the rotor resistance equals the rotor reactance. Rotor power factor is 0.707 or 70.7%. Any additional load added to the motor beyond this torque value causes the motor to stall. In the stall condition, the stator current is approximately five times the full-load value.

Performance curves of a four-pole, three-phase 440-VAC, 15-HP SCIM are shown in Figure 5–9. Note that the full-load slip is only approximately 3.5%. In a stalled condition, the rotor reactance of this type of motor is nearly five times the rotor resistance. At full load, however, the rotor reactance is much less than the rotor resistance. When the motor is running, the rotor current is determined mainly by the rotor resistance. Both torque and slip increase up to the pull-out point. Beyond the pull-out point, the torque decreases and the motor stalls. Because the change in speed from no load to full load is small, the motor torque and the power output are considered to be directly proportional.

The SCIM speed regulation is excellent, changing no more than a few percent over the range from full load to no load. How does this good speed regulation happen? As an example, let us suppose that we have an SCIM running at no load. When we load down the motor, slip increases and the frequency of the current induced into the rotor increases. The current in the rotor

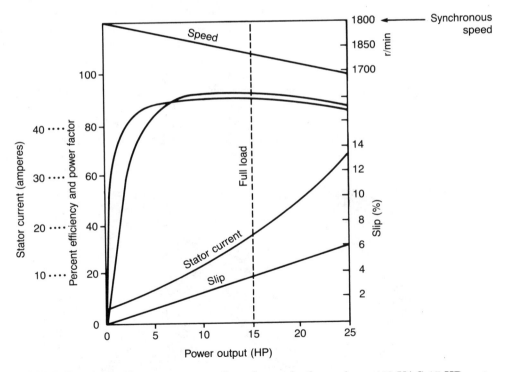

FIGURE 5–9 Performance curves for a four-pole three-phase 440-VAC 15-HP motor (squirrel cage) induction

increases as well, producing a stronger field in the rotor. The rotor torque is, therefore, increased by an amount necessary to handle the extra load. The speed stays approximately the same.

At starting, the slip in an SCIM is maximum. Large currents are induced into the rotor; these currents produce the torque needed to start the motor. As the motor picks up speed, the frequency induced into the rotor decreases, and so does the current in the rotor bars. The torque continues to increase to its maximum value. As the motor continues to accelerate, the rotor current continues to decrease until a point is reached where enough current flows to produce the torque needed to handle the motor load at a constant speed. The same type of equilibrium system is present in the SCIM motor as in the shunt-connected, wound-field DC motor discussed earlier. Increases in load cause increases in the torque needed to handle the extra load at a relatively constant speed.

5-3.6 Squirrel-Cage Induction Motor Construction

Figure 5-10 shows a typical SCIM stator construction. The SCIM stator usually has a laminated iron core. Slots are cut into the core, and the stator coil windings are placed in the slots. The coils are held in place in the core by means of sticks or wedges. Coils of large AC motors are made up of rectangular wire made to fit within the rectangular slots cut in the stator core. Smaller SCIMs use normal circular wire wound into a coil. After the coils are set into the slots, the whole stator assembly is coated with varnish or epoxy resin. The epoxy resin has two functions. It helps hold the stator coils to the core, and it prevents moisture from attacking the insulation.

As in the DC motor, the proper stator coil insulation must be chosen. The classes of insulation are the same for AC and DC motors. Classes A and B are the most commonly found, with B predominant. Class A insulation is made of cotton, silk, paper, or other organic compounds, and filled with an insulator such as varnish or resin. Class A insulation is classed as 105° insulation. Class

FIGURE 5-10 Three-phase induction motor

B insulation is more common today, especially on larger motors. Class B, rated at 130°C, is usually made of fiberglass, mica, or another inorganic insulator. Classes F and H are reserved for special applications. Class F insulation (155°) and class H insulation (180°) are made of special materials that do not break down at high temperatures.

The temperature rise is an important part of the motor's temperature characteristics. The temperature rise is based on a 40°C ambient temperature. Manufacturers determine the temperature rise of a motor from the difference between a deenergized motor and one that has been running for several hours at full load. Class B insulation allows a 90°C rise in temperature from the base temperature of 40°. For this reason, class B is called 130° insulation.

Long motor life and safe operation can be insured by choosing the proper type of motor enclosure. NEMA has standardized motor enclosures according to the method of cooling and the type of environmental protection offered. In general, there are two types of motor enclosures: open and totally enclosed. Open motor enclosures have ventilation passages that allow air to circulate freely. The free air circulation removes the heat, keeping the motor cool. Totally-enclosed motor enclosures do not allow air to flow freely from outside the motor enclosure. According to NEMA standards, each of the following types of enclosures is considered open:

TABLE 5–1   NEMA Open Enclosures

A—Dripproof
B—Splashproof
C—Semi-guarded
D—Guarded
E—Dripproof fully guarded
F—Weather protected

Open motors are the most common type of enclosure in industry. Guarded or protected enclosures limit the size and shape of the openings in these motors. Guarded motors are found in those industrial environments where parts or pieces of material may occasionally be launched into the air. Guarded motors also protect personnel from coming into contact with the motor's rotating parts.

NEMA defines the following enclosures as totally enclosed:

TABLE 5–2   NEMA Totally-Enclosed Motor Enclosures

A—Totally enclosed, non-ventilated
B—Totally enclosed, fan cooled
C—Explosion proof
D—Dust-ignition proof
E—Waterproof
F—Totally enclosed, fan cooled, guarded

Each of these types of enclosures meets a particular standard associated with protection from environmental factors and cooling. For example, a guarded machine is an open machine in which all openings giving direct access to rotating parts are limited in size by screens, baffles, or grills, which prevent accidental contact with hazardous parts. Openings that give access to the motor's rotating parts do not permit the passage of a cylindrical rod 3/4″ in diameter. The splash-proof machine is an open machine in which drops of liquid or solid particles striking the enclosure at an angle not greater than 100° from the vertical do not prevent the motor from working properly. The explosion-proof motor enclosures are designed so that if explosive gas or vapor gets into the motor and ignites, the enclosure will be able to withstand the explosion. Internal explosions in this type of motor will not cause gases or vapors outside the motor to explode. These are examples of definitions NEMA gives to motor enclosures. You can study these different types of motor enclosures in more detail by referring to the appropriate NEMA standards publications.

In discussing how torque is produced in the SCIM rotor, we have talked briefly about rotor construction. Torque is generated in bars that are shorted by end rings. This basic construction is common among all types of SCIMs. Changing the rotor design slightly can make the SCIM more suitable for certain applications. Varying rotor design can change the SCIM torque, speed, slip, and current. Based on the variation in rotor design, there are five basic types of SCIMs: NEMA designs A, B, C, D, and F. Of the five types discussed, the B, C, and D are the most commonly used. Each of these designs has different torque, speed, slip, and current characteristics. Stator construction between these SCIMs remains the same. Figure 5–11 shows the speed torque

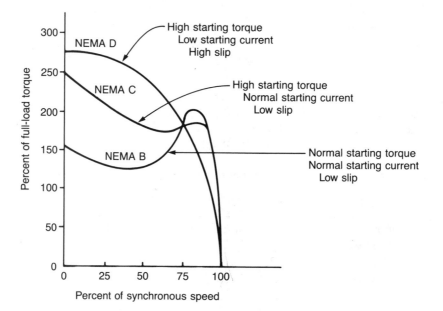

FIGURE 5–11 Speed torque curves for three induction motors

curves of these three common types of SCIMs. NEMA design B is the general purpose SCIM. It has normal starting torque, starting current, and low slip. Low rotor resistance accounts for the low slip. This motor has high efficiency. Starting current is approximately 4½ to 5 times the full-load current. Starting torque is approximately 1.5 times the rated torque; speed regulation is between 3 and 5%. This means that the speed of the motor will change only 3 to 5% from no-load to full-load. NEMA design B has a rotor construction as shown in Figure 5–12a.

NEMA B rotors are found in fans, blowers, pumps, machine tools, various types of mills, compressors (starting unloaded), crushers, chippers, and other applications needing motors that produce only a low starting torque.

NEMA design C is sometimes called the double squirrel cage rotor. This rotor design combines the characteristics of the high resistance and high reactance rotors. Note the construction of the rotor in Figure 5–12b. The bars closest to the surface have a comparatively small cross-sectional area. The other bars, set deeper into the rotor, have a larger cross-sectional area. The bars with the smaller cross-sectional area, usually made of brass, are the high-resistance bars. They have a resistance of a few tenths of an ohm. The bars embedded lower in the rotor have a resistance of a few thousandths of an ohm and are usually made of copper. At starting, the frequency of the currents induced into the rotor bars is high. The inductive reactance of the bottom, low-resistance winding is much higher than the top windings. At starting, the current in the top windings is higher than the current in the lower windings. The top windings produce most of the starting torque. The high resistance of the top bars

a. Type B-
    bars deep and narrow

b. Type C-
    two sets of
    bars

c. Type D-
    high resistance
    bars - close to
    surface

FIGURE 5–12 Three types of induction motor rotors (NEMA)

reduces the phase angle between rotor and stator currents. The power factor of the rotor is, therefore, high. This phase angle reduction produces higher starting torque. As the rotor picks up speed, the frequency of the current in the rotor decreases and slip decreases. As the rotor approaches top speed, the frequency of the induced voltage becomes low, and the inductive reactance of the bottom bars is decreased. The amount of current in the windings is now determined by their resistance rather than their reactance. Since the lower bars have a much lower resistance than the top bars, more current flows in the lower bars at close to synchronous speed.

When operating at a no-load condition, the NEMA design C SCIM behaves as if it were a single-bar NEMA B motor. Under varying load conditions, the current automatically divides between both sets of bar, depending on the slip. As slip increases, more current flows in the upper bars and less in the lower bars. NEMA design C has a starting torque of 200 to 250% of full-load torque and a slip of 2 to 5%.

NEMA C SCIMs are used in conveyors, reciprocating pumps, compressors, and crushers and pulverizers that encounter materials at starting. In general, this motor is used in applications where the motors have loads that are hard to start but must operate at a constant speed with infrequent starting and reversing.

The NEMA design D rotor is sometimes called the high slip motor. The rotor bars are placed close to the surface and have a small cross-sectional area. Starting torques are increased to 275% of full-load torque, while slip ranges between 5 and 13%. As you can see from the NEMA D curve in Figure 5–11, this motor has a higher torque at starting than the B and C designs. Because of the higher rotor resistance, the NEMA design D motor is not as efficient as the other two popular designs.

Applications for NEMA C SCIMs include punch presses, elevators, small cranes, and hoists that are very hard to start and have frequent reversals and starting.

5–3.7 Induction Motor Efficiency and Power Factor

Power Factor As stated earlier in this chapter, the power factor of an induction motor equals the power factor of the current the motor draws from the line. In the SCIM motor, data sheets usually give the motor power factor. The power factor rating is the power factor that exists when the motor meets all rated conditions, such as rated voltage, torque, and speed. Power factors for SCIMs are usually lagging. This means that the motor appears inductive to the line voltage, with stator current lagging applied voltage. The power factor (pf) of a motor is usually given as a percent. It is the ratio between true power and apparent power:

$$pf = \frac{P_T}{P_A} \times 100$$

(eq. 5–6)

where *pf* is the power factor in percent, P_T is the true power in watts, and P_A is the apparent power in kVA. Using an example, let us say we are working with a motor with an input power of 75 kVA and an output power of 56 kW. What is the power factor of the motor? Using equation 5–6, we get:

$$pf = \frac{P_T}{P_A} \times 100$$

$$= \frac{56 \text{ kW}}{75 \text{ kVA}} \times 100 = 75\%$$

Calculating the kVA input to a three-phase induction motor is not done simply by multiplying the stator voltage by the stator current. Recall that the kVA input currents and voltage will be out of phase, with current lagging voltage in the three-phase induction motor. The motor draws rated current from the line when rated voltage is applied to the stator windings. The motor then receives a certain amount of apparent power from the line supply. Since apparent power is measured in kVA, the motor input is called the kVA input. The apparent kVA input depends on the number of phases in the line, the stator voltage, and the stator current. The kVA input to a three-phase induction motor can be calculated as follows:

$$\text{kVA input} = \frac{1.732 \, V_L I_L}{1000} \qquad \text{(eq. 5–7)}$$

where V_L is the applied stator voltage and I_L is the amount of stator current drawn. An example may help at this point. A three-phase 440-VAC induction motor draws 10A from the line at full load. What is the kVA input to the motor? Using equation 5–7, we get:

$$\text{kVA input} = \frac{1.732 \, V_L I_L}{1000}$$

$$= \frac{1.732 \, (440\text{V}) \, (10\text{A})}{1000} = 7.65 \text{ kVA}$$

Efficiency   The efficiency of a motor is the ratio of the mechanical power output to the electrical power input:

$$\text{Efficiency} = \frac{P_{out}}{P_{in}} \qquad \text{(eq. 5–8)}$$

Since the output power is normally rated in HP and the input power is normally given in watts, we need to convert the HP rating of the motor to power in watts. You will recall that 1 HP is equal to 746 W. Equation 5–8 then becomes

$$\text{Efficiency} = \frac{0.746\, P_{out}\ (\text{in HP})}{P_{in}\ (\text{in W})} \qquad\qquad \textbf{(eq. 5–9)}$$

Using an example at this point, we can calculate the efficiency of a 10-HP induction motor that requires an input of 8 kW:

$$\text{Efficiency} = \frac{746\, P_{out}\ (\text{in HP})}{P_{in}\ (\text{in W})}$$

$$= \frac{(746)\,(10\ \text{HP})}{8\ \text{kW}} = .933 \text{ or } 93.3\%$$

When the rated output, efficiency, and power factor are known, these factors can be used to calculate the kVA input to the motor, regardless of how many phases the motor has. The rated kVA can be calculated thus:

$$\text{kVA input} = \frac{0.746\, P_{out}}{p_f\, n} \qquad\qquad \textbf{(eq. 5–10)}$$

where P_{out} is the output power in HP, p_f is the rated power factor as a decimal, and n is the efficiency, also expressed as a decimal. As an example, let us suppose we are working with a 10-HP induction motor with a rated power factor of 87% and a rated efficiency of 85%. What is the rated kVA input for this system? Using equation 5–10, we get:

$$\text{kVA input} = \frac{0.746\, P_{out}}{p_f\, n}$$

$$= \frac{0.746\,(10\ \text{HP})}{(0.87)\,(0.85)} = 10.1\ \text{kVA}$$

5–4 SINGLE-PHASE AC INDUCTION MOTORS

Up to this point, we have dealt primarily with the three-phase AC motor. It predominates in industry where three-phase power is readily available. Small industry and residential and commercial applications, however, do not have three-phase power as readily available. They are more likely to have single-phase AC to supply line voltage. Motors for these applications need to run with single-phase AC power. One of the advantages of single-phase AC motors is that they are less costly to make in small sizes than the three-phase machines. Single-phase AC induction motors are used in fans, refrigerators, washing machines, dryers, drills, grinders, and other machine tools. Single-phase AC motors are usually found in sizes less than 15 HP.

The single-phase induction motor (SPIM) has a squirrel cage rotor as in the three-phase machines we have studied. The SPIM has only one stator winding to run the motor, since there is only a single-phase input. The SPIM produces no starting torque, since there is no rotating field at starting. If the rotor is brought up to speed (or partially up to full-speed) by an external means, the induced currents in the rotor will cooperate with the stator fields to produce running torque. The torque produced will cause the rotor to keep running in the direction in which it was started.

Since the SPIM has no torque at starting, we must have a reliable method of starting the motor and getting it going in the right direction. The most common method of doing this is to fit the motor with an extra stator field called the start winding. This field is excited by the same stator voltage applied to the other stator winding, called the run winding. The currents in each of these windings are out of phase with each other. The phase shift is caused by applying one stator voltage across a reactive component. Single-phase AC motors are classified as either fractional or integral HP motors. They are also classified by the method used to start them. We will, in the next section, discuss several popular starting methods for SPIMs.

5–4.1   Split-Phase SPIMs

One common SPIM is the split-phase type. The split-phase SPIM has a stator core made up of slotted laminations, like the three-phase type. The windings are divided into an auxiliary or start winding and a run or main winding. Physically, these two windings are placed at 90° angles from each other. The start winding has fewer turns of smaller diameter wire than the run winding. The start winding has a higher resistance and less inductive reactance than the run winding to produce the necessary phase shift. The two windings are connected in parallel, as shown in Figure 5–13a. The single-phase line is applied across both windings since they are in parallel. The split-phase SPIM gets its name from the action of the stator during starting. The single-phase stator is split into two separate windings, start and run, that are displaced in space by 90°.

Phasors are often used when trying to explain the operation of AC circuits. A phasor is a rotating vector, shown at an instant of time. Because of the difference in inductive reactance between the two windings, the currents in the two windings are displaced electrically by approximately 15°, as seen in Figure 5–13b. The current in the start winding (I_S) lags the line voltage by approximately 30° and is less than the current in the run winding because of the greater impedance of the start winding. The current in the run winding (I_R) lags the applied voltage by approximately 45°. The total current (I_{line}) during the starting period is the sum of the two currents. The resultant stator field rotates, as shown in the two-phase rotating field example in this chapter.

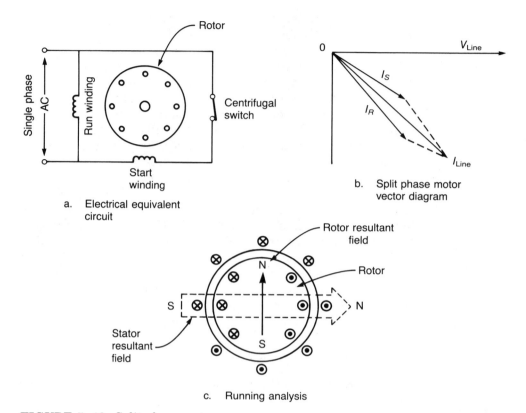

a. Electrical equivalent circuit

b. Split phase motor vector diagram

c. Running analysis

FIGURE 5–13 Split-phase motor

Starting During starting, the resultant stator field, caused by the current in the start winding and by the current in the run winding, rotates around the stator at synchronous speed. As the resultant stator field rotates, it cuts the rotor bars and induces a voltage in them. The induced rotor current lags the rotor voltage by almost 90° because of the high rotor reactance. The interaction of the stator and rotor fields produces torque that turns the rotor in the direction of the rotating field. After the rotor accelerates, the rotor voltage and current are reduced and the current becomes more in phase with the stator field.

When the rotor comes up to almost 75% of synchronous speed, a centrifugally-operated switch opens the start winding, removing it from the circuit. The motor then continues to run from the run winding alone. The rotating field is kept by the interaction of the rotor flux and the stator flux.

We assume that the stator field rotates in the CW direction at synchronous speed. Figure 5–13c shows the stator currents at the instant that the field is horizontal, extending from left to right. The left-hand rule shows that the stator currents will produce a north pole on the right side of the stator and a south pole on the left side. The motor shown is wound for two poles. Apply the left-hand rule for generators to the rotor, with the thumb pointing in the direction of the motion of the conductor with respect to the field. The direction of the induced voltage is back on the left side of the rotor and forward on the right side, as shown by the cross-dot convention. The induced rotor voltage causes a rotor current to flow. The phase angle difference, however, is very small because the slip is small. If we apply the left-hand rule for electromagnets to the rotor, we see a rotor field with a north pole at the top and a south pole at the bottom. The rotor and stator fields are, therefore, displaced by 90° mechanically and by a much smaller electrical angle. Note that the stator field does not actually rotate as in the three-phase SCIMs; it oscillates or pulsates back and forth. The resultant field (the net field produced by the pulsating stator field and the pulsating rotor field) can be considered a rotating field.

The split-phase induction motor has the same constant-speed, variable-torque characteristics as the three-phase SCIM. Most of the split-phase motors are manufactured to run on both 115 and 230 VAC. For 115-VAC operation, the stator coils are divided into two equal groups. These groups are connected in parallel. In 230-VAC split-phase motors, the groups are connected in series. The starting torque of the split-phase SCIM is approximately 150 to 200% of full-load torque. Starting current is approximately six to eight times full-load current. Split-phase SCIMs are usually found in fractional HP sizes. They are used in applications such as washers, oil burners, and ventilating fans. The direction of rotation can be changed by reversing the start winding leads. A 230-VAC split-phase SCIM is shown in Figure 5–14.

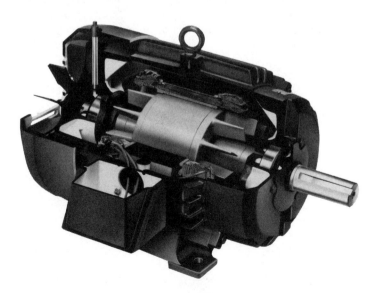

FIGURE 5–14 230-VAC split-phase SCIM

5–4.2 Capacitor Start

The capacitor start SCIM is a modified form of the split-phase starter. Like the split-phase starter, the capacitor start motor has two windings, a start winding and a run winding. The capacitor start motor has a capacitor in series with the start winding (see Figure 5–15a), a component missing from the split-phase starter. The capacitor produces a greater phase difference between the currents in the rotor and stator than the split-phase type. The start winding has many more turns of larger wire and is connected in series with a capacitor. The current in the start winding is shifted approximately 90° from the current in the run winding. Since both windings are physically located 90° from each other in space, the capacitor start motor has higher starting torque than the split-phase type. Figure 5–15b shows the relationship between currents and voltages.

Note that the current in the run winding I_S lags the stator voltage by approximately 45°, about the same amount as in the split-phase run winding. In the start winding, with the right size capacitor, the current I_S leads the current in the run winding by approximately 90°. The total current I_{line} is the sum of the other two currents. Since the stator current and voltage are nearly in phase with the stator voltage, the power factor is close to one. The starting torque, however, will be large, because the currents in start and run windings are nearly 90° out of phase. This situation corresponds to the original two-phase rotating field example in this chapter.

Starting torque can be three to four times the full load torque. Because of the high starting torque, the capacitor start SCIM finds application in loads that are difficult to start. Such applications include refrigeration units, compressors, air conditioners, pumps, conveyors, and some machine tools. The capacitor-start SCIM has the disadvantage of being noisier than the split-phase SCIM.

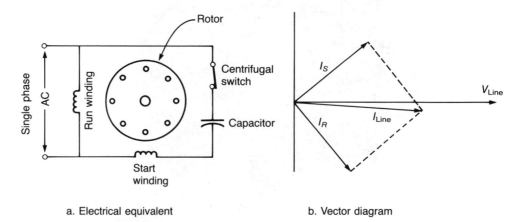

a. Electrical equivalent

b. Vector diagram

FIGURE 5–15 Capacitor start single phase AC motor

A capacitor start SCIM is shown in Figure 5–16. Note the large capacitor attached to the side of the motor. The capacitor is normally a nonpolarized electrolytic capacitor. The capacitor is made nonpolarized by connecting two electrolytic capacitors in series. Electrolytic capacitors for capacitor-start motors vary in size from 80 μf in 1/8-HP machines to 400 μf in 1-HP machines. The direction of rotation can be reversed by switching the leads to the start winding.

In applications where starting is frequent and the motor takes a long time to reach final speed, a permanent-split capacitor motor is sometimes used. In this motor, the capacitor is permanently connected to the start winding, as shown in Figure 5–17. Note that no centrifugal switch is present. The start winding is designed to be used during both starting and running.

Note that the capacitor-start SCIM in Figure 5–15a has a centrifugal switch. As in the split-phase SCIM, this switch opens the start winding after the rotor reaches approximately 75% of its full-load speed. Some larger capacitor-start SCIMs have a capacitor in series with the run winding as well as one in the start winding. Such a motor is called a capacitor-start, capacitor-run, or two-value capacitor-start motor. Figure 5–18 shows an equivalent circuit of this type of motor. Note that one capacitor, usually the larger of the two, is in series with the start winding. The other capacitor, called the run capacitor, is not connected at this time. When the motor starts, the start capacitor provides the phase-shifted current to produce the torque necessary to start the motor. When the rotor reaches 75% of full-load speed, the centrifugal switch connects the run capacitor, disconnecting the start capacitor. Note that the run winding is not disconnected from the circuit, as in the previous types of single-phase AC motors discussed.

FIGURE 5–16 Capacitor start motor

FIGURE 5–17 Permanent-split capacitor single-phase AC motor

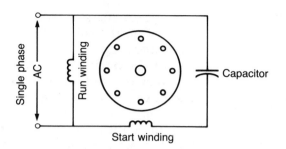

The capacitor-start, capacitor-run motor is normally found in 5- to 20-HP sizes. It combines the advantages of the high starting torque of the capacitor-start motor and the high running torque of the permanent-split capacitor design.

Troubleshooting Split-Phase AC Motors   Most of the split-phase single-phase AC motors use a centrifugal switch. In troubleshooting any system, you should be aware that moving parts are more susceptible to failure than parts that do not move. Since the centrifugal switch is a moving part, it should be checked first if a motor fails to start. After checking the centrifugal switch for proper operation, check to see if the start winding is the proper resistance. An open start winding will not permit the single-phase motor to start. Likewise, a partially-shorted start winding will not allow the motor to generate enough torque to start the motor properly. Checks for proper winding resistance can be made with an ohmmeter.

5–4.3   Shaded-Pole AC Motors

The shaded-pole single-phase AC motor is a low-cost motor that, in small sizes, generates enough torque to turn small loads. Manufactured in sizes from 1/500 to 1/4 HP, the shaded-pole motor is the simplest single-phase AC motor in construction. The rotor of the shaded-pole motor is a squirrel cage, like the split-phase versions. As seen in Figure 5–19, the stator is a salient-pole type. Salient-pole stators are constructed so that the pole pieces stick out. In fact, the

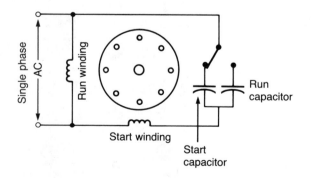

FIGURE 5–18 Capacitor-start, capacitor-run single-phase AC motor

term *salient* means projecting or jutting outwards. The projecting poles on the stator are similar to those found on DC machines, except that the entire magnetic circuit is laminated. A part of the pole piece is split, and a copper coil is placed around the smaller part of the pole. Stator windings are wrapped around the entire pole piece. As shown in the four-pole model, the four coils are connected in series across the motor terminals. The two-pole version is much simpler and less expensive to produce. When the current through the stator is increasing, the flux also increases through the pole piece. As the flux moves through the pole piece, the flux cuts the shading coil, as seen in Figure 5–20a. The flux cutting the copper conductor induces current in the copper ring. The current in the copper ring opposes the flux in that area of the pole piece. The total flux around the pole piece is concentrated on the right side, as seen in the figure. When the flux reaches its maximum value, the shading coil will not be cut. No flux will then be opposing the field around the pole piece. The flux lines will be distributed more uniformly over the entire pole piece. When the stator voltage decreases, the flux in the pole piece will be decreasing. As the main flux decreases toward zero, the induced voltage and current in the shading coil reverse their polarity. The resulting field in the coil prevents the flux from collapsing through the iron in the region of the shading coil. The resulting flux distribution, shown in Figure 5–20b, shows a total field concentrated toward

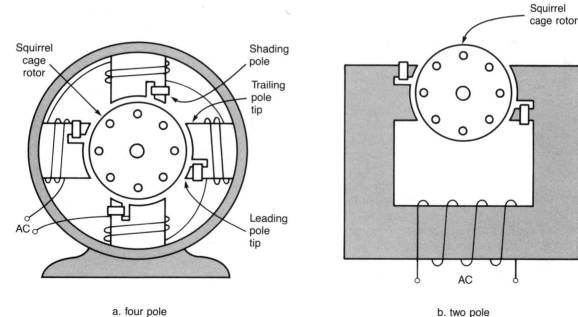

a. four pole   b. two pole

FIGURE 5–19 Shaded-pole motors

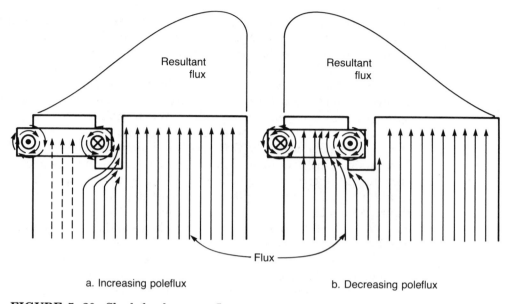

FIGURE 5–20 Shaded-pole motor flux

the left of the pole piece. The shaded pole, therefore, causes the main flux to rise first in the unshaded part of the pole and later in the shaded part. This action is equivalent to a sweeping movement of the field across the pole face in the direction of the shaded pole. The squirrel cage rotor bars are cut by the rotating field. Currents are induced in the rotor bars producing rotor flux. The rotor flux interacts with the stator flux to generate the torque needed to turn the rotor.

Most shaded-pole motors have only one edge of the pole split. The direction of rotation is, therefore, fixed and not reversible. The shaded-pole motor is similar in operation to the split-phase AC motor. It is the simplest and least expensive of all the single-phase SCIMs. Because of its simple construction and lack of moving parts, the shaded-pole AC motor has the lowest maintenance of the single-phase AC motors we have examined. This motor has several disadvantages, however. It has low efficiency, low starting torque, and is noisy. Starting torque is approximately 40 to 50% of full-load torque. The shaded-pole AC motor is found in applications such as small direct-drive fans, blowers, and pumps. In general, shaded-pole motors are used in applications where low cost is more important than a lack of efficiency.

5–5 WOUND-ROTOR INDUCTION MOTORS

Up to this point, we have discussed one particular type of induction motor, the squirrel cage induction motor. Unless we have a method of changing the frequency of the voltage applied to the squirrel cage induction motor, this motor

FIGURE 5–21   Wound-rotor induction motor

is considered a constant speed device. Some time ago, the need arose to have an AC induction motor with a variable speed, which led to the development of the wound-rotor induction motor. As shown in Figure 5–21, the wound-rotor induction motor has a stator essentially the same as a squirrel cage induction motor. The difference is in the rotor. The wound-rotor induction motor (WRIM), is basically a three-phase machine and has a rotor winding similar to

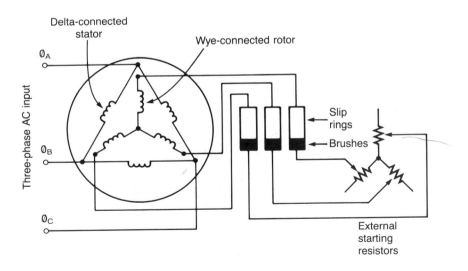

FIGURE 5–22   Wound-rotor motor system

its three-phase stator. The rotor has preformed coils placed in slots in a laminated rotor core. The windings, usually connected in wye, are brought out to slip rings on the motor shaft. Through stationary brushes in contact with the slip rings, the windings connect to three external variable resistors (Figure 5–22). These resistors provide variable resistance for changes in starting or running characteristics. By varying the resistance of the external resistors, any value of resistance can be selected to change the speed-torque characteristics. When the resistors are adjusted so that they are short circuited, the rotor resistance is minimal. Adjusted this way, the WRIM has speed-torque characteristics similar to NEMA design B with a slip of between 2 to 4%. When different values of resistance are added, a whole family of speed-torque curves can be achieved through NEMA design D and beyond. By increasing the resistance of the rotor, we can limit the starting current to a very low value. It is possible to operate the WRIM at very high values of slip. As can be seen in the speed-torque curves in Figure 5–23, each value of external resistance in the rotor circuit gives the WRIM a different speed-torque characteristic.

The WRIM was once the only form of variable speed AC machine available in large power applications. With the introduction of variable speed AC drives, the WRIM has become less popular. Many people thought erroneously that the introduction of solid-state motor drives would make the WRIM extinct. The advent of solid-state controls has found new areas of application.

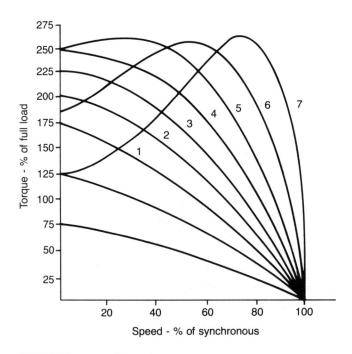

FIGURE 5–23 Wound-rotor characteristic curves—different starting resistances

While not as popular as it once was, the WRIM has been applied in areas where SCIM cannot be used. Two very important characteristics of the WRIM are its ruggedness and its resistance to high rotor currents under locked-rotor conditions. A good example of an application of a WRIM is in rock crushers. In this type of application, the rotor can be brought to a stop as the crusher jams against rock fragments. Since the WRIM has external rotor resistances, the heat built up in these resistances under low speeds and frequent starting and stopping can be dissipated externally. SCIMs quickly heat up in such applications and burn out because the heat is built up in the rotor and cannot be dissipated easily. Another application is in automobile crushers used in automobile junkyards. Many WRIMs are powered by solid-state controllers called cycloconverters, which will be discussed in Chapter 12.

CHAPTER SUMMARY

- Both synchronous and induction AC motors use the principle of the rotating field.
- The induction motor is the most commonly used motor in industry.
- The most common type of induction motor rotor is the squirrel-cage rotor.
- Currents are induced into the squirrel-cage rotor bars by electromagnetic induction.
- The difference between rotor and stator speed in an induction motor is called slip.
- The SCIM has excellent speed regulation characteristics.
- Induction motors, since they are an inductive load, cause lagging power factors when placed on line.
- Single-phase induction motors start by energizing a special stator winding, called the start winding. After the rotor has reached a certain speed, the start winding is usually disconnected.
- The shaded-pole induction motor is a salient-pole type motor.
- The wound-rotor induction motor has a rotor similar to its three-phase stator. Speed and torque are varied by adjusting the rotor resistance.

QUESTIONS AND PROBLEMS

1. How does the three-phase induction motor produce a field?
2. What are the two basic parts of the polyphase induction motor? How does power get from one part of the motor to the other? How does the induction motor create torque?
3. Find the synchronous speed of a four-pole motor with 60 Hz applied to its stator.

4. If an eight-pole induction motor has an applied frequency of 60 Hz and a slip of 5%, how fast is the rotor turning?

5. If a six-pole motor with 400 Hz applied to its stator has a rotor speed of 7600 Hz, what is the slip?

6. What is the slip of the induction motor at starting? Can the induction motor slip ever be zero? Why?

7. If a motor has a slip speed of 160 rpm and a rotor speed of 3440, what is the frequency of the supply if the motor has two poles?

8. Explain why the reactance of the rotor in the induction motor changes with slip.

9. What is the rotor frequency of an induction motor with a slip of 5% and an applied frequency of 60 Hz? What is the rotor frequency at starting?

10. A 460-VAC motor produces 150 lb · ft of torque at starting. If the starting voltage is 230 VAC, what is the starting torque?

11. To which electronic component is the induction motor compared? Why?

12. What happens to the current in the rotor bars at starting compared to the current at full load? What happens to the stator current when the motor is stalled? Why?

13. What does the class of an induction motor's insulation tell you?

14. List three types of NEMA motor enclosures.

15. What is the most common type of motor enclosure?

16. List the five NEMA rotor designs. How do they differ?

17. An induction motor is operated at a power factor of 0.8. The motor puts out 25 W of power. What is the input kVA? What is the power output in HP?

18. A motor has an input of 1000 kVA. If the output is 1.2 HP, what is
    a. the output power in watts?
    b. the power factor of the motor?

19. A three-phase 230-VAC motor draws 10 amps of stator current. If rated voltage is applied, what is the kVA input?

20. What is the efficiency of a 100-HP motor that has an input of 80 kW?

21. A 100-HP induction motor running at rated speed and delivering rated power has an efficiency of 0.9 and a power factor of 0.8. What is the kVA input to the motor?

22. What is the difference between the split-phase, capacitor-start, and shaded-pole AC motors. Describe how each starts each motor.

23. Describe how the wound-rotor induction motor differs from the squirrel-cage induction motor.

6

AC SYNCHRONOUS MOTORS

At the end of this chapter, you should be able to

- identify the parts of an integral HP synchronous motor.
- list and explain two methods used to start synchronous motors.
- define power factor and list two reasons why low, lagging power factor is not wanted.
- explain how the synchronous motor is used to correct power factor.
- define the following terms and how they relate to the synchronous motor:
  - pull-in torque
  - pull-out torque
  - normal excitation
  - underexcitation
  - overexcitation.
- describe how the synchronous motor generates torque.
- list the factors that determine the speed of a synchronous motor.
- explain why the synchronous motor cannot produce starting torque on its own.
- list three types of fractional HP synchronous motors and explain how they generate torque.

6–1 INTRODUCTION

The synchronous motor (SM) is one of the most efficient electrical motors in industry. The unique features of the SM allow efficient, economical conversion of electrical power to mechanical power. Synchronous motors are effectively applied over a wide range of speeds and loads. Because the SM can operate at a leading power factor, it is used in industry to reduce the cost of electrical power. The savings in power cost can pay for the motor in several years.

Functionally, the SM is different from the induction motor in that the SM rotor rotates at synchronous speed. The induction motor rotates at a speed slightly lower than synchronous speed. In the induction motor, the difference between rotor and stator is called slip. In the SM, the rotor rotates at synchronous speed.

6–2 SYNCHRONOUS MOTOR OPERATION

The induction motor uses the principle of the rotating field in motor operation. The increasing and decreasing currents in the stator windings cause the magnetic field to rotate, although the physical poles remain stationary. The rotor of the SM uses this same rotating field principle. In fact, an SM can be converted to an induction motor by replacing the synchronous rotor with an induction motor rotor. The primary difference between SMs and induction motors lies in the rotor. In all SMs, the rotor is accelerated until the rotor and stator are close to the same speed. At this point, the unlike poles of rotor and stator attract, and the rotor locks into step with the rotating stator field.

6–3 SYNCHRONOUS MOTOR CONSTRUCTION

Integral HP SMs, in integral HP sizes, are constructed as shown in Figure 6–1a. This type of motor is said to have ***salient poles.*** The word salient means projecting out. Salient poles are basically electromagnets, with the pole pieces projecting outward, as shown in Figure 6–1b. The pole pieces are laminated, as in the wound-field DC motors. Coils of wire are wound around the pole pieces and are connected to slip rings on the rotor shaft. Most SMs are of the salient-

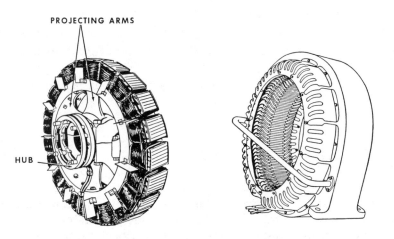

FIGURE 6–1 Integral HP synchronous motor

pole type. The salient poles on an SM are similar in construction to the poles on a wound-field DC motor. The coils are wound around the center of the pole core. In the rotor construction, the SM is similar in construction to the synchronous generator or alternator. In addition to the coil windings on the rotor, many SMs have squirrel cage bars. One of the disadvantages of the SM is the fact that it does not produce starting torque. The squirrel cage windings allow the SM to start. The squirrel cage bars are embedded in the rotor, as in the induction motor, and are shorted by end rings. These squirrel cage bars are sometimes called damper or *amortisseur windings.* There is, however, a difference between rotor construction in a squirrel cage rotor and an SM with *damper windings.* In the induction motor, the rotor bars are evenly spaced and located all around the rotor. In the SM, the damper windings are located only on the pole faces. No current flows in the damper windings. Current flows only when there is a difference between rotor and stator speeds. In the SM, this happens only during starting. As soon as the rotor turns at synchronous speed, all current in the damper windings stops. Since no current flows in the damper windings when the rotor turns at synchronous speed, no power losses occur in the damper windings, which makes the SM run more efficiently than a comparable induction motor.

The motor pole pieces are laminations punched out of metal (Figure 6–2). In larger motors, the laminations are assembled, riveted together, and wound with wire. Each pole piece is fitted into a steel or iron form called a *spider.* Each pole piece is fitted with a dovetailed end to attach to the spider (Figure 6–3a). A bar of metal called a *key* holds the pole piece to the spider. The spider on an SM is shown in Figure 6–3b. This construction technique is used in high speed SMs. In low speed SMs, the rotor is bolted to the spider. At the other end of the pole piece, the rotor is fitted with squirrel cage bars. These bars are either soldered or brazed to end pieces. The end pieces short each of the squirrel cage bars. Squirrel cage bars are made of a nonmagnetic material, such as brass or copper. As in the induction motors, the starting characteristic of the SM with damper bars is influenced by the resistance of bars. The high resistance brass bars produce different starting characteristics compared to motors using low resistance copper bars. The entire rotor assembly is shown in Figure 6–4.

Another construction feature separating the synchronous motor from the induction motor is the *exciter.* Recall that the rotating stator field induces a field in the rotor bars. Many small SMs have a permanent-magnet rotor. Large SMs have a rotor field created by an electromagnet. Current flows from an external source into the rotor windings, creating a magnetic field. The current usually comes from a source called the DC exciter.

6–3.1 Rotor Excitation

The rotor in the SM gets its magnetic field in several different methods. One method is to use a wound-field DC generator, usually mounted on the rotor shaft. A schematic diagram of this type of device is shown in Figure 6–5a. As

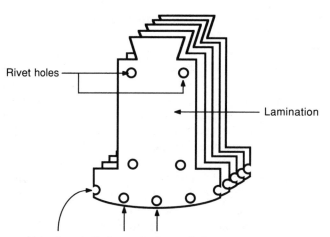

Rivet holes

Lamination

Holes in pole tip for squirrel cage bars

a. Unassembled salient pole

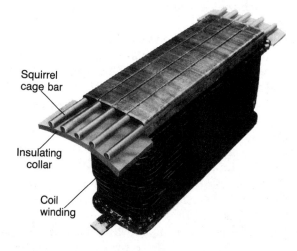

Squirrel
cage bar

Insulating
collar

Coil
winding

b. Assembled salient stator pole

FIGURE 6–2 Construction of an AC motor bolted field pole

the rotor shaft rotates, the DC generator also rotates, producing a DC voltage. The DC voltage is then fed into the rotor of the SM via brushes and slip rings. The current flowing in the rotor windings produces the field in the rotor. The end of the motor attached to the load is called the ***drive end.*** On SMs with exciters, the DC exciter is located on the end of the motor opposite the drive end. On some large SMs, the current required to generate the rotor field can be hundreds of amps. In this case, another exciter, called the *pilot exciter,* controls the main exciter. This arrangement is shown in Figure 6–5b. The exciter using the DC generator has fallen out of industry favor. Downtime and maintenance

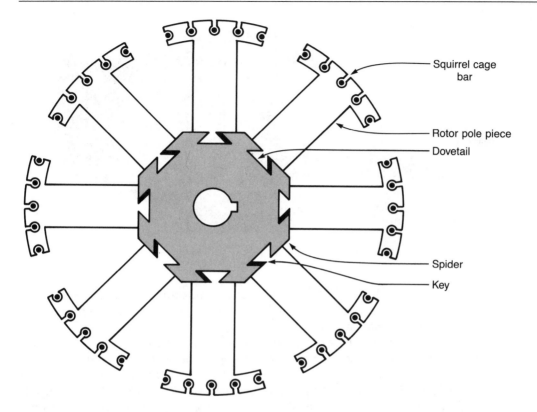

Squirrel cage
bar

Rotor pole piece

Dovetail

Spider

Key

FIGURE 6–3 Synchronous motor

FIGURE 6–4 Entire rotor assembly with fan

costs for replacing brushes and repairing commutators make DC excitation expensive. As the brushes wear, conductive dust gets into the machine, causing further maintenance problems.

Today, brushless DC excitation systems are more popular because they avoid the problems associated with the DC generator. One type of brushless excitation is shown in Figure 6–6a. The exciter is a three-phase alternator instead of a DC generator. The rectifier changes the AC to DC. The DC voltage is then sent to the rotor of the SM. Again, since the excitation currents are high, a pilot exciter (Figure 6–6b) provides a control at a lower value of current and voltage.

Modern excitation methods sometimes get rid of the DC and AC generator altogether. This type of excitation is called *static excitation.* The AC is rectified by one or more high-power semiconductors called SCRs (Figure 6–7). The output of the SCRs goes directly to the rotor of the SM. SCRs can be turned on by an external signal. They are simple, fast, and can carry currents in the thousands of amps. The SCR is discussed in greater detail in Chapter 9. Static excitation has great cost advantages over other types of excitation.

6–3.2 Starting the Synchronous Motor

The stator windings of the three-phase SM and the three-phase induction motor are practically the same. Application of three-phase AC power to the stator of a three-phase induction motor produces a rotating field. At starting, the rotor is stationary and the DC excitation voltage is zero. If the poles were excited with DC, a magnetic field would be created. By looking at Figure 6–8, we can see that no starting torque is produced when the rotor is excited with DC. At time t_0, the sweeping stator field is aligned as shown (Figure 6–8a). No net torque is produced on the rotor. Time t_1 finds the stator field in the position shown in Figure 6–8b. The torque created tends to move the rotor CCW. Before

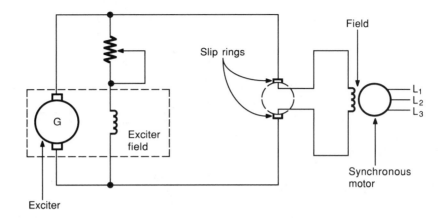

a. DC generator excitation

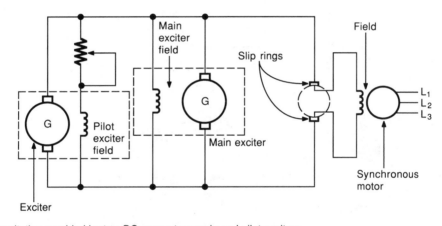

b. excitation provided by two DC generators-main and pilot exciters

FIGURE 6–5 DC generator excitation

the rotor has a chance to move significantly, the rotor is in the position shown in Figure 6–8c. No torque is generated in the rotor at this position. Finally, the stator field sweeps to the position shown in Figure 6–8d. Torque is generated, but this time in the CW direction. During the entire cycle, the torque was first CW, then CCW, averaging zero. The stator fields move too quickly for the rotor fields to lock onto the stator fields. All that happens is motor vibration and, finally, overheating.

The SM needs a special system to start the motor. One obvious way to start the motor is to slow the rotating field down. The rotor then has a chance to lock onto the stator field. This method requires a source of variable frequency AC. Until recently, this was not an acceptable method of starting SMs.

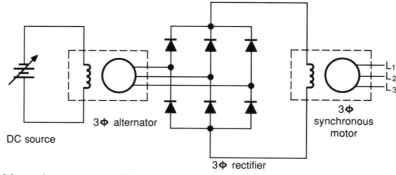

a. 3Φ synchronous motor with brushless excitation

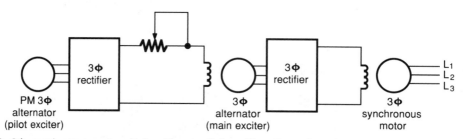

b. 3Φ synchronous motor with brushless excitation with a pilot exciter

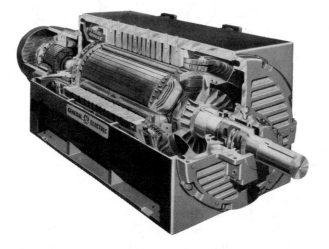

FIGURE 6–6 Synchronous motor brushless excitation

No inexpensive means of providing variable frequency AC existed. With the introduction of circuits that can provide variable frequency AC, this type of starting is used more often.

Another way to start SMs uses windings embedded in the rotor. These embedded windings are called damper or amortisseur windings. If the SM has

Fixed-frequency
fixed-voltage
AC

Phase-controlled
SCRs

Variable DC

FIGURE 6–7 Static DC excitation

damper windings, the motor starts as an induction motor. The rotor is, therefore, driven to almost synchronous speed by the squirrel cage bars in the rotor. The DC excitation voltage is not applied until the motor has accelerated to almost synchronous speed. When the rotor DC excitation voltage is applied, the rotor becomes magnetized. The rotor is constructed so that one of the salient poles is a north pole and the adjacent pole is a south pole. North and south poles alternate around the salient pole rotor. As an example, when a north pole on the stator passes over a particular pole on the rotor, DC power is applied. The south pole of the rotor is attracted to the north pole of the stator. The rotor then locks into step with the stator field. The salient pole rotor produces torque only after locking into step with the rotating stator field. The speed of the rotor then is identical to the stator synchronous speed. This is why the motor is

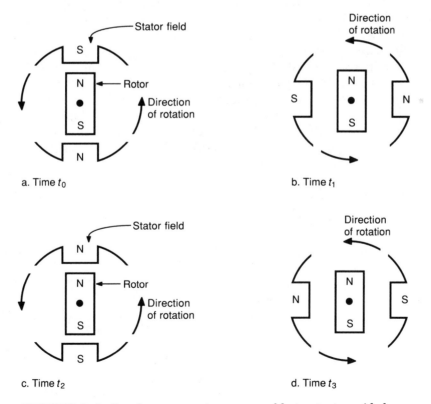

a. Time t_0

b. Time t_1

c. Time t_2

d. Time t_3

FIGURE 6–8 Synchronous motors are unable to start unaided

called a synchronous motor. The speed of the SM depends only on the frequency of the applied stator voltage.

A final method of starting synchronous motors uses an external system to drive the synchronous motor rotor to near synchronous speed. These starting motors are called ***pony motors.*** The pony motor can be an AC induction motor or a DC wound-field motor.

Applying Excitation Voltage   The SM is classified as a constant-speed machine. As long as the line frequency does not change, the rotor rotates at a constant synchronous speed. The DC excitation voltage must be applied at the correct time and with the correct polarity. Figure 6–9 presents an undesirable situation. What if power were applied when two north poles are created as shown? Instead of the rotor pulling into step, the opposite will happen. The rotor will slow down suddenly. When this happens, the air gap flux is reduced and a large amount of current rushes into the stator from the line. If DC rotor excitation continues to be applied, large fluctuations in stator currents occur as north and south rotor poles pass under the stator. The best time to apply DC excitation voltage is when a rotor pole of opposite polarity passes under a stator pole. This brings up an important point about SM construction. An SM should have the same number of stator and rotor poles.

Another important point concerns the torque produced by the motor when it pulls into synchronization. Obviously, there is a certain attractive force that pulls the rotor into synchronization with the stator field. If the force produced by this attraction is not large enough, the rotor will never be pulled into lock. The amount of torque needed to pull the rotor into synchronization is called the ***pull-in torque.*** Once the rotor is synchronized with the stator, the torque load on the rotor can be increased. If the load torque is too great, the magnetic tie between the rotor and stator is broken. The motor quickly comes to a stop when this tie is broken. The torque at which the rotor pulls out of synchronization with the stator is called the ***pull-out torque.*** Increasing DC

FIGURE 6–9   Field application

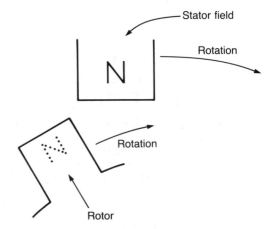

excitation current increases the attractive force between the rotor and stator fields. Increasing the excitation can then increase the pull-out torque. Further increases in the load torque without an excitation increase will reach a point where the rotor pulls out of synchronization with the stator.

Load Effects in Synchronous Motors The power factor of an induction motor depends on the load and varies with it. The power factor of an SM can be one or less than one, either leading or lagging. In the induction motor, the power factor can only be lagging.

When a load is applied to the rotor, it causes the poles to be pulled a certain number of mechanical degrees behind their no-load position, shown in Figure 6–10a. The loaded rotor pulls back to the position in Figure 6–10b. The CEMF induced into the stator is that same amount of electrical degrees behind the applied voltage. The resultant voltage in the stator causes the stator current to lag behind the applied voltage. In this condition, with the stator current lagging behind the stator voltage, the power factor is lagging. The SM then appears inductive to the line. If the load is increased further, the rotor poles are pulled further behind the stator poles. This causes the stator current to lag further behind stator voltage. The power factor decreases, and the stator current increases. We should point out here that the load cannot continually be increased. A point will be reached where the flux link between the rotor and stator will be broken. This happens when the rotor pole is about halfway between the stator pole it is attached to and the one behind it. The torque produced at the point where the motor loses synchronization with the load is called the pull-out torque.

Because the speed is constant, if we decrease the rotor excitation, the CEMF decreases. The resultant stator voltage, therefore, increases. The stator current increases and lags the applied voltage V_A by a greater amount. On the other hand, if we increase the rotor field excitation until the current I_S is in phase with the applied voltage, the power factor is one. The power factor will be one, however, for this particular value of load. For one particular load value,

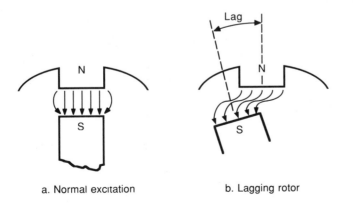

a. Normal excitation   b. Lagging rotor

FIGURE 6–10 Rotor-stator position with load

the stator current and resultant voltage are minimum when the power factor is unity. If excitation is increased past this point (when power factor is unity), stator current increases and leads the applied voltage. The motor presents a capacitive reactance to the line and is said to have a leading power factor. If the power factor is leading, the motor is said to be overexcited. For a particular load, the power factor is controlled by the amount of excitation in the rotor. A weak rotor field produces a lagging motor power factor. On the other hand, a strong rotor field produces a leading power factor. When the stator current is in phase with the applied voltage, the power factor is one. Synchronous motors are designed by manufacturers to be operated at rated HP at either unity power factor or leading power factor. Synchronous motors come in two values of leading power factors, 0.9 and 0.8.

Synchronous motor curves, which indicate the changes of current for a constant load and varied rotor excitation, are called *V curves* and are shown in Figure 6–11. The corresponding changes in power factor are also shown.

Before moving on to the SM as a power factor correction device, we need

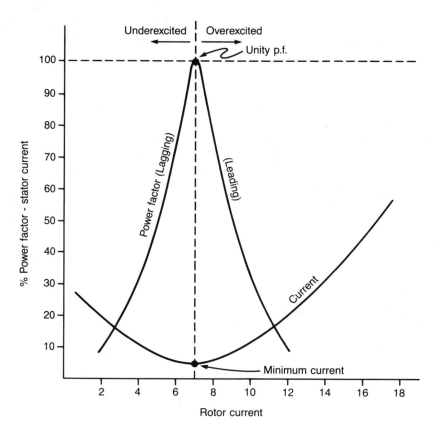

FIGURE 6–11 V-curves for a 15-kVA synchronous motor

to mention the torques associated with SM operation. An SM has many important torques that relate to how well the motor starts, accelerates, and pulls into step. First, the ***starting torque*** (or breakaway torque) is the torque produced at starting when the speed is zero. The ***accelerating torque*** is the torque produced between starting and pull-in speed. The ability of the motor to synchronize is defined by the term pull-in torque. The ***pull-in torque*** is the maximum amount of load torque that can be synchronized with a specified amount of load inertia. This parameter is sometimes called the actual pull-in torque. The nominal pull-in torque is the torque developed at 95% of synchronous speed. The ***nominal pull-in torque*** is often used to compare SMs. The ***synchronous torque*** is the torque developed during synchronous operation. Finally, the ***pull-out torque*** is the maximum torque developed by the motor for one minute before it pulls out of step due to an overload.

Power Factor Correction Synchronous motors are used in certain systems to change the power factor of the system. Industrial power users can have

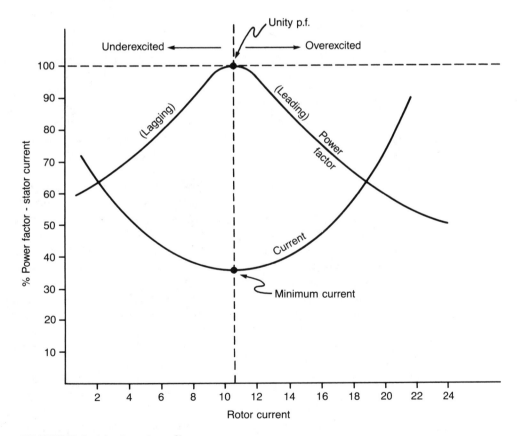

FIGURE 6–11 *(continued)*

many reasons for wanting to change the power factor of the system. Large numbers of induction motors and transformers make the power factor of an industrial location low and lagging. Several important problems arise when a power system has a low, lagging power factor. Low, lagging power factors lower the supply voltage. A lowered supply voltage can cause motors to overload and other electronics equipment to work improperly. Lower power factor reduces the ability of the system to carry power. The kVA rating of a piece of equipment determines the amount of kVA it can carry, that is, its capacity. When power factor is low, larger generators, transformers, and switches must be used for each kilowatt of load. The power factor of the load determines the power handling capability of transformers. Transformers are rated in kVA. Their kilowatt-carrying capacity depends directly on the power factor of the load. At low power factors, the effective capacity of the transformer is reduced. Reduced power factor makes it necessary to have larger power distribution lines and increases the voltage drop in those lines. Power factor also has an effect on the cost of electricity. Power company kilowatt-hour meters register the energy used by industrial loads. A low, lagging power factor means that the kilowatt-hour meter registers the use of energy, with no useful work done by the reactive current. Most power companies assess penalties for low, lagging power factor and give incentives for high power factor.

The best way to deal with low, lagging power factor is to prevent it from happening. In the case of induction motors, a lightly-loaded motor has a high proportion of lagging, reactive kVA. The first step in preventing low, lagging power factor is to make sure all induction motors are loaded as fully as possible. Excessively high voltages can also decrease power factors. The second step, then, is to make sure line voltages are correct. Capacitors can also be used to improve the power factor. Finally, the SM can be used to adjust the power factor.

Power factor is adjusted by adjusting the excitation current of the SMs attached to the line. If operated without a load, the SM power factor can be adjusted to a value of 10% leading. When operated in this condition, the motor is called a synchronous condenser. It is similar in operation to a capacitor (or condenser) because it makes current lead voltage. The synchronous condenser takes only enough true power from the line to supply its losses. At the same time, it supplies a high leading power factor. The high leading power factor cancels out the lagging power factor from such inductive loads as induction motors. The entire power factor of the line is improved. With enough SM excitation, the line power factor can be adjusted to one. When the SM is used strictly for adjusting power factor, it is called a *synchronous condenser.* Synchronous condensers are built to operate without a mechanical load. It makes sense to use SMs for all appropriate applications, such as in compressors. The SM in this type of application can be adjusted so that it has unity power factor. It will not contribute to the lagging power factor problem, like an induction motor. It adds to a plant load without requiring reactive kVA. When it is partially loaded, it can operate at a leading power factor, helping boost the power factor of a site. Synchronous motors, then, do double duty. They can drive a load economically, efficiently, and precisely. They also improve the power factor of the system.

6–4 FRACTIONAL HP SYNCHRONOUS MOTORS

The majority of the synchronous motors in industry are fractional HP, rather than integral HP, machines. Most fractional HP SMs do not use slip rings and brushes, as do the integral HP machines. Three specific types of fractional HP machines do not use slip rings and brushes: the reluctance, hysteresis, and permanent-magnet synchronous motors. These motors are usually very small and do not produce very much torque.

6–4.1 Reluctance Synchronous Motor

The reluctance synchronous motor is very similar to the squirrel-cage induction motor. If grooves were cut into the rotor of the SCIM motor, you would have essentially the rotor of a reluctance SM. The grooves cut into the rotor form salient poles on the rotor (Figure 6–12). Recall that salient poles jut or stick out of the rotor toward the stator. The rotor in Figure 6–12 is called the cloverleaf rotor. The squirrel cage bars in the rotor make this machine start as if it were an induction motor. The rotor accelerates to almost synchronous speed.

The starting torque of the motor depends on the type of material used in the construction of the squirrel cage bars and their position in the rotor. When

FIGURE 6–12 Reluctance synchronous motor rotor structure

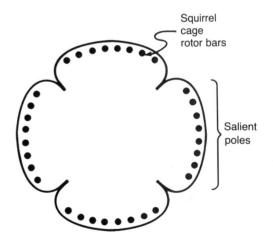

the rotor approaches synchronous speed, the salient poles of the rotor are attracted by the stator poles. At synchronous speed, the salient poles pass slowly by the rotating stator poles. The rotating stator field pulls the salient poles of the rotor. At some point, the salient poles become magnetized by the slowly passing stator field. At a rotor speed below approximately 95% of synchronous speed, the rotor has too much inertia. The pull of the stator does not generate enough torque. The rotor poles will not lock in step with the stator field. Above 95% of synchronous speed, the magnetism induced in the rotor by the stator field is much greater than the field induced in the squirrel cage bars. As synchronous speed approaches, slip and current induced in the squirrel cage bars will drop off. The field in the squirrel cage bars approaches zero. The pull of the stator generates torque in the rotor and pulls the rotor into step with the stator fields.

This type of SM has about 1/3 to 1/2 of the horsepower rating of comparably sized induction motors. The reluctance SM must, therefore, be two to three times larger than an induction motor of the same rating. The reluctance SM has poor efficiency, and power factor is worse than a comparable induction motor. Since the reluctance SM has no external excitation, power factor cannot be adjusted. Despite these disadvantages, the reluctance SM is used in many constant-speed applications, such as tape transports and turntables, recording instruments, and synchronous timers. Reluctance SMs are rarely found in sizes greater than 10 HP.

6–4.2 Hysteresis Synchronous Motors

The hysteresis SM uses the principle of magnetizing metal in its rotor. C. Steinmetz showed as early as 1908 that the magnetic hysteresis principle could be used to drive an AC motor. It was not until the 1940s that this idea resulted in a practical AC motor. The stator of a hysteresis motor is similar to the conventional induction motor stator with slots for coils cut into a laminated core. The rotor is made of a single piece of hard, heat-treated steel. The rotor is a perfectly round cylinder with no salient poles like the reluctance SM. A cross-sectional view of the rotor is shown in Figure 6–13a. A side view of this rotor can be seen in Figure 6–13b. When a rotating stator field moves past this rotor, the crossbars present a low reluctance path for the magnetic flux. Poles are induced into the steel with the magnetic north and south poles aligning along the lines of the crossbars. The poles that are formed lock into synchronization with the rotating stator field. Below synchronous speed, the motor gets its torque from the hysteresis effect. As the stator field moves across the rotor, the magnetic domains in the rotor tend to follow along with the stator magnetic fields. The flux created by the stator field cannot follow the stator field perfectly because the rotor material resists the magnetizing force of the stator fields. A certain amount of power is expended in getting the rotor fields to rotate. The power exerted in changing the position of the domains is called

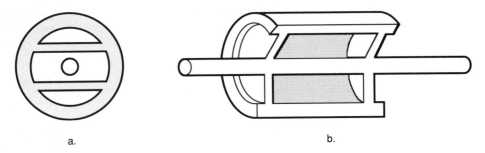

a.   b.

FIGURE 6–13 Hysteresis synchronous motors

hysteresis loss. The sweeping stator fields induce eddy currents in the rotor material, adding to the torque generated by the hysteresis effect. When the rotor reaches synchronous speed, the stator field does not move relative to the rotor. All torque produced by eddy currents and hysteresis stops. The rotor, however, remains magnetized and locks into step with the stator field.

Hysteresis synchronous motors come in sizes of only fractions of 1 HP, around 20 W and under. Because the hysteresis SM produces torque all the way up to full speed, it finds application in driving loads with high inertia, such as in tape transports. It is much more commonly used in this application than the reluctance SM. It has efficiencies that are close to induction motors of comparable size.

Because the motor produces torque by hysteresis and eddy currents, which have relatively low power drains, the hysteresis motor has low starting currents when compared to induction motors of the same size. These motors are, therefore, used in applications where starting currents must be kept low. Since the hysteresis rotor is a smooth, nonsalient construction, its operation is smooth and quiet, which makes it ideal for tape transports and turntables. Variations in turntable and tape speeds, called flutter, can be reduced to a very low value.

6–4.3 Permanent-Magnet Synchronous Motors

As in DC motors, permanent magnets are also used in the rotors of synchronous motors (Figure 6–14). Recall that, in the integral sizes of synchronous motors, a field was created by an electromagnet, with current flowing from an excitation source through brushes and *slip rings.* The only difference here is that the rotor is a permanent magnet rather than an electromagnet. The permanent-magnet SM has the same difficulty starting as does the larger synchronous motor with slip rings. The permanent-magnet SM usually has squirrel cage bars built into the rotor. As with the reluctance SM, the rotor starts as an induction motor. As the rotor approaches synchronous speed, the rotor permanent magnets lock into step with the stator fields. It is important that this

FIGURE 6–14 Rotor of a permanent-magnet synchronous motor

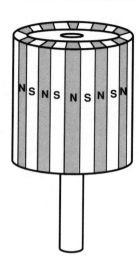

Figure 6-14

motor's rotor have the same number of permanent magnet poles as there are stator poles.

Because of the strong permanent magnet fields in the rotor, this motor has a relatively high pull-in and pull-out torque. The permanent-magnet SM has high power factor and efficiency, rivaling that of the polyphase induction motor. It can be found in sizes ranging up to 2 HP (approximately 1.5 kW).

6–5 CONTROLLING SYNCHRONOUS MOTORS

We have already stated that the integral HP SM needs a source of DC called the exciter. In high speed SMs, the exciter can be a DC generator mounted on the rotor shaft. The controller for the SM has requirements that the induction motor controller does not have. For example, the SM controller must have a method of applying the DC excitation at the right time to pull the rotor into step with the stator. In general, the SM controller for large SMs must meet six requirements.

First, the SM controller must have a method of applying power to start and of removing power to stop. This requirement is usually met by a high-power switch, which will be discussed in Chapter 8. Typically, a start switch applies power to a high-power relay that, in turn, applies power to the stator of the motor.

Second, the motor stator windings should be protected from the damage that can be caused by short circuits. This kind of protection is provided by a semiconductor or conventional fuse or by a circuit breaker. The fuse or breaker is designed to trip when a certain predetermined current level is reached. It removes power by introducing an open circuit in the power line or from the relay that is supplying the power to the motor.

Third, the SM controller must be able to protect the motor against over-loads. An overload condition causes higher than normal currents to flow in the motor. These higher currents are not enough to trip the fuse or circuit breaker. Overloads are caused by several factors, such as an excessive mechanical load or low line voltage. Whatever the cause, overloads can damage the motor by causing it to overheat. High temperatures reduce the life of the stator winding insulation and shorten the life of the motor. In extreme cases, the conductors can melt or cause the motor to catch fire. Controllers must then have a method of detecting overloads and alerting the operator and shutting down the equipment.

The three requirements mentioned above are necessary in every motor, not just synchronous motors. The rest of the requirements apply particularly to the SM. The fourth requirement is that the controller must protect the damper windings from damage. Damper windings are especially liable to damage when the motor is stalled or started and stopped too much. Remember that the SM is not designed to run as an induction motor, only to start as an induction motor. Heavy current flow in the damper windings can cause permanent heat damage, since the SM rotor is not designed to dissipate a lot of heat from the damper windings. This type of protection is usually provided by some type of **_thermal overload relay_** in parallel with a mechanical relay. Both of these devices are in series with the excitation circuit. The exciter windings are usually shorted out during SM starting. The thermal relay usually has the same type of heating characteristics as the damper windings. As seen in Figure 6–15, the thermal relay is in parallel with a coil. When the motor starts, most of the current flows through the thermal relay, since the coil has a high reactance to the AC current when slip is one. As the motor picks up speed, more current flows through the coil and less through the thermal relay. If the motor slows down and runs as an induction motor for any length of time, the thermal relay will heat up and trip when the damper windings heat up.

Fifth, the motor must be protected against the high currents that occur at

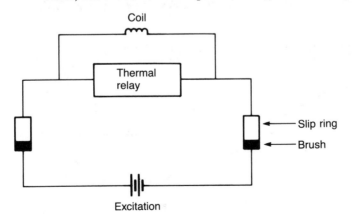

FIGURE 6–15 Damper winding protection

pull-out. Generally, these currents are not high enough to trip fuses and circuit breakers and can cause damage before the overload circuitry can react. If the field remains excited when the motor pulls out of synchronous operation, high pulsing currents will flow and the power factor will be lagging. In addition to the motor, the load can also be damaged at pull-out. Damage to the load can result from torque pulsations, which can be greater than three times rated torque. The rotor excitation must be removed when the rotor breaks out of synchronous operation. One device used to protect the motor system from damage at pull-out is the ***power-factor relay (PFR).*** The PFR has a current and voltage coils and monitors the power factor of the motor system. If the motor pulls out of synchronous operation, the PFR removes the excitation. The motor then runs as an induction motor. Many solid-state motor protectors (discussed in more detail in Chapter 7) allow power factor to be monitored and the excitation to be removed if necessary.

Last, the controller should be able to apply the excitation at the right time and remove it when the motor is turned off. A device capable of accomplishing these functions is called the angle application relay. There are a number of different types of angle application relays that sense different parameters in the motor. When the controller connects the synchronous motor to the line, a voltage is induced into the rotor. The frequency of that voltage is equal to the slip frequency, the difference between rotor and stator speed. When the motor is stalled, the slip frequency is equal to the line frequency, usually 60 Hz. As the motor picks up speed, accelerated by the damper windings, the slip frequency decreases. We can see this graphically illustrated in Figure 6–16. The top synchrogram shows the 60-Hz supply. The middle part shows the stator current. The lower curve shows the current induced into the rotor (field) winding. During the starting interval, the field winding is shorted through a

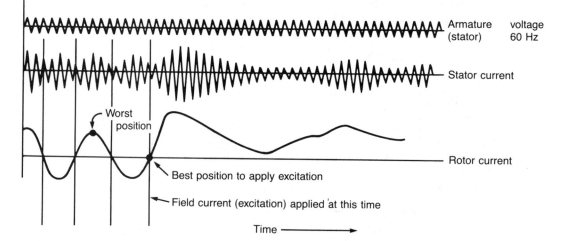

FIGURE 6–16 Application of synchronous motor excitation

resistor, known as the field discharge resistor. The best and worst times to apply the rotor excitation are noted on the synchrogram. The best time is when the rotor is close to the appropriate stator pole. If excitation is applied at this time, the rotor pulls in and synchronizes smoothly and powerfully with a minimum of line disturbance. The worst time is when the rotor is far away from the appropriate stator pole.

A PFR is shown in Figure 6–17. At the moment of starting, the armature of the relay pulls into position C, opening the contacts that apply excitation to the rotor. Relay coil B is in series with the rotor excitation winding. The high AC resulting from the large slip at starting causes the relay to pull closed at position C. A constant DC is also applied to the relay coil A, polarizing the relay. Coils A and B are wound so that the flux in the entire relay during the positive half cycles is the difference between the AC and DC fluxes. During negative half cycles, the resultant relay flux is the sum of the AC and DC fluxes. As the rotor speed increases, the slip frequency decreases. The reactance of coil B decreases, causing a reduction in the current in coil B. At approximately 95 to 96% of synchronous speed, the resultant flux is not strong enough to hold the relay armature. The relay armature pulls out to position D, closing the rotor excitation circuit. Because of the polarization, the resultant flux during the negative half cycles is always greater than that of the flux during positive half cycles. This guarantees that the relay will drop out during a positive half cycle. This action, plus small delays in the relay, applies the excitation at the best time.

The point in the starting cycle at which the rotor excitation is applied has a great effect on the amount of pull-in torque the motor generates. The maxi-

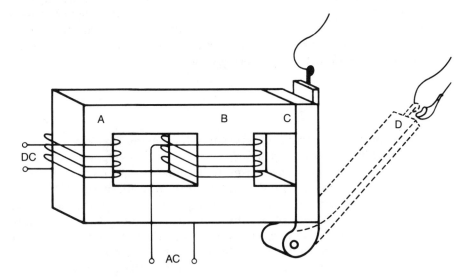

FIGURE 6–17 Power-factor relay

mum load that the motor can pull into synchronous operation varies with the point where the excitation is applied. The pull-in torque is a function of the point where the excitation is applied. According to Figure 6–17, the pull-in torque can vary as much as 35% between the places where the excitation is applied. The difference between the best and worst places to apply excitation changes with the synchronous motor. The difference, however, can be as much as 50%. The oscillations of the rotor current after synchronizing are due to the large flywheel effect of the mechanical load. In the starting method shown in the figure, an angle application relay senses the change in frequency of the induced voltage to determine the best time to apply the rotor excitation. Some angle application relays sense the variations in field current, while some sense the change in power factor. Most often, the angle application relay senses the induced rotor current.

6–6 SYNCHRONOUS MOTOR APPLICATIONS

Synchronous motors are used in two types of applications. First, they are applied where their power factor correction and efficiency are needed. Although the power factor issue is an important one, it is not the only factor considered when looking into a motor for a particular application. After considering the extra expense in the exciter, synchronous motors are cost competitive. As a rule of thumb, the SM is less expensive than the induction motor in sizes greater than 250 HP and speeds under 1200 rpm. In small motor applications (10 to 200 HP), the induction motor is cheaper. In these low power ratings, another requirement sometimes arises that dictates the choice of a synchronous motor. This brings us to the second general area of SM application. Some motors must run at exactly the same speeds and in synchronism with each other. An example of such an application is in the manufacture of man-made fibers. All the machines in the process line must be turning at exactly the same speed or the quality of the product will suffer.

6–7 MOTOR EFFICIENCY

Efficient motor operation has been of increasing concern to industry since the early 1970s. Historically, energy in the USA was cheap and available. Only recently have energy costs risen to become a major part of the cost involved in producing a product. The Federal Energy Administration (FEA) recently sponsored a study on electric motor usage in the USA that found that electric motors consume 64% of all the electric power used in this country. Of that power, 43% is consumed by the industrial sector. Specifically, motors rated 1 to 125 HP consume 26% of all the generated electric power. The FEA report stated that electric power quadrupled in the 1970s and was expected to triple in the 1980s. Conservation of energy resources is not the only issue here. The more energy it takes to manufacture a product, the more the manufacturer must charge for that product. This factor is one of several that drive up the cost of our goods and keep prices from being competitive. Reducing energy costs increases productivity and reduces operating costs. An example may help at this point. Let us suppose that we pay $3000 for a 100-HP AC induction motor. If the motor runs continuously, as in a pump application, the annual operating cost for this motor, based on a conservative price of $0.05/kWh and an efficiency of 90% will cost approximately $30,000 to operate for one year. This cost is ten times the cost of the motor. Even small improvements in motor efficiency will reap large savings in reduced operating costs.

A motor's efficiency is a measure of how effectively the motor converts electrical energy to mechanical energy. We can define efficiency as the ratio of power out to power in. The difference between the electrical power we put in and the mechanical power we get out is called a motor's losses. This difference is illustrated graphically in Figure 6–18. In this system, a 50-HP motor is coupled to a load, a pump in this case. The total power input is 41.2 kW. Of that input power, 37.3 kW is converted to mechanical power to drive the pump. The motor consumes 3.9 kW of power in making this conversion. The 3.9 kW is called the motor's losses. Losses can be loosely grouped into two areas: mechanical and electrical.

Every AC motor has five losses that determine its efficiency. The energy lost in the motor is converted to heat that is dissipated by fans or other cooling methods. Let us take, for example, a NEMA type B, 10-HP induction motor, operating at 1750 rpm. The largest losses are electrical and come from the rotor and stator I^2R losses. Current flowing through the rotor and stator windings causes this loss. The power loss is proportional to the square of the current flowing through the windings times the resistance of the windings. These losses are 435 W for the stator loss and 215 W for the rotor loss in our motor example. The next electrical energy loss is called iron loss. Iron losses are caused by hysteresis and eddy currents in the cores of the rotor and stator.

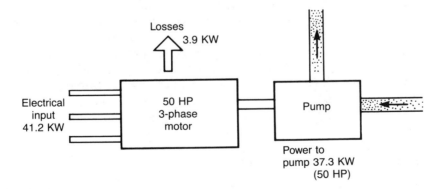

FIGURE 6–18 Electrical-mechanical system

These losses are reduced by laminating the cores of these structures. Iron loss is also reduced by using a grade of steel for the core that is less susceptible to this loss. Iron loss for our example motor will be approximately 300 W. The next electrical loss is called the stray-load loss. This loss is caused by high frequency flux pulsations in the rotor and stator cores. It amounts to approximately 200 W of loss in our example motor. Finally, the mechanical losses from friction and windage are approximately 20 W in our motor. Motor manufacturers must reduce these losses to improve motor efficiency.

6–8 TROUBLESHOOTING AC MOTORS

The AC motor, whether synchronous or induction, is a simple device. Troubleshooting the AC motor is not a difficult task. Before attempting troubleshooting on any motor, you should become familiar with all available information on the unit. This information can take the form of an equipment history record or a technical manual. If an equipment history record is not available, you should make a thorough, systematic inspection of the unit. Visual inspection can offer much information and can suggest possible causes of the trouble.

6–8.1 Visual Inspection

Inspection should begin by deenergizing the equipment to be inspected. It is important that the circuit breaker be tagged by the person who will be doing the maintenance. The purpose of tagging the breaker (called tagging out) is to prevent a motor from being accidentally energized. In the next step, excessive dirt, grease, and oil should be removed from the surfaces of the unit (commonly called cleaning). With the unit clean, inspect leads for burned, cracked, or unserviceable insulation where it enters the machine. Observe the end bells for cracks and mismatch of mating surfaces. Check the bearings for wear and looseness. Disconnect the motor from the load and try to turn the rotor by hand. If it does not turn freely, inspect the unit for a bent shaft, misalignment of end bells, loose or frozen bearings, a loose pole piece, or foreign objects in the motor. Use caution here and wear canvas gloves to protect the hands from the sharp edges of the rotor keyway when turning the rotor shaft. Make SURE that the unit is deenergized before doing any of these maintenance procedures.

Measure and record the air gap between rotor and stator, if it is accessible. Checks of the air gap can give a clue to possible bearing wear or misalignment. For large open machines, make air gap measurements at four equally spaced points around the rotor. The air gap measurements should be within +/− 5% of each other.

6–8.2 Operational Test

Problems encountered in the visual inspection can require further maintenance procedures. If the visual inspection reveals nothing unusual, it is time for an operational test. Make sure that the motor is loaded properly before making any operational test. First, note and record all *nameplate* data. A nameplate from a 100-HP motor is shown in Figure 6–19. According to the National Electric Code, the following information should appear on the nameplate of a polyphase AC motor:

1. the maker's name

2. rated voltage and current

3. rated frequency (usually 60 Hz) and number of phases

4. rated full-load speed, at rated frequency and voltage

5. rated temperature rise or the class of insulation and rated ambient temperature

6. time rating

7. rated power if more than 1/8 HP

8. code letter indicating locked rotor input

9. NEMA frame classification code letter.

In special application motors, other information, as specified by Article 530 of the NEC, may be needed. Check all terminal connections and jumpers. Make sure that the connections are tight and correctly made. At this point, check the quality of the insulation in the stator by checking the resistance between the stator coils and ground or from one set of coils to another. In the synchronous motor, check to see that the brushes are making contact with the commutator. Connect the motor to the power source, making sure that the source is deenergized. Make sure that the unit is connected to a proper load. Connect the test equipment (ammeters, voltmeters, oscilloscopes, power factor meters, etc.). Start the unit, observing all the steps and safety precautions in the manual.

If the motor fails to start, check the applied voltage. Since we have already ruled out mechanical problems, such as a bent shaft and locked bearings, the problem is most likely electrical. If the correct voltage is applied to the motor, we can assume that the problem is within the motor. If the voltage is not present, the problem may be in the motor control circuits (called the con-

FIGURE 6–19 Nameplate from a 100-HP AC motor (courtesy of Reliance Electric)

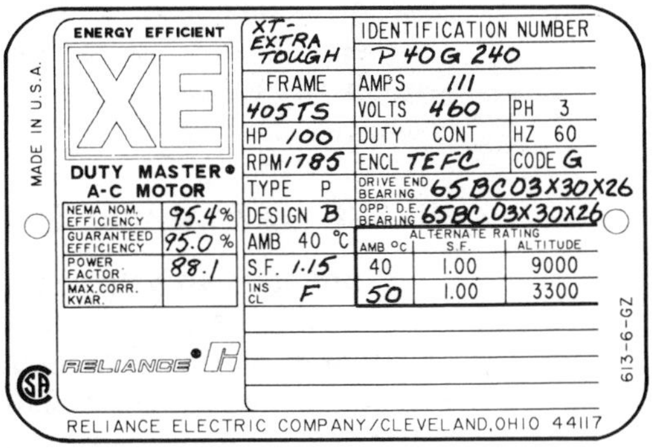

troller). Check for blown fuses or circuit breakers and other possible sources of an open circuit condition. The problem may also be too low a voltage applied to the motor. If this is the problem, look at the controller or the input power source to the controller. Another possible cause for the failure of an AC motor to start is an open or shorted stator. This condition can easily be checked with an ohmmeter. Be sure to deenergize the unit before making an ohmmeter check.

If the motor runs slowly, the problem may again be in the line voltage. The line voltage may be too low to allow the machine to develop the right amount of starting torque. We have not ruled out the load as a possible problem. The load may have increased significantly, causing the motor to turn more slowly. The load should be examined to see if this has occurred.

If the motor overheats, there are many possible causes. The motor may simply be overloaded. The load needs to be examined to see if it has increased the torque demand on the motor. The load may have to be decreased. If the load cannot be decreased, a new and more powerful motor may need to be installed. Overheating can be caused by losing a phase in a three-phase system or by a decrease in voltage of one of the phases. Phase voltages should be checked carefully. The motor should have been checked for proper ventilation during the visual inspection. This is also a possible cause of overheating.

In the synchronous motor specifically, the following symptoms or problems can appear:

1. rotor fails to pull into step.
2. no excitation is present.
3. the rotor pulls out of step.

Possible causes for the rotor not pulling into step could be: a) no excitation, b) open rotor coils, c) the exciter has no output voltage, or d) the switch that applies the excitation could be faulty. The motor may also be overloaded. What do you think might cause lack of excitation? The rotor coil and the slip rings may be open or grounded. The exciter may not have an output voltage. The rotor pulling out of step can have several causes. The rotor can pull out of synchronous operation if the exciter voltage or the stator voltage is too low. Remember that the exciter produces the field in the rotor. The cables that supply the voltage to the stator may have an intermittent short circuit, which removes the power briefly.

6–8.3 Diagnostic Tests

Certain tests can help you to locate problems quickly and efficiently. It may be helpful at this point to discuss testing procedures and results of tests. One common problem in AC motors is in the stator windings. Stator windings can be shorted, grounded, or burned open. The presence of a shorted or open winding is easily tested with an ohmmeter.

Opens in stator windings can be caused by vibration or a poor connection. Whatever the cause, neither induction nor synchronous motors will start with a winding open. Both types of motors will run if the open occurs while they are operating and are lightly loaded. Motors running with one winding can cause an overload and destroy the motor due to overheating. Note the wye-connected winding in Figure 6–20a. Placing an ohmmeter between phase A and phase B indicates low resistance. The ohmmeter placed between phase A and phase C and between phase B and phase 3 will indicate a high resistance. The phase in common, phase C, has the open circuit. The same strategy would be used in troubleshooting the delta-connected winding in Figure 6–20b. In the delta, phase AB will read twice the AC resistance, but will not read infinite ohms. Shorted windings are checked in the same way. The indication is, however, a lower-than-normal resistance.

Another problem commonly found with stators is the grounding of a winding due to insulation breakdown. The insulation resistance can be checked by a megger. The megger is placed between the stator lead and ground and should produce a reading in the hundreds of megohms. Another method is the voltmeter method, shown in Figure 6–21. To perform this check, connect a voltmeter to one side of the line through a fuse. Connect the other lead of the voltmeter to one of the stator leads. Connect the case of the machine to the grounded side of the power source. If the voltmeter indicates a voltage near or equal to the source voltage, the winding is grounded. A reading of a lower voltage indicates a high resistance ground. A meter reading of 0 V indicates no ground on the winding. In a three-phase stator, each lead must be checked individually. This check can also be done with an ohmmeter but is better done with a voltmeter or megger. The ohmmeter may not have enough voltage in its source to break down the winding insulation resistance. As in all tests, exercise caution and follow appropriate safety procedures.

Another test that can be made on a motor, especially a three-phase motor,

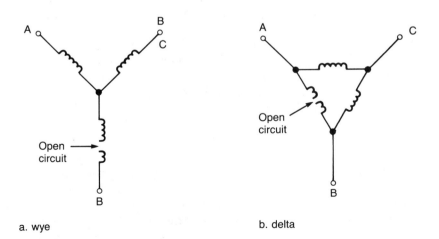

a. wye   b. delta

FIGURE 6–20 Checking for open circuits in three-phase AC circuits

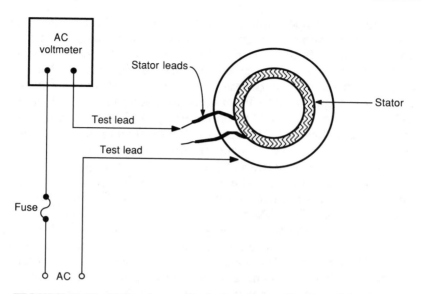

FIGURE 6–21　Voltmeter method of checking for grounds

is shown in Figure 6–22. It uses a device called a clamp-on ammeter, shown in Figure 6–22a. The clamp-on ammeter should be clamped around each of the three-phase leads to the motor, one at a time, as shown in Figure 6–22b. It is important that the ammeter be placed around only one lead at a time. If placed around two or more leads, the fields created by the out-of-phase currents can cancel, giving a false reading. Under loaded conditions, each phase should draw approximately equal currents. A phase imbalance (that is, one phase drawing more or less current than it should) indicates a problem in the motor that should be studied further.

Induction motors present a special problem in checking the squirrel cage bars, since they have no external electrical connection. Squirrel cage rotor bars can be checked with the following procedure. First, connect the motor to a power supply with current and voltage regulating ability. Provide a means to rotate the rotor mechanically. Small induction motors can be rotated by hand. Next, connect an ammeter in one phase of the stator (a clamp-on ammeter will do). Apply a reduced voltage to the stator winding on which the ammeter is connected. The voltage applied to the stator winding should be 25% or less of the rated stator voltage. Turn the rotor slowly. Observe the variations in stator current. Any changes greater than 3% indicate an open or high resistance squirrel cage winding. An additional check can be made by turning the rotor one complete revolution and counting the ammeter pulses. If the ammeter pulses are caused by open or high resistance bars, the number of pulses will be equal to the number of poles in the stator winding.

On an induction motor, it will be necessary to check the speed of the rotor. Slip calculations require a precise reading of the rotor speed. Speed can be measured on small machines by a hand-held tachometer, as shown in Figure

FIGURE 6–22 Clamp-on ammeter (courtesy of Simpson Electric Company)

6–23a. With larger motors (where it would be dangerous) or on small motors (where the shaft is not accessible), non-contact tachometers are used. A non-contact, photoelectric tachometer is shown in Figure 6–23b. Light reflected back to the photosensor in the probe from a reflective strip on the motor shaft is used to make the speed measurement. The speed of the shaft will then be proportional to the pulses of light reflected back to the sensor.

Whatever the problem in the AC motor, a good grasp of the theory of

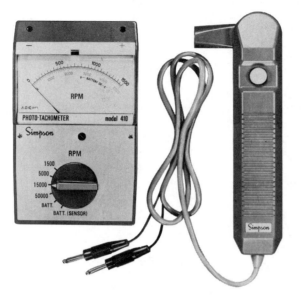

FIGURE 6–23 Photo-tachometer (courtesy of Simpson Electric Company)

operation of the motor will suggest troubleshooting strategies. The trouble-shooting task should proceed in a systematic manner. We have not discussed all the troubles you are likely to encounter in troubleshooting an AC motor. You have seen many problems that are likely to develop in motors. Every troubleshooting situation is different, since all motors, controllers, and loads are different.

6–9 UNIVERSAL MOTORS

The ***universal motor*** gets its name from its ability to operate in DC or in AC circuits. It is normally found in fractional HP sizes. The direction of rotation of the universal motor does not depend on the polarity of the supply. If a universal motor is connected to an AC source of power, it will develop torque in one direction only.

Electrically, the universal motor is a series-connected field DC motor. The diagram in Figure 6–24 shows the basic electrical construction of the universal motor. Using the left-hand rule for coils, we can see that magnetic fields are created with the polarities shown. The instantaneous polarities of both field and armature oppose each other and torque is developed. Now let us reverse the polarity of the applied voltage. Using the left-hand rule for coils, we can see that the magnetic fields created in both the armature and field still oppose each other. This occurs because reversing the voltage reverses the cur-

FIGURE 6–24 Universal motor—electrical equivalent circuit

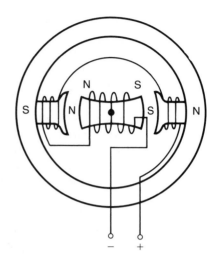

rent in both the armature and the field. The applied AC voltage causes these reversals to take place continuously at a rate of 60 times per second.

A simple DC series motor does not operate well on AC. Hysteresis and eddy current losses that occur in the unlaminated parts of the motor reduce the motor's efficiency and cause heat losses. In addition, the alternating current creates large currents in the coils that are short-circuited by the brushes. This causes excessive arcing and sparking at the commutator. Finally, the inductance of the field and armature windings causes a very low, lagging power factor. The universal motor is specially constructed to deal with these problems. The eddy current losses are reduced by laminating the field poles, yoke, and armature. The hysteresis losses are minimized by the use of high permeability silicon steel for the laminations. The high inductance of the field is lowered by using shallow pole pieces, fewer turns of wire, and reducing the air gap. The inductance of the armature is lowered by using compensating windings in series with the armature.

Like the DC series motor, the universal motor is used in applications where starting torque is high. The universal motor is basically a variable speed machine. Recall that the induction motor and synchronous motors are constant speed machines. Universal motors have similar operating characteristics to series DC motors. Speed and load are inversely proportional. When load decreases, speed increases, and vice versa. No-load speeds are high but generally not high enough to damage the motor, as in the series motor. The speed of a particular universal motor depends on the applied voltage and the physical design of the motor. Most universal motors are manufactured for use at speeds in excess of 3600 rpm. Many universal motors operate at speeds up to 12,000 rpm and can go as high as 20,000 rpm. The average speed for machine tools is between 3600 and 7500 rpm. The speed of the universal motor is directly proportional to changes in voltage. For example, if the applied voltage is lowered by 5%, the speed decreases by approximately that same amount. For small motors, speed control is achieved by placing a variable resistance in series with the applied voltage. Increases in the resistance increase the voltage drop across the resistor, lowering the voltage to the motor and decreasing its speed.

Because of its high operating speeds, the universal motor has the highest horsepower-per-pound ratio of any DC motor. A typical 1/2-HP 1750-rpm induction motor weighs approximately 30 pounds. A 1/2-HP 19,000-rpm universal motor weighs approximately 2 1/2 pounds. The efficiency of universal motors is not as good as many other types of AC motors. Small DC motors may have an efficiency of 30%. Larger motors have efficiencies up to 75%.

Because of its brushes and commutator, the universal motor has a relatively short operating life. Universal motors used in appliances require brush replacement after 300 to 1000 hours of use. If the motor is specially built, the brush life can be extended to approximately 5000 hours. Because of the expense, applications involving the universal motor should be carefully studied.

CHAPTER SUMMARY

- The synchronous motor (SM) uses the principle of the field, as does the induction motor.
- Most SM rotors use salient poles.
- SMs require special devices to develop enough torque to start them.
- Damper or amortisseur windings are special squirrel cage windings used to start SMs.
- SMs can appear inductive, resistive, or capacitive, depending on the amount of excitation.
- If an SM is overloaded and pull-out torque is exceeded, the motor will come to a stop.
- The SM is a constant-speed device, unlike the induction motor.
- Fractional HP SMs use PM rotors or the hysteresis or reluctance principle to generate the rotor field.
- Application of the excitation in an integral HP SM must be done at the proper time.
- The efficiency of a motor is a measure of how effectively the motor converts electrical energy to mechanical energy.

QUESTIONS AND PROBLEMS

1. List and explain two methods used to start synchronous motors.
2. Discuss the construction of the synchronous motor. How does it differ from the construction of the induction motor?

3. Define power factor and list two reasons why low, lagging power factor is not wanted.

4. Explain how the synchronous motor is used to correct power factor in an industrial site.

5. Define the following terms and how they relate to the synchronous motor:
   a) pull-in torque
   b) pull-out torque
   c) normal excitation
   d) underexcitation
   e) overexcitation.

6. Describe how the synchronous motor generates torque. How is the operation of the synchronous motor different from the operation of the induction motor?

7. List the factors that determine the speed of a synchronous motor. What is the most practical way to change the speed of a synchronous motor?

8. Why is the synchronous motor unable to produce starting torque on its own? How are synchronous motors started?

9. List three types of fractional HP synchronous motors and explain how they generate torque.

10. List five things that a synchronous motor controller must be able to do.

11. Explain why it is important to apply the excitation at the right time in the motor's starting cycle.

12. What is meant when we say that a synchronous motor is overexcited and underexcited?

13. What is a synchronous condenser and for what purpose is it used? Can it be used to drive a mechanical load?

14. List some applications of fractional HP synchronous motors.

7

MOTOR CONTROL DEVICES

At the end of this chapter, you should be able to

- describe the difference between full-voltage and reduced-voltage starters.
- describe the operation of a solenoid.
- contrast the electromagnetic and solid-state relays.
- calculate inrush and sealed current in an electromagnetic relay.
- define and give examples of pilot devices.
- contrast overcurrent and overload conditions in a motor.

7–1 INTRODUCTION

All motors, regardless of size or type, need a form of control. At its simplest level, control is needed to start the motor, accelerate it to its running speed, and stop it. The simplest control device is the *motor starter.* Other applications can require more complex circuitry to do more complex tasks. We may want to run the motor at several speeds, start it and stop it at regular intervals, and change its direction. In these cases, we would need a more complex control circuitry system. Such a system is called a *motor controller.* This chapter will discuss those electrical devices used widely in industry for the control of motors. Electronic controller devices will be discussed in Chapter 10.

Types of Controllers Motor controllers can be classified as either manual or magnetic. As its name implies, the *manual controller* is controlled directly by the operator, usually from only one position. All functions, including acceleration, starting, stopping, and reversing, are done by hand. The manual controller is limited in size and capacity to the smaller motors and motors that can be operated locally.

Magnetic controllers control power to the motor by relay action. The magnetic controller performs all basic functions by electromagnets. Small control circuit currents energize coils that open and close power circuits to the motor. The control circuit coil is energized manually by a switch or automatically by a limit, float, or pressure switch. After the pushbutton (or other pilot device) is switched, the magnetic controller automatically performs all its functions. Magnetic control circuits provide an important advantage over manual controllers, the ability to operate from remote locations. With manual control, the controller must be located in a place easily accessible to the operator. With magnetic control, the pushbutton stations or other pilot devices can be operated anywhere, with connections to the coil provided by control wiring.

Controllers can also be classified as full-voltage or reduced-voltage, depending on which voltage is applied to the motor terminals at starting. *Full-voltage controllers* put the full line voltage across the motor. Also called *across-the-line starting,* this method draws the full inrush current at starting as well as providing the maximum starting torque. *Reduced-voltage controllers* apply less than the full line voltage across the motor terminals at starting. As the motor is accelerating or after it has reached full-load speed, the line voltage is increased to the full amount.

7–2 MOTOR STARTERS

Manual motor starters are used to start and stop a motor and to provide overload protection. The motor starter is used to control equipment that is not frequently turned on and off and where automatic control is not necessary. These starters are also used widely as a way to disconnect the motor from the line.

A manual motor starter is basically an ON-OFF switch with overload protection in the form of *overload relays.* The manual starter is operated by hand with a toggle switch or a pushbutton, as shown in Figure 7–1. The switch is then operated by a mechanical linkage from the button or handle. Manual starters are used on fans, blowers, compressors, pumps, small machine tools, and conveyors. Because of their simple construction, they cost less than other types of motor starters and do not have the annoying AC hum caused by the coils of AC magnetic starters. Manual starters, usually two-pole or three-pole switches, remain energized until the stop button or lever is actuated or until the overload relay trips. Manual starters are used in single-phase motors up to 3.7 kW (about 5 HP) at 240 V. They are used on three-phase motors up to 10 kW (about 15 HP) at voltages up to 600 V.

7–2.1 Reversing Starters

Industrial application often requires that a motor's direction of rotation be reversed. Three-phase motors can be reversed by reversing the connections of any two of the three stator connections. The *reversing starter,* as shown in

FIGURE 7–1 Manual motor starter (courtesy of Allen-Bradley Company)

Figure 7–2a, is actually two separate starters wired together. When the F contacts close, lines L1, L2, and L3 are connected to terminals T1, T2, and T3, respectively. A special mechanical interlock prevents both F and R contacts from being closed at the same time. When the F contacts are opened and the R contacts are closed, L1 is connected to terminal T3 and L3 is connected to T1. The motor should now run in the opposite direction.

Figure 7–2b shows the reversing starter with two overload relay contacts, an overload relay, and a mechanical interlock to prevent both forward and reverse contacts from being closed together.

7–3 SOLENOIDS

A *solenoid* is an electromagnetic device that converts electrical energy to mechanical energy. Other devices, such as relays, use this mechanical motion created by the solenoid to do work. In the case of the relay, the solenoid's motion is used to close a set of relay contacts. The diagram in Figure 7–3 shows a cutaway diagram of a solenoid. Note that the solenoid is an electromagnet. You will recall that an electromagnet is a coil of wire with current flowing through it, wrapped around an iron core. The difference between this electromagnetic solenoid and other types of electromagnets lies in the core construction. Note that the core is divided into two pieces, one stationary and the other free to move. The movable part of the core, sometimes called the *plunger,* is kept away from the stationary part by a spring.

When current is allowed to flow through the coiled conductor, a magnetic

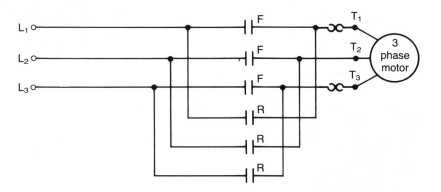

FIGURE 7–2 Reversing starter (courtesy of Allen-Bradley Company)

field is created in the pole pieces, with north and south poles created as shown
in Figure 7–3. Note that the north pole in the movable part of the core and the
south pole in the stationary part will attract, moving the right part of the core
inward, as shown in Figure 7–3b. When current stops flowing in the coil, the
attractive force disappears, and the movable core goes back to its previous
state (Figure 7–3a). The strength of pull on the plunger depends on several
factors. The strength of pull depends on the distance between the plunger and
the stationary core piece. The more distance between them, the weaker the
attraction. For short distances, the pulling strength is strong. As the distance
increases, the strength of the attraction drops off steeply. In general, the force
or pull of a solenoid is inversely proportional to the square of the distance
between the pole pieces. The force depends on the magnetic flux density generated when current passes through the coil windings. The force is also affected
by other interrelated factors, such as the length of the iron path, the magnetic
saturation properties of the solenoid case and plunger, and the area and shape
of the pole pieces.

FIGURE 7–3 Cutaway diagram of a solenoid

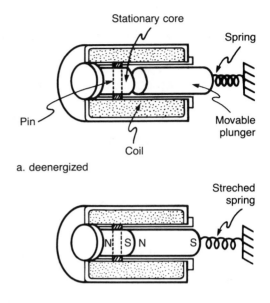

a. deenergized

b. energized

When the solenoid is used as a control device, the current to the coil is usually energized by another type of switching device. One of the advantages of the solenoid is isolation, both mechanical and electrical. The solenoid can provide control at a considerable distance from the controlling device. The electrical control circuit can also be independent of the electrical circuit controlled. This isolation adds to operator safety, since the control circuit can be low-current and low-voltage, while the controlled circuit can be high-current and high-voltage.

7–4 RELAYS

Electromechanical relays are the workhorses of industry. Today, relays can be grouped into two broad classifications: solid-state and electromechanical. The basic idea behind the relay is the concept of an electrically controlled switch, which is used to control the power to high current/high voltage loads or to isolate a control circuit from a load. Both the solid-state relay and the electromechanical relay can accomplish these tasks.

7–4.1 Electromechanical Relays

The electromechanical relay (EMR) uses a principle similar to the principle of the solenoid just described. The solenoid, however, has a movable plunger, while the relay does not. Unlike the solenoid, the relay has switch connections, called contacts, that open and close when the relay is actuated. As a generalization, most relays are actuated by heat (thermal relay) or by current flowing in an electromagnet (electromagnetic relay). In motor circuits, a relay is used either to protect the motor or to control it. Protection is achieved by disconnecting the motor from the line voltage if the motor overheats or is drawing damaging short-circuit currents. Relays can be used to control a motor's starting, stopping, direction, or speed. Before we talk about specific relays, let's review the basics of relay operation.

Construction Figure 7–4 shows three popular types of relays. When current flows through the coil in each of these relays, the armature moves to its "closed" position.

The coil, armature, and assembly make up the magnetic circuit. When current flows through the coil, it creates a field in the stationary part of the magnetic circuit, called the iron magnet assembly. This assembly supports the coil, which is wound around it. The *armature* is the moving part of the magnetic circuit. When the coil current creates a magnetic field, the armature is attracted to its "closed" position. By closing, the magnetic circuit is completed in much the same way as an electrical circuit. To provide maximum pull to

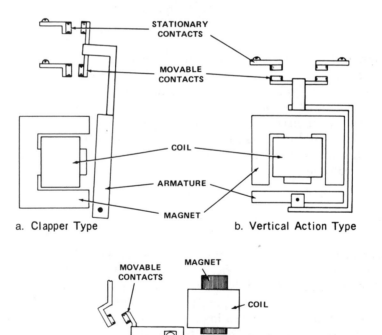

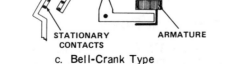

FIGURE 7–4 Three types of relays (courtesy of Square D)

close the contacts and to help insure quiet operation, the faces of the armature and the magnetic assembly are smoothed to a close tolerance.

Note that the same principle applies in each of the circuits just discussed (Figure 7–4). In the clapper type (Figure 7–4a), often used in low power relays, the armature is hinged. As it pivots to close under magnetic force, the movable contacts close against the stationary contacts. In the vertical action device (Figure 7–4b), the armature and contacts move upwards in a straight line. The bell-crank lever (Figure 7–4c) device transforms the vertical action of the armature motion into a horizontal contact motion. The momentum of the armature is not transmitted to the contacts, reducing **contact bounce** that occurs when the contacts come together and extending contact life. Since the contacts are made of a metal with elasticity, the contacts will not stay together on contact. In addition to bouncing being annoying, each time the contacts separate, arcing and oxidation occur. Separation reduces the life of the contacts.

Shading Coils Until now, we have assumed that the relay coils were supplied with DC. In many cases, such as control applications, this is true. However, in large relays (called contactors) and in such specialized relays as motor starters, AC may be supplying the coil. From our study of AC, we know that when the current passes through zero, any flux created by that field will also be zero. The armature attraction will then stop, opening the contacts. A shading coil, which is a single coil of copper or aluminum mounted on the armature assembly or on the armature (Figure 7–5), is used to prevent this from happening. The flux in the magnetic assembly alternates, inducing currents in the shading coil. These currents cause flux of their own that is out of phase with the flux in the main assembly. The shading coil flux produces a pulling force that is also out of phase with the main assembly flux. The shading coil exerts a pulling force on the armature when the main assembly flux is zero. This extra pulling force keeps the armature closed when the main flux goes to zero. Without the shading coil, the armature would open each time the flux fell to zero, which would occur approximately 120 times per second. The time delay principle used to prevent relay chattering is also used in time-delay applications.

Coil Specifications When the coil is deenergized and the armature is in the open position, there is a large air gap in the magnetic circuit. At this point, the coil impedance is low, due to the large air gap. When the coil is energized, it draws a large current. As the armature moves closer to the magnet assembly, the air gap is progressively reduced. This reduction in air gap causes the coil impedance to increase to a maximum when the armature is totally closed or sealed. The final current is called the **sealed current.** The high current that flows when the coil is first energized is called the **inrush current.** The inrush current is usually between six to ten times the sealed current, the ratio varying with individual designs.

Data on the coil are usually given in volt-amperes (VA). For example, let's say we are working with a coil rated at 600 VA inrush and 60 VA sealed.

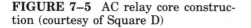

FIGURE 7–5 AC relay core construction (courtesy of Square D)

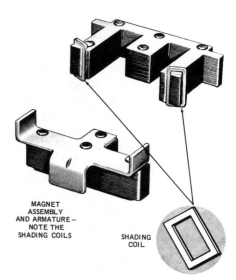

MAGNET
ASSEMBLY
AND ARMATURE –
NOTE THE
SHADING COILS

SHADING
COIL

Using a 120-V coil, we can calculate the inrush current by dividing the VA
rating by the operating voltage:

$$\text{Inrush Current} = \frac{600\text{VA}}{120\text{V}} = 5 \text{ A} \qquad \textbf{(eq. 7–1)}$$

The sealed current can be calculated in the same way using the sealed VA
rating:

$$\text{Sealed Current} = \frac{60\text{VA}}{120\text{V}} = 0.5 \text{ A} \qquad \textbf{(eq. 7–2)}$$

What sealed current and inrush current will the same relay draw with a 480-V
coil?

$$\text{Sealed Current} = \frac{60\text{VA}}{480\text{V}} = 0.125 \text{ A}$$

$$\text{Inrush Current} = \frac{600\text{VA}}{480\text{V}} = 1.25 \text{ A}$$

Two other important parameters are the pick-up and drop-out voltages.
The ***pick-up voltage*** is the minimum coil voltage that will cause the armature
to start to move. The ***drop-out voltage*** is the voltage at which the armature
unseats from the magnetic assembly. The pick-up voltage is normally much
larger than the drop-out voltage due to the air gap. The presence of the air gap

when the coil is not energized increases the voltage (and current) needed to pull the armature in. When the armature is pulled in, less flux is needed to maintain the position of the armature since no air gap is present.

NEMA standards require that all electromagnetic relay devices operate properly with low- or high-coil voltages, ranging from a high of 110% to a low of 85% of rated-coil voltage. In meeting this requirement, the coil designer ensures that the coil will withstand voltages up to 10% over rated voltage and that the armature will pick up and seal in, even though the voltage may drop to 15% under the rated voltage.

7–4.2 Solid-State Relays

The National Association of Relay Manufacturers defines the *solid-state relay (SSR)* as "a relay with isolated input and output whose functions are achieved by means of electronic components and without the use of moving parts." SSRs come in sizes ranging from 2 to 40 amps, with both AC and DC control voltages. Figure 7–6 shows basic SSR operation. A high voltage on the control line will turn on an LED. The light from the diode turns on a switch, usually a phototransistor or light-activated SCR (LASCR). The solid-state power switch applies power to the load the next time the source voltage crosses zero volts. This part of the circuit uses a detector called a zero-voltage switch. When the control voltage is removed, the power to the load is disconnected the next time the load current passes through zero. Switching the power to the load only when the load current goes through the zero point reduces the generation of electromagnetic interference (EMI), a recurrent problem with EMRs. Unlike the SSR, the EMR has a useful life that is dependent on the number of actuations the relay performs. The SSR has an extremely long life because of its lack of moving parts. Another advantage of the SSR over the EMR is the SSR's resistance to shock and vibration. EMRs do have some advantages; they include multipole and multithrow operation, ability to withstand such surges as those encountered in motor starting, and resistance to triggering by transient voltages. The advantages of both types of relays must be carefully weighed by the engineer before an intelligent choice can be made. It is unwise to say,

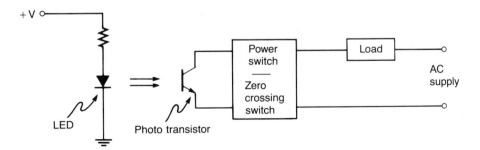

FIGURE 7–6 Functional diagram of a solid-state relay

categorically, that SSRs are replacing EMRs in all applications. EMRs will be popular in industry for the foreseeable future.

Relays vs Contactors   A relay is primarily used in control circuits drawing less than 15 amps at 600 V. Consequently, they are often called control relays. Relays can have as many as 10 to 12 poles in normally-open and normally-closed configurations. A *contactor,* on the other hand, is a high-power relay, repeatedly switching high-current loads. The contactor has heavy-duty contacts that are usually open, with anywhere from one to five poles. Normally neither contactors nor relays have overload protection built-in. They can not be used for motor-starting or controlling without adding overload protection.

7–5 MAGNETIC MOTOR STARTER

A magnetic motor starter is simply a contactor with overload protection built-in. When the proper voltage is applied to the coil of the contactor, the normally-open contacts close, applying power to the motor terminals. The contacts must therefore be able to carry full-load current. The diagrams in Figure 7–7 show a magnetic starter used to start three-phase AC motors. A photograph of the starter is shown in Figure 7–7a. Note that the AC power is connected to the top set of terminals. The motor windings are connected to the bottom set of

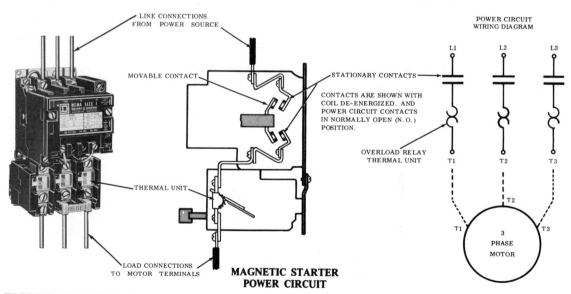

FIGURE 7–7 Full-voltage magnetic starter

starter terminals. The simplified diagram in Figure 7–7b illustrates the overload protection at the bottom of the figure and the contacts at the top. You will notice that the contacts are open, implying that the starter is shown in the deenergized condition. The wiring diagram represented in Figure 7–7c shows the normally-open contacts, the overload relays, and the connection to the motor terminals.

7–6 TIME-DELAY RELAYS

The *time-delay relay (TDR)* is a control relay in which the contacts open or close at a preset time interval after the coil is energized or deenergized. A delay after energization is called an *on-delay.* A delay after the coil has been deenergized is called an *off-delay.* Large machine lubrication is an example of an on-delay application. In some large motors, a small oil pump must pump lubricating oil to the motor bearings *before* the motor starts. The on-delay relay will prevent the motor from being started or turned on until the pump has enough time to lubricate the bearings.

Timers are based on three main technologies: pneumatic, electromechanical, or electronic.

7–6.1 Pneumatic Timers

The three most commonly used timers are electropneumatic, hydraulic-pneumatic, and mercury timers. In electropneumatic timers, an input voltage energizes a solenoid that puts force on a diaphragm within a sealed air cylinder (Figure 7–8). The diaphragm then moves at a rate determined by length of the flow path created by the diaphragm. When the diaphragm reaches a certain point, it actuates a snap-acting switch connected to the relay contacts. When power is removed from the solenoid, a one-way valve exhausts the chamber quickly, resetting the device.

Hydraulic-magnetic timers have a solenoid similar to the electropneumatic timer shown in Figure 7–9. When the solenoid coil is energized, a movable core is pulled into the solenoid. Since the core is filled with fluid, the movable core cannot move immediately into the solenoid. It must first displace the fluid that surrounds it and resists its movement. After a time delay, the core reaches the end of its travel and touches a pole piece. The resulting reduction in air-gap reluctance increases the field strength, pulling down the armature and actuating a switch. When power is taken away from the solenoid, the spring returns the movable core to its original position. A one-way valve lets the core move back quickly through the fluid.

The mercury timer is similar to the hydraulic timer. The time delay is determined by the speed at which a gas passes through a porous ceramic plug. The switch uses mercury to complete the electric circuit.

FIGURE 7–8 Electropneumatic timer (courtesy of Machine Design, May 31, 1984)

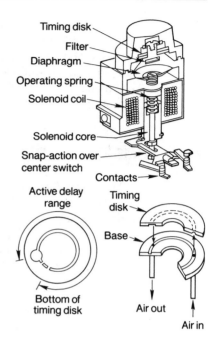

7–6.2 Electromechanical Timers

Electromechanical timers can be further broken down into two classifications: motor-driven and thermal. In the motor-driven timer (Figure 7–10), a synchronous motor is locked on the input AC line. The motor is connected to switch contacts through a mechanical linkage and a clutch-solenoid mechanism. When the relay is actuated, the clutch is pulled in by the solenoid and

FIGURE 7–9 Hydraulic-magnetic timer (courtesy of Machine Design, May 31, 1984)

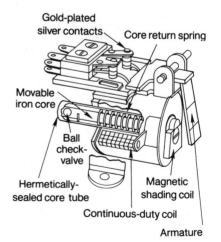

FIGURE 7–10 Motor-driven timer (courtesy of Eagle Signal Controls Division, Gulf and Western Mfg. Co.)

linked to the mechanical linkage. At the end of the time delay, the motor closes or opens a set of switch contacts.

The thermal timer classification can be subdivided into bimetallic and expansion timers. Bimetallic timers, shown in Figure 7–11, use the principle of differential expansion of metals. Two bimetal strips are rigidly supported on a frame, with a heating coil wound around one strip. When current flows through the heating coil, the bimetal strip heats and bends. When it bends to a predetermined point, the switch contacts close. The unit compensates for ambient temperature changes by making both bimetal strips with identical coefficients of expansion. Both will move identical amounts in the same direction with ambient temperature changes.

In the expansion timer, a heating coil is wound around a stainless steel rod. When heated, the rod lengthens and closes a switch attached to it (Figure 7–12). The adjusting arm is used to vary the time delay by moving the adjusting arm. Moving the arm so that the contacts are further away will increase the time it takes for the contacts to close.

7–6.3 Electronic Timers

Some electronic timers use the principle of a charging capacitor to give a time delay. The diagram in Figure 7–13 shows a capacitor charging through an RC network at the start of the timing cycle. The time the capacitor takes to charge

FIGURE 7–11 Bimetallic timer

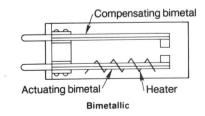

Compensating bimetal

Actuating bimetal   Heater

Bimetallic

FIGURE 7–12 Expansion timer

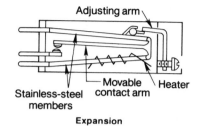

Adjusting arm

Movable contact arm

Heater

Stainless-steel members

Expansion

up to a preset threshold is the time delay interval. When the capacitor voltage reaches the threshold level, the voltage turns on a solid-state switch (such as a thyristor, transistor or FET) that applies power to a relay coil.

The other principle used in electronic timers is the principle of the time base. The time base can be an internal clock or oscillator, or it can be the line voltage (Figure 7–14). In either case, the counters count clock pulses or zero-voltage crossings in the AC line. The decoder output will go high after a preset number of pulses are counted. The high voltage is amplified and used to turn on a relay coil. The time delay is varied by changing the number of counts the decoder needs to put out a high voltage.

7–7 PILOT DEVICES

A *pilot device* is a switch actuated by a non-electrical means. Operation of a pilot device often actuates a controller and, therefore, a motor. Pilot devices are classified as either *maintained contact* or *momentary contact* devices. A maintained contact device is one that, when actuated, causes a set of contacts to open or close. These contacts will remain open or closed until a reverse actuation occurs. The thermostat is a good example of this type of device. The thermostat actuates when the temperature goes below some predetermined point. As long as the temperature remains below that actuation point, the thermostat contacts remain in their previous position. In momentary contact devices, actuation causes normally-closed contacts to open and normally-open

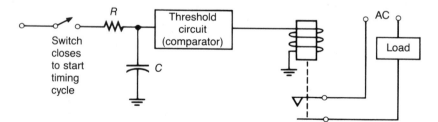

FIGURE 7–13 Electronic timer—RC type

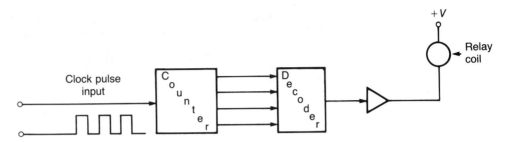

FIGURE 7-14 Electronic timer—time base type

contacts to close. These contacts remain in position as long as the device is actuated. A good example is a pushbutton, which actuates the contacts when the button is pushed. When the button is released, the contacts go back to their previous position.

7-7.1 Pushbuttons

The pushbutton is one of the most frequently used pilot devices in industry. Pushbuttons, available from many manufacturers with from one to seven buttons, can be momentary contact, maintained contact, or both, as in Figure 7-15. Pushbuttons are normally mounted on walls or directly on the machines they are controlling. It is good practice for pilot devices to be mounted where the effects of starting can be seen. Pushbuttons can also be mounted in a panel with other pushbuttons or pilot devices. Pilot devices are normally mounted in sheet metal enclosures; when necessary, they can be mounted in dust-tight or water-tight containers, such as aboard ships.

Modern technology has combined the old concept of the pushbutton with the idea of computer control in the programmable pushbutton, shown in Figure 7-16. This innovation permits the operators to have a compact control panel with all necessary information clearly displayed without the need for a keyboard. The programmable pushbutton replaces many pushbuttons and indicating instruments at a relatively low price.

FIGURE 7-15 Pushbutton switches (courtesy of Allen-Bradley Company)

FIGURE 7-16 Programmable pushbutton switch (courtesy of MICRO SWITCH)

The programmable pushbutton was originally developed for aircraft and space vehicles to save on space and training time. The programmable pushbutton, shown in Figure 7–16, is called the Programmable Display Pushbutton (PDP) and was developed by Microswitch Corporation. This single illuminated pushbutton can perform the functions of large pushbutton panels and indicating instruments, and eliminate large and cumbersome operator manuals.

The PDP allows communication between the operator and the control computer using the traditional pushbutton rather than a computer keyboard. Operators receive information about the system from an illuminated display panel on the pushbutton. The pushbutton allows the operator to gain information about the system (a process called interrogation in process control terminology) as well as control the system. A single pushbutton can give information on the speed of a number of different motors. One of the difficulties with manual motor starters is the training needed by operators to start the motor properly. These pushbuttons can be programmed to lead the operator through the proper startup sequence. In emergencies, the unit can be programmed to instruct the operator on the proper course of action. Programming can also include the instruction to disregard incorrect operator commands.

Float Switches The *float switch,* shown in Figure 7–17, is a device used to sense the level of a liquid in a tank. It controls the pump used to fill or empty the tank. When the liquid level reaches a predetermined minimum, the float switch closes, turning on a pump that fills the tank. By reversing the operation, the float switch empties the tank. The switch can be used to directly control a motor. Generally, the switch contacts are used to control a magnetic motor controller.

7–7.2 Limit Switches

A *limit switch* limits the travel of moving parts driven by a motor-driven machine or other apparatus. It has either normally-open or normally-closed contacts (or both) and is actuated by contact with a moving part or machine. The most popular limit switch is the lever type, shown in Figure 7–18. The contacts are opened or closed by a lever that projects out from the main body of the switch. On the end of this lever is a roller. The moving part of the machine comes in contact with the roller, which moves the lever to a position where the contacts in the switch either open or close. Many times lever actuation will open the circuit to the coil of a magnetic motor controller, which disconnects

FIGURE 7–17 Float switches (courtesy of Allen-Bradley Company)

the motor from the line. Typical applications of the lever-type limit switch in motor control are limit switches at the top and bottom of elevators, motor-driven doors, and windows. When the lever-type switch is used in a door, some part of the door will contact the lever when the door is almost fully open. The limit switch contacts then open, and the motor is stopped. To close the door, a pushbutton switch is energized turning on the motor. When the door is fully closed, the lever of another limit switch is actuated, stopping the motor.

Photoswitches The *photoswitch,* shown in Figure 7–19, replaces the limit switch in many applications. It is more desirable than a limit switch because of its lack of moving parts, which reduces the need for maintenance. The photoswitch has a light sensor with control circuitry and a relay. When the light from a light source strikes the sensor, the relay is energized, closing or opening one or more sets of relay contacts. In some cases the light source is included in the photoswitch package and is called a retroreflective photoswitch.

Photoelectric switches, with operating distances of a few millimeters to 1000 ft, are used to turn on and off a motor that opens and closes a door. They are also used on automatic dumb waiters and on automatic drinking fountains.

Figure 7–20 illustrates the many industrial uses of photoswitches. Many are directly related to control of motors.

7–7.3 Proximity Switches

As its name implies, the proximity switch detects the presence or nearness of a metallic object, either ferrous (containing iron) or nonferrous. Both the photoswitch and the proximity switch are classified as noncontact, or no-touch, controls because they do not need the touch of an object to actuate them. The limit

FIGURE 7-18 Limit switches (courtesy of Square D)

switch, however, operates on contact only. Noncontact switches are preferable to contact devices because the number of moving parts is reduced, thus reducing wear and maintenance costs.

One of the most common proximity switches is the ***eddy current killed oscillator (ECKO),*** shown in Figure 7-21. The end of the ECKO probe incorporates the coil from an oscillator as part of its circuitry. The oscillator has just enough feedback to keep it oscillating. The oscillating sine wave is converted to DC by the integrator, which provides an input to a comparator. When a metal is placed close to the probe, eddy currents are induced into the surface of the metal. The induction of the eddy currents takes energy away from the oscillator, causing oscillations to die out. The integrator, detecting the change in voltage, trips the comparator, which changes the state of the switch. Recall that an integrator is a device that performs the mathematical function of integration. The output of an integrator is proportional to the length of time an input is present. The comparator is a circuit that compares two electrical signals, giving an output when one signal is above or below a preset value.

Figure 7-22 shows several ECKO proximity switches. These switches can be hermetically sealed, protecting them from harsh industrial environments.

7-8 PROTECTIVE DEVICES

Now that we have discussed basic starters, relays, and pilot devices, we will examine devices used to protect a motor from damage. Section 430 of the National Electrical Code lists detailed requirements of motor protection. The in-

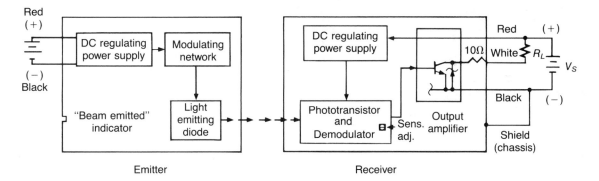

FIGURE 7–19 Photo and functional block diagram of a photoswitch (courtesy of MICRO SWITCH)

tent of the code is ". . . to protect the motors, the motor control apparatus, and the branch circuit conductors against excessive heating due to motor overloads or failure to start."

Motors can be damaged or their useful life reduced when they draw current only slightly higher than their full-load current rating times the service factor. *Full-load current* is defined as the current required to produce *full-load torque* at the rated speed. The service factor, given to each motor by the manufacturer, is the amount of extra power the motor can develop without the insulation being damaged or deteriorating. The service factor is a margin of safety.

Fill Level Control
Two light source-photoreceiver pairs are used to keep hopper fill level between high and low limits.

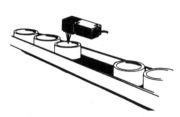

Conveyor Control
A retroreflective control is the most common form of scanning across a conveyor. Applications such as collision control, product flow, jam-up etc.

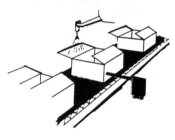

Food Processing
A polarized photoelectric control is ideal when detecting shiny objects. These units respond only to retroreflective targets, and not the light reflected by the objects.

Labeling
A convergent beam photoelectric control with time delay assures the conveyor holds enough parts for a later operation. The inherent short range of the control insures that the conveyor belt will not be detected.

Package Handling
A diffuse scan photoelectric control is used to detect the light reflected from the object in this application. The control detects the light reflected off the box, turning On and Off the gluing machine.

FIGURE 7–20 Industrial uses of switches

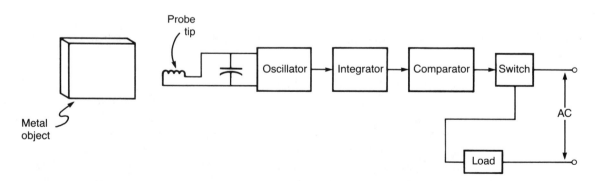

FIGURE 7–21 Functional block diagram of an ECKO-type proximity sensor

FIGURE 7–22 ECKO switches (courtesy of MICRO SWITCH)

For example, if a manufacturer of a 100-W motor gives it a service factor of 1.2, the motor can develop 120 W without damage.

Motors can also be damaged by a sudden high current for a short time, such as in short-circuit conditions. All currents greater than full-load current, whether they are high and of short duration or only slightly above full-load current, are classified as overcurrents.

In addition to causing motor damage, short-circuit currents can also damage ancillary wiring. When current flows through a conductor, it generates heat proportional to the square of the current flow. If current doubles due to a malfunction, the heat generated quadruples. Enough heat will cause the conductor's insulation to break down or, possibly, to start a fire. A protective device must then be installed to match the current-carrying capability (or ampacity) of the conductor. When too much current flows, the protective device will open the circuit.

7–8.1 Overcurrent Protection

The function of the overcurrent protective device is to protect the motor and associated circuitry from short circuits and grounds. The two devices commonly used to prevent overcurrents are the thermal magnetic *circuit breaker* and the *fuse*. Both of these devices should be capable of carrying the starting current of the motor without disconnecting the motor from the line voltage. These devices are set to break the circuit between 250% and 400% of full-load current. The National Electrical Code requires the overcurrent protection to provide a means of disconnecting the motor and associated circuitry from the line while providing an overcurrent protection device to clear short-circuit faults. The circuit breaker accomplishes both functions. When overcurrent protection is accomplished by fuses, a disconnect switch is usually provided. Figure 7–23 shows both the fuse and the circuit breaker.

7–8.2 Overload Protection

In most motors, the current drawn by the motor depends on the mechanical load. The amount of load can be anywhere from a *no-load* situation to full load. The motor has no built-in protection against a load that is too great. When the

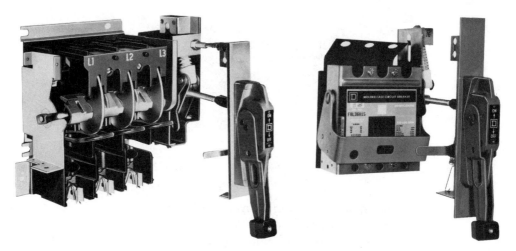

FIGURE 7–23 Overcurrent protective devices (courtesy of Square D)

mechanical load is greater than the torque rating of the motor, the motor will draw current that will exceed the full-load current. This condition is called an *overload.* The maximum overload occurs when the rotor or armature is stalled or at a standstill, a condition referred to as a locked rotor condition. Although overloads are usually mechanical, they can be electrical. An electrical overload can occur when a polyphase motor is run with only one phase connected or when the line voltage is low.

Why is an overload dangerous? An overload in a motor causes a rise in the temperature of the motor windings. The term *overload* is something of a misnomer. The function of motor overload protection is to protect the motor from over-temperature resulting from the overcurrent that flows under overload conditions. Too high a temperature can damage the winding insulation and the motor's lubricating fluids. Every overload will damage the motor by reducing the life of the insulation. Relatively small overloads over a long period of time can be as dangerous as a large overload accomplished quickly. The effects are cumulative. One conclusion we can draw about the size of the overload is that the larger the overload, the faster the temperature will rise to the danger point. The relationship between current and time is an inversely proportional one. The higher the current, the shorter the time before the motor will be damaged. The relationship between the size of the overload and the time it takes the motor to reach a damaging condition is illustrated in the curve in Figure 7–24. With a 300% overload on this particular motor, a potentially damaging condition would occur after only three minutes.

Thinking about this curve and its relationship to motor overload protection, the ideal protection circuit is a device with sensing properties that respond in a similar manner to the motor heating curve. This device would take away power from the circuit when full-load current is exceeded. The operation

FIGURE 7–24 Motor heating curve

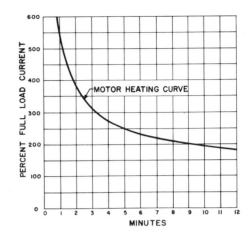

of this sensing device should respond in such a way that nondamaging over-loads, such as those present when the motor starts, are ignored.

7–8.3 Overload Relays

The most common way to prevent motor overloads is to use the overload relay. This relay holds in during motor start-up and acceleration, yet it opens when small or large overloads are encountered during normal motor operation. It should be stressed that overload protection is not the same as short-circuit protection. Short-circuit protection is provided by fuses and circuit-breakers.

Overload relays are composed of two basic parts. The sensing mechanism detects the overload, which causes the circuit-breaker to directly or indirectly open the circuit. Two basic types of overload relays are commonly used today: the thermal and the magnetic. The thermal overload relay uses the principle of heat rise due to excess current as the triggering mechanism.

Thermal Overload Relays Two types of thermal overload relays are used in motor control systems. In the melting alloy overload relay, the motor current passes through a small heating coil, shown in Figure 7–25. When too much current passes through the heater coil, the heat generated melts a special solder, called a *eutectic alloy,* in a container. Because of its construction, this device is also called a solder pot overload relay. Within the solder pot is a ratchet wheel device that is held in place by the solidified solder. When over-load current flows through the heating coil, the solder melts and the ratchet wheel is free to turn (Figure 7–26). The special eutectic alloy changes rapidly from solid to a liquid at a specific temperature. The spring then pushes the contacts open and the relay trips. Response to different overload conditions is achieved by placing different heating elements in the overload relay. After

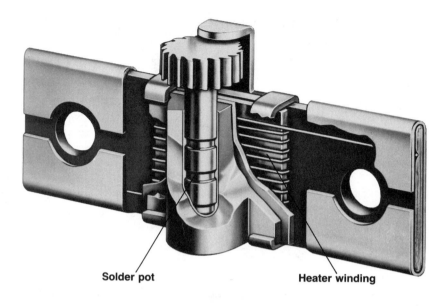

Solder pot   Heater winding

FIGURE 7–25 Solder pot overload relay (courtesy of Square D)

tripping due to an overload, this relay must usually be reset by hand. A front-panel reset button is provided for easy relay resetting. The heating elements are rated in amperes of motor full-load current.

The other common thermal overload relay uses the bimetallic strip and the principle of differential expansion of metals. A bimetal strip is shown in Figure 7–27. Let us say that metal A on the left side of the strip has a tendency to expand more when heated than metal B on the right. If the two metals are joined together so that they cannot slip and heat is applied, the strip will move towards the right, in the direction indicated in the diagram.

An overload relay using the bimetal strip is shown in Figure 7–28. It has a heater coil, similar in function to the one in the melting alloy relay. The

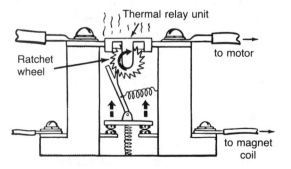

FIGURE 7–26 Melting alloy overload relay (courtesy of Square D)

FIGURE 7–27 Principle of differential expansion of a bimetal strip

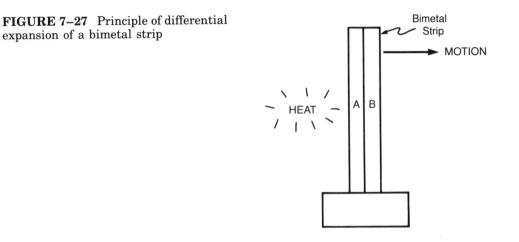

heater, when carrying an overload, heats the bimetal strip, which causes the strip to move. The moving strip pushes against the breaker contacts, causing the contacts to open. As in the melting alloy device, different heaters cause the relay to trip with different overload conditions. Depending on the device, the bimetallic strip relay can be either manually or automatically reset. Automatic reset is done by a spring-loaded mechanism on the relay contacts. When the bimetal strip cools down, it releases pressure on the relay contacts, which are then driven closed by the spring mechanism. Automatically resetting relays should be used with caution. If the relay automatically resets without taking away the cause of the overload, the relay will trip again soon after resetting. This cycle will repeat until the motor burns out due to the overheating caused by the inrush current at starting. Another reason for caution with automatic resetting relates to the safety of maintenance personnel. The motor can start at any time while the maintenance personnel are trying to diagnose the problem, a situation that can place personnel in danger.

Both types of overload relays are designed to open when the actuating

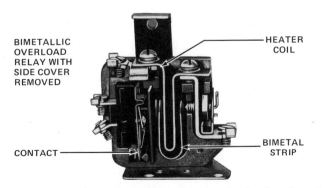

FIGURE 7–28 Bimetallic strip overload relay (courtesy of Square D)

FIGURE 7–29 Relationship between motor heating time and percentage of full-load current (courtesy of Square D)

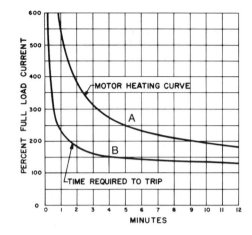

temperature is approximately equal to the temperature of the overloaded motor. When the temperature of the motor increases, the heating coil temperature should also increase. The graph in Figure 7–29 shows the relationship between the time it takes for the motor to heat up and the percentage of full-load current. Notice that curve A shows the motor heating relationship, while curve B shows the overload relay trip curve. The overload relay will always trip before the motor overheats.

Although most overload relays using bimetal strips have a built-in series heater, some relays are mounted directly on the motor itself. The heat generated in the motor is then used to trip the overload relay. An example of this type of overload switch is illustrated in Figure 7–30. This type of bimetal switch has a ***snap-action*** characteristic that allows it to remain closed until it reaches a specific temperature. At this temperature, it snaps open, opening the switch contacts and removing power from the motor.

FIGURE 7–30 Snap-acting bimetallic switch (courtesy of Machine Design, May 31, 1984)

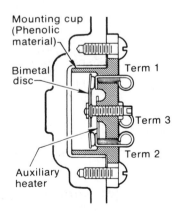

Magnetic Overload Relays Magnetic overload relays are often used when it is not convenient to locate the overload protection near the motor. This type of overload relay responds to a magnetic field made by the motor current. The magnetic overload relay has a movable core inside a coil through which the motor current passes (Figure 7–31). The core moves inside a sealed, nonmagnetic tube. Normal motor current passing through the coil is not enough to cause the core to move. When enough current passes through the coil, the core moves left, pulling the armature right. The movement of the core is slowed by its movement through a thick viscous liquid in the tube. Some relays use a piston working in an oil-filled dash pot mounted on the coil. This provides a controlled time delay, making it difficult to trip the relay under starting conditions when current is high. The piston-dash pot system is similar in construction to a shock absorber. The tripping time is varied by uncovering more holes in the piston. The more holes that are uncovered, the faster the piston rises, lowering the amount of delay between the time an overload is applied and the relay trips.

Alarms The relay contacts of an overload relay are normally in the closed position. Current is then allowed to flow through the motor. When an overload occurs, the relay contacts open, preventing current from flowing through the relay. It is sometimes necessary for an alarm to indicate when a motor overload has occurred. This is usually accomplished by a special set of contacts, called alarm contacts, that close when the relay trips. The closed relay contacts complete an alarm circuit, which indicates that an overload has occurred.

Simulation Protectors In our previous discussions, we have concentrated on current and temperature sensing devices. Although these devices work well in most cases, they have some limitations. Electronic overload protection circuits

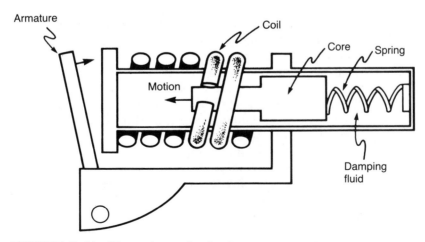

FIGURE 7–31 Magnetic overload relay

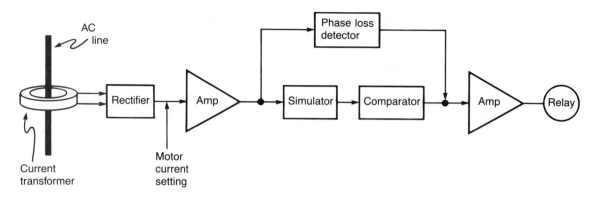

FIGURE 7–32 Simulation protectors

use sensors that monitor current in motor leads to overcome some of these limitations. Hall effect devices are sometimes used. Current is sensed (Figure 7–32), converted to a voltage, amplified, and fed to a circuit that simulates the rise and fall of the motor's temperature. If the temperature rises too quickly, as in the case of an overload, the network drives an amplifier that trips a motor shutoff relay. AC circuits can also have a phase-loss detection circuit. Losing one or more phases will also trip the shutoff relay.

7–8.4 Ground-Fault Protection

Large motors and expensive specialty motors require more sophisticated protection than simple overload and overcurrent protection. Ground-fault protection adds an important factor to such motors. A *ground fault* is a current imbalance between hot and neutral leads of a line supplying power. It occurs when a current-carrying conductor accidentally touches a grounded conducting material. Depending on the nature of the materials, the current can be small or large. As an example, ground-fault protection would prevent damage to the motor if a stator winding were to short-circuit to the grounded case. The power supply would also be protected. Low resistance ground faults causing high current will usually blow a fuse or trip a circuit breaker. Higher resistance ground faults may not draw enough current to trip a circuit breaker or fuse. Although the current may not be enough to trip short-circuit protective devices, it could be enough to injure or kill a human. A fire could also be caused by these smaller currents without opening overcurrent protective devices.

A simple circuit demonstrating a ground-fault protection circuit can be found in Figure 7–33. It employs a current transformer (called a toroidal current transformer because of its doughnut shape) and a ground-fault relay in a three-phase motor. During normal operation, all three conductors will be carrying nearly equal currents. The flux in the transformer is close to zero, which produces little current in the secondary winding. When a ground-fault occurs,

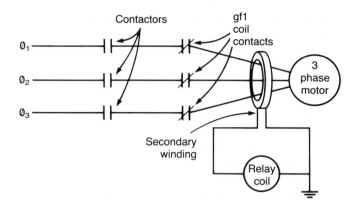

FIGURE 7-33 Ground-fault protection circuit using one current transformer

the motor phase currents become unbalanced, which produces flux in the secondary winding, tripping the ground-fault relay. The relay normally disconnects the motor from the line. This method works well with small and medium sized motors. It is not as convenient with large motors due to the difficulty in getting large leads through the small current transformer.

Another method of ground-fault protection that works better on large motors is illustrated in Figure 7-34. This method uses a sensitive relay in the grounded neutral lead of three current transformer secondary leads. In normal operation, the phase currents are equal and no current flows in the neutral

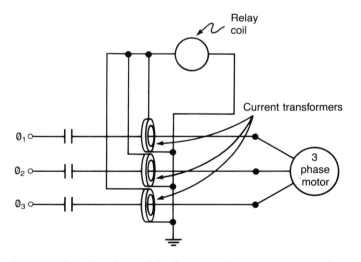

FIGURE 7-34 Ground-fault protection circuit using three current transformers

lead. When a ground fault occurs, an imbalance causes current to flow in the neutral lead, which trips the ground-fault relay. Again, power is usually removed from the motor, protecting the motor, power supply, and associated leads and circuitry.

7–8.5 Solid-State Motor Protectors (SSMPs)

We have seen how electromechanical devices can be used to provide most types of protection necessary in a motor system. Providing a wide range of protection methods would require a large number of individual relays, taking up valuable space. All of the protection features we have discussed can be grouped together in a single solid-state motor protector. This saves space, installation time, and money, depending on the protection measures required.

Overloads are detected either by embedding temperature sensors in the windings or by using current transformers to monitor current. A few SSMPs use both temperature sensors and current transformers and make decisions on whether an overload is occurring by evaluating the information from both sensors together. Most SSMPs have some common features. They all have a means of setting the trip point based on the manufacturer's data for the partic-

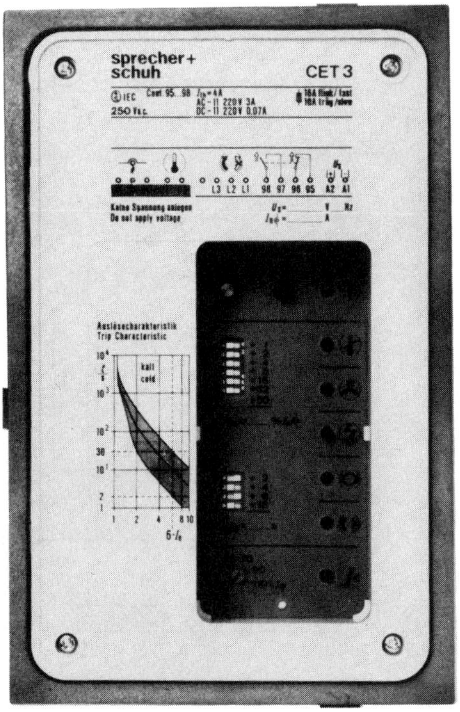

FIGURE 7–35 Solid-state motor protector (courtesy of Sprecher & Schuh)

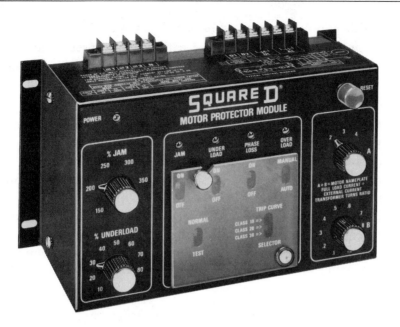

FIGURE 7–36 Solid-state motor protector (courtesy of Square D)

ular motor used. All SSMPs are relatively immune to ambient temperature changes, a problem with other overload protection devices. All have programmability built in, making it easy for operators and installation personnel to adjust the SSMP to respond to a particular motor's characteristics.

Several companies manufacture these protectors, one of which is shown in Figure 7–35. This SSMP, manufactured by Sprecher & Schuh, uses a current transformer to provide overload protection. This device also provides protection against losing phases in a three-phase motor, ground faults, phase reversal, underloads, and motor stall. A Square D Company version of the SSMP is shown in Figure 7–36.

CHAPTER SUMMARY

- Motor controllers can be broken down into two basic groups: manual and magnetic.
- Manual controllers are operated by hand; magnetic controllers use a solenoid.
- The solenoid principle is used in the electromagnetic relay.
- The solid-state relay is made up of only electronic devices and has no moving parts.

- EMRs can be operated by either DC or AC.
- A motor starter is a contactor with overload protection built in.
- Time-delay relays are used for timing purposes in the control of motors.
- The most popular time-delay relays work on pneumatic, electromechanical, and electronic principles.
- Pilot devices are another class of devices used to control motors.
- Pilot devices include pushbuttons, float switches, limit switches, photoswitches, and proximity switches.
- Controlling motors is important; equally important is the protection of motors.
- Motors can be damaged by overcurrents, overvoltages, and overloads.
- The most modern development in motor protection is the solid-state motor protector. In this device, many of the most common damage-producing factors are sensed and appropriate action is taken by the SSMP.

QUESTIONS AND PROBLEMS

1. Describe the difference between the following:
   a. a manual and a magnetic starter
   b. a relay, a solenoid, and a contactor
   c. a full-voltage, and a reduced-voltage starter
2. List and describe the three types of EMRs.
3. Describe the operation of an EMR.
4. Identify the basic parts of an EMR.
5. A coil is rated at 500-VA inrush and 40-VA sealed. If the coil is a 240 V coil, what is
   a. the inrush current?
   b. the sealed current?
6. Describe the operation of the pneumatic, electromechanical, and electronic timers.
7. What is the purpose of an overload relay? Describe the operation of one type of overload relay.
8. Explain the purpose of the mechanical interlock on a forward reverse motor controller.
9. Contrast the normal pushbutton and the programmable pushbutton.
10. Identify and describe the operation of the following pilot devices:
    a. limit switch
    b. photoswitch

    c. proximity switch

    d. pushbutton

    e. float switch.

11. Contrast an overload and an overcurrent condition.

12. Define a ground fault. Draw and explain a circuit used to detect and protect against a ground fault.

8

ELECTRICAL MOTOR STARTING CIRCUITS

At the end of this chapter, you should be able to

- describe the operation of AC and DC motor controllers.
- explain the difference between a ladder diagram and a schematic diagram.
- differentiate between the following circuits:
  - two-wire vs three-wire
  - full-voltage starters vs reduced-voltage starters.
- list three methods of reduced-voltage starting.
- contrast dynamic braking and plugging.
- list two characteristics that accompany the following motor fault conditions:
  - overvoltage
  - undervoltage.
- list three environmental conditions that can cause faults in motors or control circuits.

8–1 INTRODUCTION

Every motor in industry needs a type of control. In this chapter, we will examine the ways in which different kinds of motors are started, using the electrical devices discussed in the last chapter. A motor starting circuit is defined by NEMA as "an electrical controller for accelerating a motor from rest to normal speed." Motor starters range in complexity from simple across-the-line starters to more complicated circuits that actually modify the motor's acceleration characteristics. Although the motor starter will be discussed in isolation from other types of circuits, you should realize that the motor starter is only a small part of a more complicated control system. The system may contain multiple pushbutton stations, such pilot devices as the limit switch or thermostat, and sensing devices. Today, computers are an integral part of industrial process control, ultimately feeding signals to motor controllers to start, stop, or adjust a motor's mechanical output.

8–2 INDUSTRIAL LADDER DIAGRAMS

A technician must be able to read the different types of industrial diagrams. Before we study motor starting circuits, we need to gain some skill in reading a schematic diagram used to represent control circuits. Note the complexity of the diagram in Figure 8–1. Although the function is very simple, the schematic diagram is very hard to follow. A simpler diagram of this circuit is found in Figure 8–2. The diagram in Figure 8–2 has several names: simplified diagram, ladder diagram, or control diagram. As you can see, it is much easier to understand. When we push the start button, current flows through the relay coil, marked M. When the coil is energized, all relay contacts change state; i.e., normally-open contacts close and normally-closed contacts open. Before the start button is pushed and with the relay in a deenergized condition, the green lamp is off and the red lamp is on. After the button is pushed, the green lamp goes on and the red lamp goes off. Note also that after the start button is pushed and the relay energizes, the relay remains energized even after the start button is released. This condition occurs because current flows through

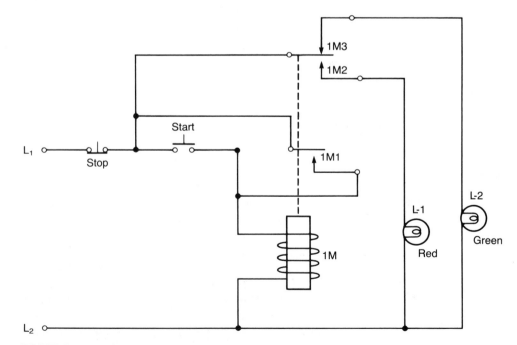

FIGURE 8–1 Schematic diagram of a control circuit

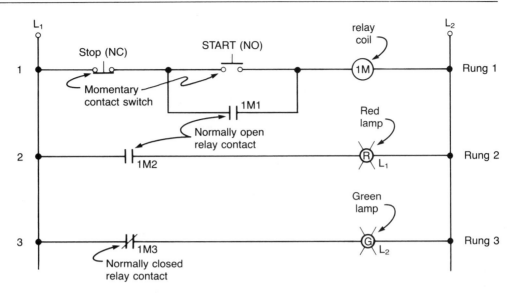

FIGURE 8–2 Industrial ladder diagram

contact *1M1* when the start button is pushed. When released, current continues to flow through the relay coil and contacts *1M1* even after the start button is released. The relay is said to be in a latched condition, and this relay circuit is called a latching relay configuration.

Industrial ladder diagrams usually adhere to certain drawing practices or standards. Note that the diagram looks somewhat like a ladder with rungs running sideways held together by vertical bars on each end. The vertical bars are the line voltage or power busses, labeled *L1* and *L2*. The numbers down the left side of the diagram vertically are the rung numbers. Loads (such as relays, coils, solenoids, timers, and pilot devices) are normally located on the right-hand side of the diagram, with only one load per line. Note that every rung of the ladder *must* have a load. The lack of a load could cause a short circuit when power is applied to such a rung. Relay contacts associated with a relay coil are labeled with the same letter designation. Relay contacts associated with the *1M* coil are labeled *1M1, 1M2,* and *1M3,* two normally-open and one normally-closed contact. Another coil, when added, may be named *2M.* Thus, its contacts would be *2M1, 2M2,* etc.

A motor can be added to our simple control circuit, as in Figure 8–3. When we close the start switch, power is applied to the left side of *4CR, 5CR,* and *6CR,* provided that the fuses are operating properly. Since these switches are normally open, no power is applied to the motor until these relay contacts are closed. As stated above, when the start button is pressed, all normally open switches close (including *1M4, 1M5,* and *1M6)*, power is applied to the motor, and the green light comes on. Note the overload relay contacts *10L, 20L,* and *30L* in the motor circuit. If an overload occurs, the overload relay contacts will open, disconnecting the motor from the line.

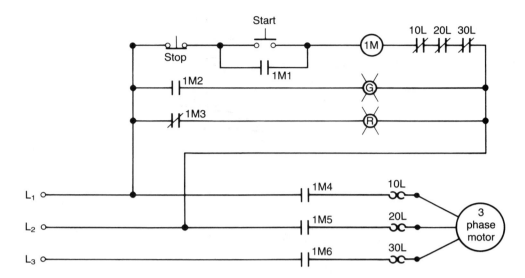

FIGURE 8–3 Ladder diagram shown controlling the three-phase motor

TABLE 8–1 Abbreviations in ladder diagrams

Abbreviation	Description
AT	Autotransformers
B	Braking relay
CB	Circuit breaker
CON	Contactor
CR	Control relay
CT	Current transformer
CTR	Counter
D	Diode
F,FOR,FWD	Forward magnet coil or contacts
FS,FLS	Float switch
L1,L2,L3	Incoming line voltages
LS	Limit switch
LT	Pilot light
M	Magnet coil (relay or motor starter)
MTR	Motor
OL	Overload
R,REV	Reverse magnet coil or contacts
RES	Resistor
S	Switch
SCR	Silicon-controlled rectifier
T	Transformer
TR,TD,TDR	Timer, time delay relay
1,2,3,4,	Terminal points or connections
T1,T2,T3	Motor terminal connections

Table 8–1 shows abbreviations commonly used in industrial ladder diagrams. Schematic diagrams of different electrical devices are shown in Figure 8–4.

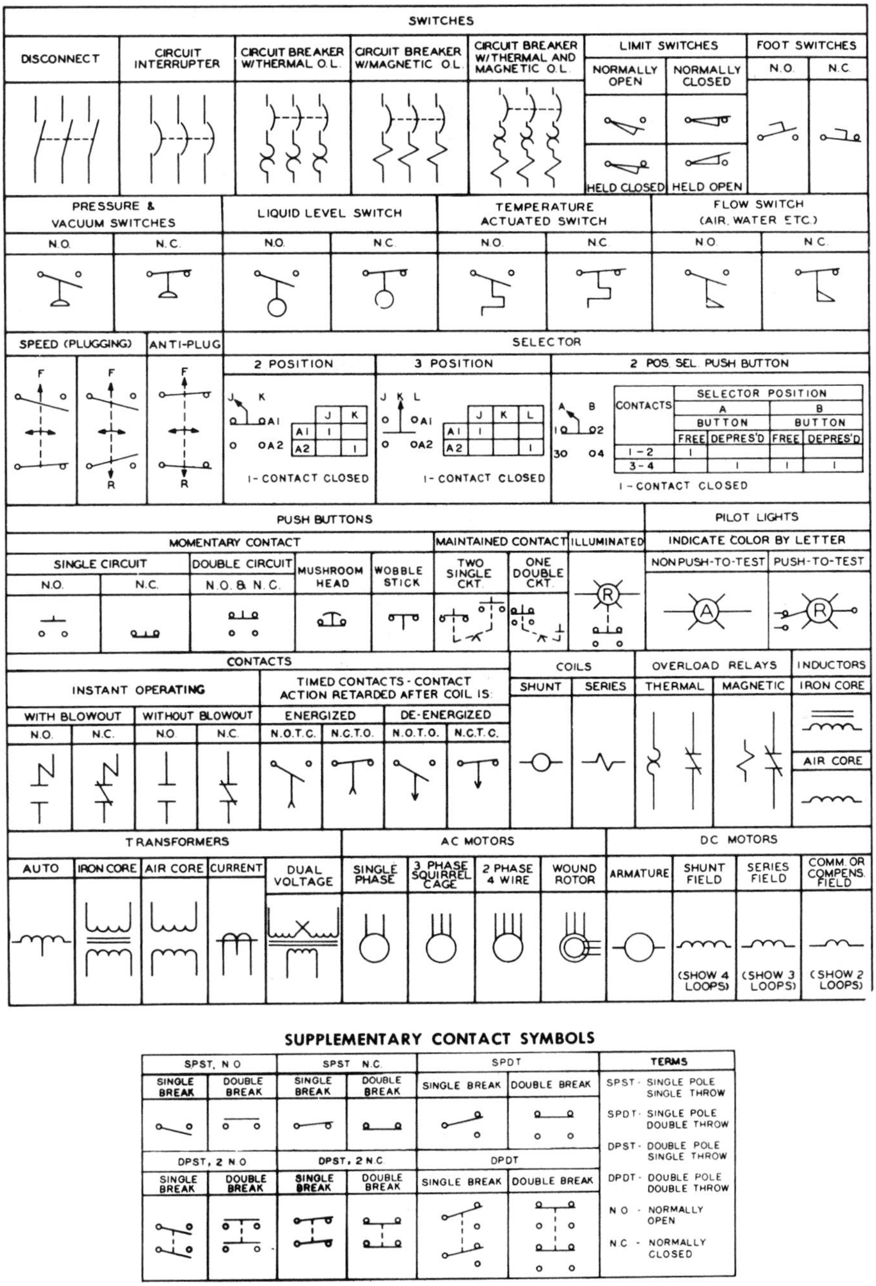

SWITCHES								
DISCONNECT	CIRCUIT INTERRUPTER	CIRCUIT BREAKER W/THERMAL O.L.	CIRCUIT BREAKER W/MAGNETIC O.L.	CIRCUIT BREAKER W/THERMAL AND MAGNETIC O.L.	LIMIT SWITCHES		FOOT SWITCHES	
					NORMALLY OPEN	NORMALLY CLOSED	N.O.	N.C.
					HELD CLOSED	HELD OPEN		

PRESSURE & VACUUM SWITCHES		LIQUID LEVEL SWITCH		TEMPERATURE ACTUATED SWITCH		FLOW SWITCH (AIR, WATER ETC.)	
N.O.	N.C.	N.O.	N.C.	N.O.	N.C.	N.O.	N.C.

SPEED (PLUGGING)	ANTI-PLUG	SELECTOR		
F / R	F / R	2 POSITION	3 POSITION	2 POS. SEL. PUSH BUTTON
		1-CONTACT CLOSED	1-CONTACT CLOSED	1-CONTACT CLOSED

PUSH BUTTONS							PILOT LIGHTS	
MOMENTARY CONTACT					MAINTAINED CONTACT	ILLUMINATED	INDICATE COLOR BY LETTER	
SINGLE CIRCUIT		DOUBLE CIRCUIT	MUSHROOM HEAD	WOBBLE STICK	TWO SINGLE CKT.	ONE DOUBLE CKT.	NON PUSH-TO-TEST	PUSH-TO-TEST
N.O.	N.C.	N.O. & N.C.						

CONTACTS								COILS		OVERLOAD RELAYS		INDUCTORS
INSTANT OPERATING				TIMED CONTACTS - CONTACT ACTION RETARDED AFTER COIL IS:				SHUNT	SERIES	THERMAL	MAGNETIC	IRON CORE
WITH BLOWOUT		WITHOUT BLOWOUT		ENERGIZED		DE-ENERGIZED						
N.O.	N.C.	N.O.	N.C.	N.O.T.C.	N.C.T.O.	N.O.T.O.	N.C.T.C.					AIR CORE

TRANSFORMERS					AC MOTORS				DC MOTORS			
AUTO	IRON CORE	AIR CORE	CURRENT	DUAL VOLTAGE	SINGLE PHASE	3 PHASE SQUIRREL CAGE	2 PHASE 4 WIRE	WOUND ROTOR	ARMATURE	SHUNT FIELD	SERIES FIELD	COMM. OR COMPENS. FIELD
										(SHOW 4 LOOPS)	(SHOW 3 LOOPS)	(SHOW 2 LOOPS)

SUPPLEMENTARY CONTACT SYMBOLS

SPST, N.O.		SPST N.C.		SPDT		TERMS
SINGLE BREAK	DOUBLE BREAK	SINGLE BREAK	DOUBLE BREAK	SINGLE BREAK	DOUBLE BREAK	SPST - SINGLE POLE SINGLE THROW
						SPDT - SINGLE POLE DOUBLE THROW
DPST, 2 N.O.		DPST, 2 N.C.		DPDT		DPST - DOUBLE POLE SINGLE THROW
SINGLE BREAK	DOUBLE BREAK	SINGLE BREAK	DOUBLE BREAK	SINGLE BREAK	DOUBLE BREAK	DPDT - DOUBLE POLE DOUBLE THROW
						N.O. - NORMALLY OPEN
						N.C. - NORMALLY CLOSED

FIGURE 8–4 Standard elementary diagram symbols (courtesy of Square D)

8–3   CIRCUIT CONFIGURATIONS

This section will examine the different circuit configurations commonly found in electrical motor controls in industry. These circuits are broken down into two basic classifications: two-wire and three-wire circuits.

8–3.1   Two-Wire Circuits

The two-wire control circuit, illustrated in Figure 8–5a, shows two wires connecting the maintained contact control device, in this case a limit switch, in series with the starter *1M*. When the limit switch contacts close, they energize the coil electrical circuit connecting the three-phase motor to the line voltage. If the limit switch contacts were to open, power would be disconnected from the motor. Two-wire control circuits provide a type of circuit protection called low voltage or undervoltage release, or ***undervoltage protection.*** This type of pro-

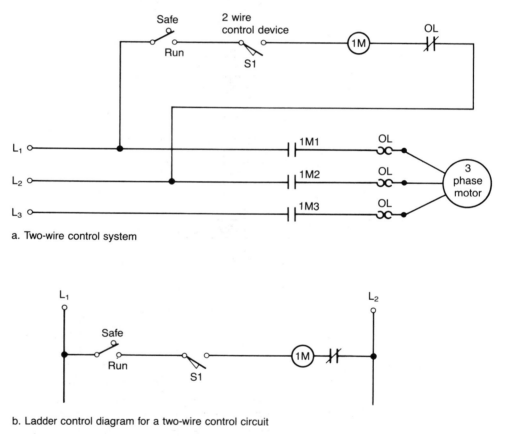

a. Two-wire control system

b. Ladder control diagram for a two-wire control circuit

FIGURE 8–5 Two-wire control circuit

tection is necessary because a motor can be damaged by running it with too low a line voltage. Low voltage release is defined as the action of a device that operates on the reduction or loss of power to the main circuit, but does not prevent the reconnection of power to the main circuit when voltage returns. This circuit will stop the motor under a low-voltage or no-voltage condition but will automatically resume when proper power is again restored. The term *two-wire control* gets its name from the basic circuit where only two wires are required to connect the pilot device (limit, pressure, photo switch) to the starter. The diagram in Figure 8–5b shows the ladder control diagram as it is normally drawn in industry.

If the line voltage in Figure 8–5 goes to zero or decreases below the seal-in voltage of the coil, the closed contacts applying power to the motor will open. The motor will then stop. If or when power is restored to a level high enough to energize the coil, the contacts will close. Power will then be applied to the motor, and it will start. Note the presence of the SAFE/RUN switch. Its purpose is to act as an emergency off switch in case the system needs service or in case of danger to equipment or personnel.

Two-wire control is often used with pumps, fans, compressors, and similar equipment where it is important that the machinery remain in operation as long as there is voltage on the line. It is also used where unexpected restarting of the machinery will not be dangerous to maintenance personnel or operators. Sometimes automatic restarting presents the possibility of danger to personnel or damage to the machinery or to the work in progress. In these cases, two-wire control should not be used. In some applications, the automatic restarting of the motor after a power failure or brown-out is not desired for other reasons. Some processes require that motors be started in a certain sequence. This sequence will not be observed if all the motors come on at once after a power failure.

8–3.2 Three-Wire Circuits

Three-wire control circuits use momentary contact start-stop buttons and a latching relay configuration with the motor starter, as shown in Figure 8–6. When the start button is pressed, power is applied to the motor through *1M1, 1M2,* and *1M3*. When the motor starter coil energizes, the starter button can be released since the closed relay contacts provide an alternate path around the reopened start button. If we press the normally-closed stop button, the circuit to the coil will open, causing the starter to open the power circuit to the motor. The starter will also be deenergized if an overload occurs. If a power failure or brown-out occurs, the starter will also deenergize provided that, in the case of the brown-out, the voltage fell below the relay coil drop-out voltage. When the starter drops out, the latching relay contact opens, preventing the motor from starting when power is restored. In this configuration, the motor will not start when voltage returns unless the starter is again depressed. This circuit is called three-wire control because three wires are connected to the starter: one

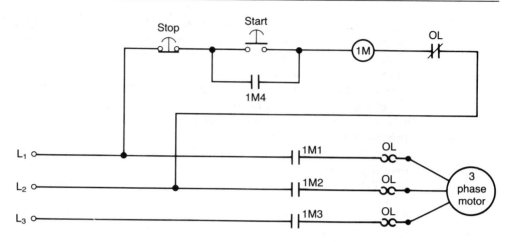

FIGURE 8–6 Three-wire control circuit

from the pushbutton, one from the latching relay contact, and one from the other side of the line.

As with two-wire control, three-wire control provides undervoltage or low voltage protection. This type of protection should be provided when the unexpected starting of the motor, after voltage failure, may present a danger to personnel, equipment, or processes.

8–4 TYPICAL CONTROL CIRCUIT DIAGRAMS

In this section, we will consider several motor control circuits popular in industry. Most of these control circuits can be broken down into two basic types: full-voltage starters and reduced-voltage starters.

8–4.1 Full-Voltage Starting

Because of its simplicity and low cost, full-voltage starting is one of the most widely used methods of starting induction motors. It also offers fast acceleration and maximum starting torque. An example of full-voltage or across-the-line starting is shown in Figure 8–6. Note that when the START pushbutton is pressed, full-voltage is applied to the motor. This type of control is present in both manual and magnetic starters. It does have two possible disadvantages. First, it develops very high starting torques that can damage some loads or driving systems such as belts or gears. If this is the case, another starting method with lower starting torque must be used. Second, this starting method causes the largest line disturbance. A large motor coming on line can actually cause a decrease in the power line voltage Such a decrease can have unwanted effects on circuits powered by the same line. For example, a power dip can cause relays on the line to drop out, causing possible equipment or process damage. In addition to possible damage to systems, power companies often set restrictions on motor starting currents in industry. Applications that cannot use full-voltage starting for either or both of these reasons often use reduced-voltage starting.

8–4.2 Reduced-Voltage Starting

Most electric utilities have a limitation on the current motors can draw from the line on starting. In most areas, local power companies have specific restrictions on starting inrush currents. These regulations vary depending on the type of system, its capacity location, frequency of starting operations, and other loading. Power company restrictions are usually on the maximum current in a single starting step or increment or on the total current drawn in a starting operation. Power fluctuations are a particular problem on 208-V lines with combined power and lighting loads. Excessive motor starting currents will cause lights to flicker and can interrupt service. Although far less frequent than current limitations, there are occasional installations where across-the-line, full-voltage starting can cause mechanical damage. Certain loads may need a more gentle start as well as smooth acceleration to a particular speed. One way to meet these requirements is to reduce the voltage applied to the stator. A starter that uses this method is called a reduced-voltage starter.

Autotransformer Starting current can be reduced while still keeping torque high by use of the *autotransformer,* shown in Figure 8–7a. The autotransformer starter energizes the motor across the secondary taps of an autotransformer. The current drawn by the motor varies directly with the applied voltage. The developed torque varies with the square of the voltage. Therefore, the percentage of full-voltage at the tap determines the starting characteristic of the motor. Since the motor draws its current from the line through the autotransformer, line current varies with normal motor current as the square of the percent voltage at the tap. Figure 8–7a illustrates a type of autotransformer starting called open transition. As you can see in the diagram, taps are available so that personnel can choose a starting voltage of either 50, 65, or 80% of full-line voltage. When the operator presses the *START* button, current flows through the *1M* coil, closing the normally open *1M1* relay contacts (Figure 8–7b). This is the familiar latching relay configuration. Current also flows through the normally closed *1TDR1* and *2CR1* contacts, energizing the coil of *1CR*. All the normally open *1CR* contacts, *1CR2, 1CR3, 1CR4, 1CR5* and *1CR6,* (shown in Figure 8–7a) close. The time-delay relay *1TDR* is set to change state after the motor reaches a certain speed. When the relay times out, *1TDR1* opens and *1TDR2* closes. This action deenergizes the *1CR* coil, opening *1CR2, 1CR3, 1CR4, 1CR5* and *1CR6* contacts. At the same time the *1CR* contacts open, the *2CR* contacts (*2CR2, 2CR3, 2CR4*) close. Examination of

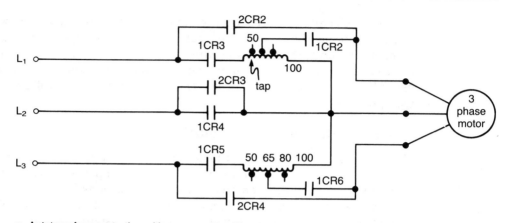

a. Autotransformer starting with an open transition

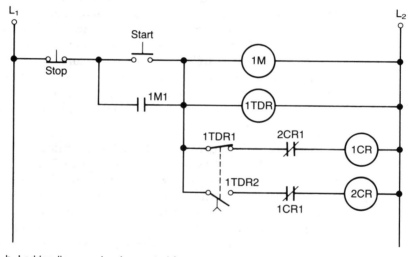

b. Ladder diagram showing control for open transition starter

FIGURE 8–7 Open transition autotransformer starting circuit

this circuit reveals that when *1CR2, 1CR3, 1CR4, 1CR5,* and *1CR6* close, the motor is actually disconnected from the line just prior to the closing of the main circuit contactor. This action may produce a high transient current and line surges. These undesirable effects can be prevented by using the Korndorfer starting circuit, a type of closed transition circuit, shown in Figure 8–8a. This circuit operates in two steps. First, *M1, 1CR2, M3,* and *1CR3* close. Next, *1CR2* and *1CR3* open, completing the second stage of the transfer. Note that part of the autotransformer winding stays in series with the stator, limiting the inrush of stator current. In the last step, the main contactor is energized, closing contacts *2CR2* and *2CR3*. The control circuit for the closed transition starter is shown in Figure 8–8b. The motor always has current flowing

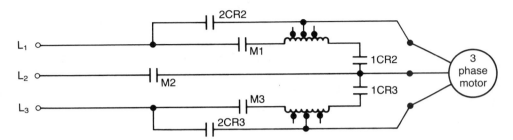

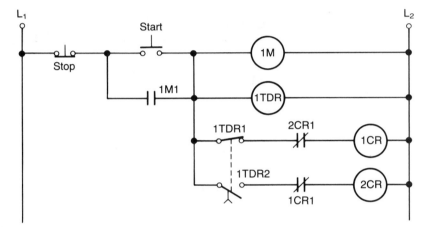

a. Autotransformer starting with closed transition

b. Ladder diagram for closed transition starter

FIGURE 8–8 Closed transition autotransformer starting circuit

through the stator, eliminating surges and producing a smooth torque output. After the motor has reached full speed, all the transformer windings are usually disconnected. Stage stepping can be done manually but more often is accomplished automatically by using time-delay relays. The current-torque ratio in the autotransformer starter depends on which of the three taps is selected. The 50% tap gives 25% of across-the-line inrush current and 25% of starting torque. Moving to the 65% tap pulls current and torque to approximately 42%; 80% tapping allows current and torque to reach 65%. Note that both of these diagrams, like many others in this chapter, do not have overload heaters and contacts. Overload protection is important but has been omitted from these diagrams to keep the diagrams as simple as possible.

Primary Resistor Starting Primary resistor starting employs a resistor connected in series with each stator winding to reduce the voltage to the motor windings. Any voltage dropped across the resistor reduces the amount of voltage presented to the windings. Since stator current is also reduced, this reduction in voltage produces a low torque. Because of power losses in the resistors, this method is less efficient than the autotransformer. An example of a resist-

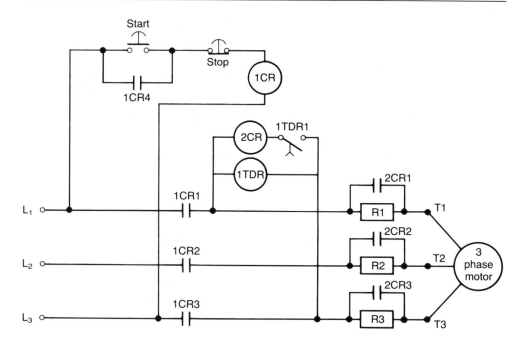

FIGURE 8–9 Primary resistor starting

ance starter motor is found in Figure 8–9. When the starter pushbutton is pressed, voltage is applied through the resistors *R1, R2,* and *R3.* Note that *2CR2* contacts will remain open until the time-delay relay times out. When the time-delay relay times out, current will flow through the *2CR* coil, closing the three normally-open *2CR2* contacts and shorting out the resistors. When the resistors are shorted out, the full voltage is applied to the stator windings. An advantage of primary resistor starting is the smooth acceleration produced. Smoothness of acceleration results from the natural reduction in current as the motor accelerates, reducing the voltage across the resistors and applying more voltage to the stator windings. The torque produced by the motor starts out low and increases as the motor picks up speed. There is no loss in speed or torque as there is at transfer in autotransformer starting methods since the resistors are shorted out at a preset speed. This starting method is generally used in systems where the starting torque can be less than 50% of full-load torque. It is the most commonly used type of reduced-voltage starting because of its simplicity.

Primary Reactor Starting Reactor starting uses reactors in series with the windings instead of resistors. The operation of this starting circuit is very similar to the resistance starter. It is normally used on larger motors with voltages greater than 600 V. Although efficiency and starting torque are better than the resistor starter, the starting power factor is poorer, due to the high reactance of the reactor. The power factor does, however, improve as the motor picks up speed.

Part-Winding Starting Part-winding starting, shown in Figure 8–10, is generally regarded as the most cost effective means of achieving low-voltage starting. The connection is very simple. The motor used in this type of starting must have two or more sets of parallel-connected stator windings and should be specially designed to use this method of starting. Starting is achieved by first switching in the start winding and, after a suitable time delay, connecting the run winding.

On startup, with one section of winding across the line, current drawn will equal about 60% of full across-the-line currents. Torque characteristics are better if the stator is designed for part-winding starting, but standard and dual-voltage motors are sometimes used.

Although there are cost and simplicity advantages, part-winding starting does have its disadvantages. The starting current runs between 60 to 80% of normal current while starting torque is between 40 to 48% of normal torque. Torque efficiency (the ratio of starting current to starting torque) is low.

Wye-Delta Starting In this starting method, the stator is started in the wye configuration (Figure 8–11). Since the line voltage is reduced by a factor of the square root of three, the stator will have a reduced voltage applied. The reduced voltage produces a reduced torque at starting. After the motor accelerates, a time delay relay is normally used to switch the stator voltage to delta configuration. This applies full-line voltage to the stator fields. As in the part-winding motor, the wye-delta method should use a specially designed motor.

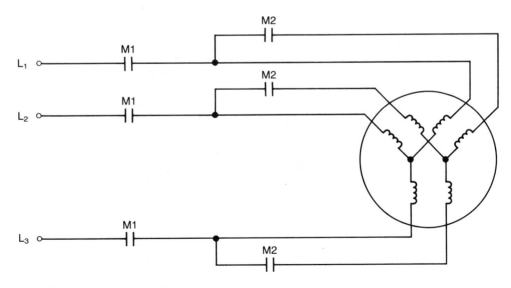

FIGURE 8–10 Part-winding starter

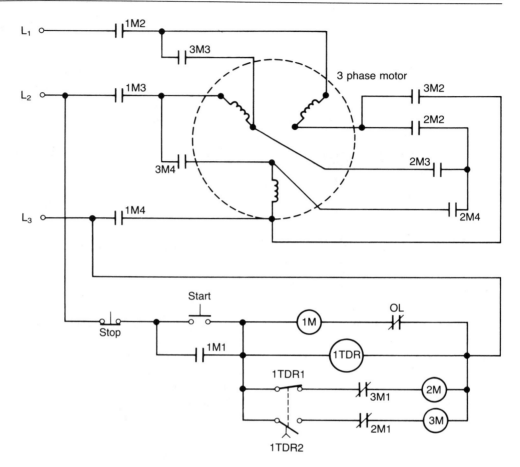

FIGURE 8–11 Wye-delta reduced-voltage starter

Although it suffers the disadvantage of low-starting torque (one third of full-load torque), the wye-delta method gives better torque behavior than the part-winding method. When connected wye, the winding voltage is approximately 58% of full-value, with starting torque and current at approximately 33% of across-the-line levels.

The biggest obstacle to this starting method is industry preference for wye-connected windings and dual-voltage machines. The wye-delta arrangement means that the motor will run delta, making dual-voltage service impossible.

Reversing Some industrial applications require that a motor be started in either direction. This application uses two contactors, as shown in Figure 8–12. One contactor and its associated contacts will be used to run the motor in one direction. The other will be used to run the motor in the opposite direction. A mechanical interlock prevents both contactors from being energized at the

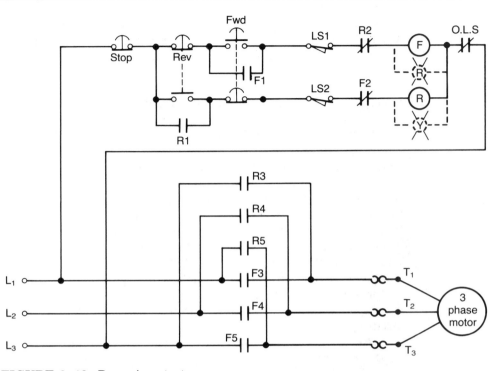

FIGURE 8–12   Reversing starter

same time. Lines *L1, L2,* and *L3* are connected to the motor terminals *T1, T2,* and *T3.* Note the overload relay contacts in lines *L1* and *L3.* Limit switch *LS1* is used for forward operation and limit switch *LS2* for reverse operation. The limit switches are used to stop the motor at a certain point in the motor operation.

The control device for this circuit is the three-button momentary contact pushbutton marked *FWD-REV-STOP.* When the *FWD* button is pushed, current flows through coil *F,* which closes the *F* contacts, moving the motor in the forward direction. Pressing the reverse button interrupts current to the forward coil, deenergizing it, and opening the forward contacts. Pushing the reverse button allows current to flow through the *R* coil, energizing it, and closing the *R* contacts, reversing the direction of the motor. Note that the forward button cannot be pressed without the reverse circuit opening and vice versa. This prevents the motor from getting mixed signals. The stop button, when pressed, will interrupt current in either forward or reverse circuits, stopping the motor.

Certain applications require that the operator know in which direction the motor is turning. By adding pilot lights in the position shown in Figure 8–12 (dotted lines), we can see that when the motor is turning in the forward direction, the red light will be on and the yellow light off. The red light turns on when power is applied to the forward contactor coil. Since the pilot light and the contactor coil are in parallel, they receive the same voltage. When running in the reverse direction, the yellow light will be on and the red light off.

8–5 STARTING SYNCHRONOUS MOTORS

Synchronous motor starters are similar to three-phase induction motor starters. Full-voltage and reduced voltage methods are common in starting circuits. Reduced-voltage uses the primary resistor and reactor, wye-delta, part-winding, and autotransformer configurations. Synchronous motors differ from induction motors in their lack of starting torque. External devices, such as other motors, are used to start the synchronous motor. Internally, the synchronous motor rotor may have embedded squirrel cage bars that allow it to start as an induction motor. Whatever method is used to bring the rotor up to speed, each must accelerate the motor to 92 to 98% of synchronous speed. DC rotor excitation is applied when the rotor reaches this speed and the rotor is lagging the stator field slightly. Sensing and application of DC power can be accomplished by a special relay called a synchronizing relay or by electronic sensors and circuitry.

Reversing the synchronous motor can be accomplished in the same way as the induction motor, reversing one of the stator connections. Improperly reversing stator connections can lead to motor and control circuit damage. Do not attempt this action without checking with an instructor or supervisor.

8–6 STARTING DC MOTORS

DC motors under 750 to 1500 W (approximately 1-2 HP) do not need any special starting circuits. They can be started by full-voltage across-the-line starters, similar to those used in induction and synchronous AC motors. Larger motors, however, can suffer damage to windings due to high inrush currents at starting. You may recall that when a DC motor is started, CEMF is zero and armature currents are high. DC motor starting can be divided into two basic categories: variable resistance starting and reduced voltage starting.

8–6.1 Variable Resistance

In this type of starting, different values of resistance are placed in the armature circuit to limit the inrush current. As the motor accelerates, the resistance values are reduced and finally eliminated. The reduction of resistance is accomplished in several ways, two of which are CEMF and definite-time methods.

CEMF Starting In CEMF starting, the CEMF, which is directly proportional to speed, is used to switch different amounts of resistance in the circuit. When the start button is pushed (Figure 8–13), all normally-open M contacts close,

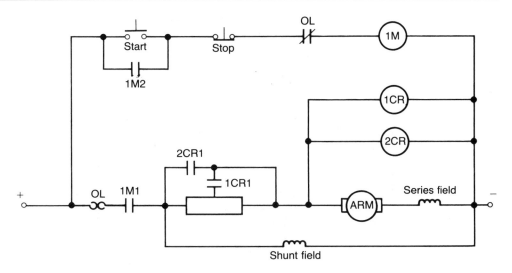

FIGURE 8–13 CEMF starting a DC motor

applying power to the motor through the full accelerating resistance. At the first instant of starting, CEMF is zero and armature current is limited by the full accelerating resistance. Since the CEMF and armature resistance are small, the voltage developed across the armature is small.

Note that the coils of relays *1CR* and *2CR* are in parallel with the motor armature. Relays *1CR* and *2CR* are not standard control relays. *CR2* has a higher pickup voltage than *CR1*. The low armature voltage keeps both coils deenergized. As the motor accelerates, both CEMF and armature voltage increase. When the motor speed is approximately 50% of full-load speed, the voltage across the armature is sufficient to energize *1CR* coil. The normally open *1CR1* contact closes, shorting out part of the accelerating resistance. As the motor continues to accelerate, the CEMF increases, making the armature voltage higher. At approximately 75% of full-load speed, the armature voltage trips *2CR*, closing the *2CR1* contact, which shorts out the remaining acceleration resistance. One disadvantage of this system is poor low-voltage behavior. At low line voltages, the armature voltage may never reach the level needed to energize the second contactor *2CR*. Part of the accelerating resistance would remain in the circuit permanently, causing it to overheat and possibly burn out. The same circumstances would occur if a heavy load were placed on the motor. The motor may not reach the speed and CEMF needed to energize the second relay.

Definite Time   Unlike the CEMF motor starting circuit, the definite-time starter operates independently of the load variations. The diagram in Figure 8–14 illustrates a constant speed DC starting circuit using the definite-time method. Note that relay *1TDR* and *2TDR* are time-delay relays, normally open, timed closed, or ON delay. When the start button is pressed, current

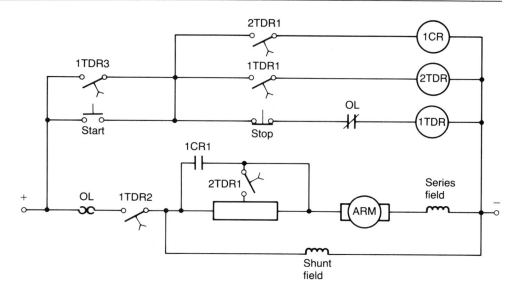

FIGURE 8–14 Definite time starter—DC motor

flows through the *1TDR* coil, closing the normally-open *1TDR2* contacts and
starting the ON delay timer function. Note that the DC power is applied across
the complete acceleration resistance. The motor will now accelerate until the
1TDR relay times out and closes contact *1TDR1*. Closing this contact applies
power to the *2TDR* coil that starts its ON delay sequence. Closing *1TDR1*
shorts out part of the acceleration resistor. The motor continues to accelerate
and, after a delay, *2TDR* times out, changing the state of its relay contacts.
Contact *2TDR1* closes, applying power to *1CR*, which energizes. Contact *1CR1*
then closes, shorting out the remainder of the acceleration resistance.

8–7 JOGGING

Jogging or ***inching*** is defined by NEMA as the momentary operation of a motor
from rest for the purpose of accomplishing small movements of the driven
machine. Usually, jogging is a momentary, or on-off operation of a drive. Often
the motor does not come up to full speed. One method of jogging is shown in the
circuit in Figure 8–15. Selector switch *A* opens the latching relay contact, the
circuit that parallels the start button. Jogging is accomplished by pressing the
start button, which applies power to the motor only as long as it is pushed. If
the start button is pressed and released quickly, machines can be moved a
fractional part of an inch. Applications for inching circuits include moving
machines to certain positions for inspection or repair and threading material
through machines. Placing the selector switch in the *A1* position will disable
the jogging function.

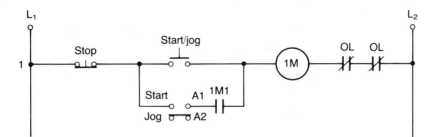

FIGURE 8–15 Combination starting/jogging control

Jogging may use a control relay as in Figure 8–16. Note the two relays, one a motor starter labeled *2M* and the control relay labeled *1M*. When we press the start button, the *1M* coil is energized, closing the two normally-open *1M* contacts, *1M1* and *1M2*. Closing *1M2* will energize the starter coil that applies power to the motor. The normally open *1M1* and *M1* contacts establish a latching relay configuration only when the start button is pushed and both *2M* and *1M* relay coils energize. Pressing the jog button energizes the *2M* starter relay coil, but since the *1M* contacts are open, no latching takes place. The motor will run only as long as the jogging button is pushed.

Sometimes it is necessary to jog a motor in both directions. A circuit that

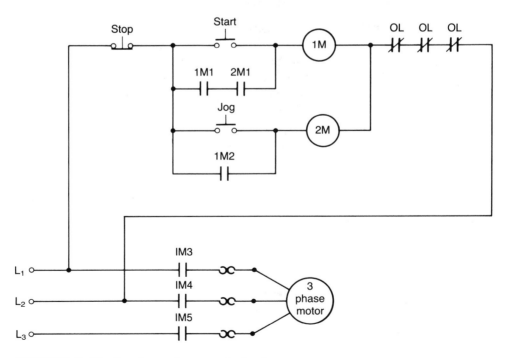

FIGURE 8–16 Jogging using a control relay

can accomplish this is found in Figure 8–17. This circuit allows jogging the motor in either direction, whether it is rotating or at a standstill. Pressing the *start-fwd* button energizes the *1F* forward coil. Current through the *1F* coil changes the state of all contacts associated with it. Closing *1F2* contact energizes the *1CR* coil, which closes the *1CR* contact. Closing *1CR* completes the latching circuit around the *start-fwd* button. When this button is released, current will continue to flow through the *1F* coil. Pressing the *jog-fwd* button interrupts current in the *1CR* coil, opening *1CR1* and the start holding circuit. The motor will now receive power only as long as the *jog-fwd* button is depressed.

8–8 BRAKING MOTORS

Electric motors need ways to bring the rotor to a stop upon command. This process is known as **braking** the motor. In some cases, the motor can be disconnected and allowed to merely coast to a stop. Many motor braking systems do nothing more than this. Other applications require that the motor be stopped quickly. In these cases, special braking circuits and methods are needed. Generally, motor braking systems can be classified into two groups: internal and external.

8–8.1 External Braking

External braking is accomplished in both AC and DC motors with spring-set friction brakes, such as those found in an automobile. Also, as in the automobile, braking is done with either a drum-shoe or disk-type brake. A drum-shoe brake is illustrated in Figure 8–18. In normal operation, the brake is held open by a solenoid. When power is taken away from the solenoid, the brakes close

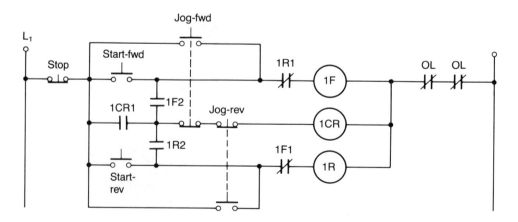

FIGURE 8–17 Jogging in both directions

Drum mounted on
motor shaft

Brake shoes

Brake and wheel

FIGURE 8–18 Mechanical brake (courtesy of Square D)

over the rotor bringing it to a stop. These magnetic brakes, as they are called, are rated in maximum torque, which must be equal to or greater than the full-load torque of the motor. For example, a motor's full load torque is equal to 100 lb · ft. A proper size for the magnetic brake would be 100 lb · ft or greater.

Electrically, the brake is connected across the line, as in Figure 8–19. The power that runs the motor keeps the brake off by energizing the solenoid.

Brake
solenoid coil

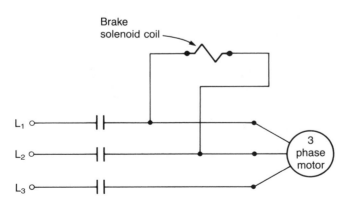

L_1

L_2

L_3

3 phase motor

FIGURE 8–19 Electrical connection for a mechanical brake

When power is taken away from the motor, power is also taken away from the magnetic brake, which causes the brake to close, stopping the motor by friction. Note that the motor will stop regardless of how power is removed, whether by the operator or by a power failure. The operation of the brake to stop the motor during a power failure can be an important safety feature. As in any mechanical brake, adjustments will be necessary from time to time to keep the brake operating properly.

8–8.2 Internal Braking

Internal braking is accomplished by currents flowing within the motor's own windings. Unlike the mechanical brakes, power is required for these systems to function properly. They suffer the added disadvantage of creating more heat in the motor, which can lead to an overload situation.

Dynamic Braking Originally used only for DC motors, *dynamic braking* is also used today in induction and synchronous motors. In dynamic braking, the motor operates as a generator, dissipating the mechanical energy in the armature through a resistor across it. A dynamic braking arrangement is illustrated in Figure 8–20. This is a diagram for a shunt DC motor, since the field is in parallel with the motor armature. Speed control is accomplished by adjusting *R1*. Note that, when the motor is to be stopped, contact *1M3* closes, placing a resistor across the motor armature. Since the shunt field remains energized and no power is applied to the armature, the motor is acting as a generator. If *R2* is small, a great amount of current will flow, generating a large counter-torque that will effectively brake the motor. The power to the field should be

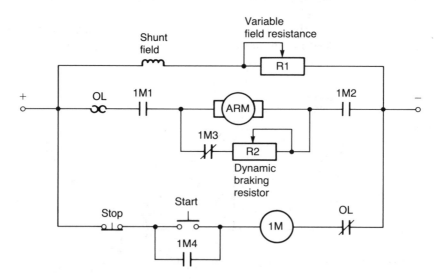

FIGURE 8–20 Dynamic braking of a shunt-connected DC motor

disconnected after the motor has stopped. Adjusting the size of the resistor $R2$ will change the amount of current drawn, and, therefore, control the speed at which the motor stops. This method of braking can be used with equally successful results with permanent-magnet DC motors.

Synchronous motors can also use dynamic braking as a method of stopping the motor. An example is shown in Figure 8–21. The DC excitation voltage to the stator remains connected when the braking is applied. Braking is accomplished by opening contacts $CR1$ to $CR3$ and closing $BR1$ through $BR3$. The stator fields are now producing a voltage, since the energized rotor is inducing voltage in the stator coils. The power produced by the stator is then dissipated in the resistors.

Dynamic braking is becoming more popular as a braking method in induction motors. Generally, dynamic braking in induction motors is accomplished by disconnecting AC power to the stator and connecting a DC power source, as in the polyphase motor in Figure 8–22. In this application, DC power is connected to two of the three windings of the polyphase motor. This method of braking will work with the stator connected in wye or delta configuration. When the stop button is pushed, contacts $CR1$ through $CR3$ open and contacts $BR1$ and $BR2$ close. The DC is applied to two stator coils that are cut by the induction motor rotor, producing the countertorque necessary to stop the motor. An interlocking system must accompany this design to keep BR and CR contacts from being closed at the same time. A method is also needed to disconnect the DC supply from the stator coils after the rotor has stopped. This can be accomplished by a friction switch or a time-delay relay. One advantage of this type of braking is that it does not reverse the motor.

Single-phase motors can be dynamically braked as well as polyphase motors. Figure 8–23a illustrates the type of system that would be used in a shaded-pole motor. When the switch is in the *run* position, AC power is deliv-

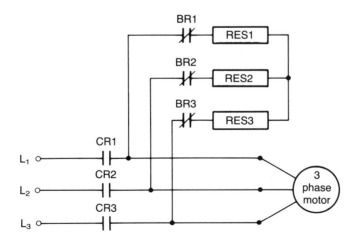

FIGURE 8–21 Dynamic braking of a synchronous motor

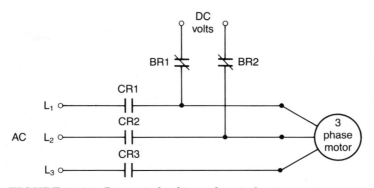

FIGURE 8–22 Dynamic braking of an induction motor

ered to the rotor through electromagnetic induction. When the switch is moved to the stop position, the AC is disconnected and the DC connected to the stator. Heavy rotor currents flow, producing the countertorque to stop the motor. A split-phase dynamic braking circuit is illustrated in Figure 8–23b. Its operation is identical to the description of the shaded-pole motor.

The source of DC for this method of braking varies from batteries to filtered power supplies. Batteries are generally less expensive than filtered power supplies but must be changed often and do not supply high currents.

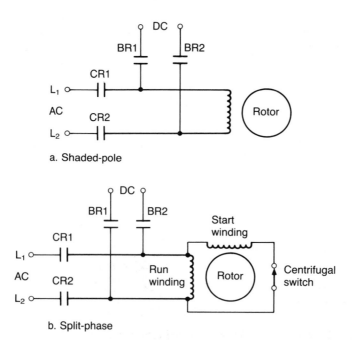

a. Shaded-pole

b. Split-phase

FIGURE 8–23 Dynamic braking of single-phase induction motors

Plugging *Plugging* is defined by NEMA as a system of braking in which the motor connections are reversed so that the motor develops countertorque, thus exerting a retarding force. Used in both AC and DC motors, it reverses the power connection to some of the motor coils. This connection produces torque in the direction opposite to that which the rotor is exerting as it turns. In this scheme, shown in Figure 8–24, when the start button is pushed, power is applied to the motor, which turns the rotor. The turning rotor in the forward direction closes the normally-open plugging switch contact *1SW1*. The plugging relay is a centrifugal switch with contacts in series with the reversing coil *1R*. Mounted on the motor, the plugging relay contacts close, allowing the motor to run forward when the start button is pressed. When the motor is plugged to a stop, the switch opens before the motor has a chance to reverse. When the stop button is pushed, current is interrupted in the *F* coil, deenergizing it. This changes the state of all the *F* relay contacts. Contact *1F2* closes, completing the circuit in run *2*, which energizes the reverse contactor through the plugging switch *1SW1*. This action reverses the motor connections; the rotor tries to run in the opposite direction and the motor comes to a stop. When the rotor nears zero rpm, the plugging switch, a centrifugal switch much like the one used to disconnect the start winding in the single-phase motor, opens and disconnects the reversing contactor. If this switch is not present or malfunctions, the motor will run in the opposite direction. The reversing contactor is used only to plug the motor to a stop.

8–9 TROUBLESHOOTING

Inevitably, motors and their control circuitry will malfunction. No circuit or system is immune to either internal or external damage. When motors and their associated control circuitry do not perform as they should, they must be repaired by maintenance personnel.

Before discussing troubleshooting, it is necessary to mention the types of faults and possible causes.

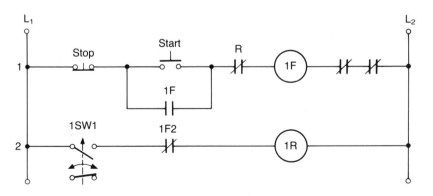

FIGURE 8–24 Control circuit for braking a motor by plugging

8–9.1 Fault Conditions

Those involved in the maintenance of rotating electrical machinery and controls see certain patterns in the faults that occur. The voltages applied to the machines can change or the load on the machines can change.

Voltage Variations Supply voltages can be either too high or too low. Too high a voltage is called an overvoltage. In AC motors, an overvoltage will cause increased stator currents, which, in turn, will cause increased power losses in the form of heat. The power losses come primarily from copper losses and core losses. The increased core losses result from the decreased power factor of the motor. Increased losses in the form of heat can cause the motor temperature to rise, possibly causing an overload condition. Overvoltages can also cause excessive starting and full-load torques. Overvoltage in DC motors will cause the motor to increase speed, a possibly undesirable consequence of an overvoltage. Overvoltages will also adversely affect the magnetic coils in contactors, relays, solenoids, and starters. As in the motor, increased voltages will cause increased currents, possibly in excess of manufacturer's recommendations. Overvoltages in relay coils will cause an increased strength in the magnetic field generated by the coil. The stronger magnetic field will exert a greater pull on the armature, possibly causing impact damage over time. Contact bounce will also be increased in a coil with an overvoltage.

A condition where the line voltage is too low is called an undervoltage condition. In AC motors, an undervoltage condition will reduce starting torque, full-load torque, and the overload capacity of the motor, ultimately causing overheating. In DC motors, the motor can run slower with an undervoltage. Too low an applied voltage also has unwanted effects on magnetic coils. Coils can pick up but fail to seal, a condition that will cause greater current flow in the coil leading to an overheating condition. In addition, low voltages can cause relay chattering that can, in turn, lead to excessive armature wear. What constitutes an overvoltage or undervoltage? Generally, motor systems can tolerate a 10% variation in voltage. Better motor performance will be realized the closer a motor operates to its rated voltage.

The other voltage discrepancy deals with the phases in a three-phase motor system. First, the phases can become unbalanced. An unbalanced voltage condition exists when the phase-to-phase voltages are not equal, as they should be. Unequal phase voltages, when applied to a motor, will produce unequal currents that can cause an overload condition in the motor, eventually causing failure of the motor. Most polyphase motors can tolerate only a 5% variation in the phase-to-phase voltage. A lower phase voltage can be caused by loading on the line, especially if it is used to supply single-phase power for lighting, heating, etc.

Second, and more extreme, is the loss of one or two of the three phases. In some cases, the motor will run on one phase if lightly loaded. The motor will rapidly overheat, however, causing eventual motor failure.

Loading Variations A motor that is jogged or started and stopped often in a short period of time can overheat. This condition is often not detected by overload protection devices, since the protection devices have lower thermal mass than the motors. This means that protective devices can cool off more quickly than the motor can; therefore, they will not heat up to the same temperature as the motor.

8–9.2 Determining Symptoms

Before troubleshooting can proceed, the technician must be fully aware of the problem. The technician should personally observe the problem and talk with the equipment operators. Never neglect the assessment of the operator concerning the root of the problem. The technician should have a firm grasp of the circuit operation before attempting to diagnose the problem. An understanding of the operation of the system can be gained by careful study of circuit ladder and wiring diagrams and by review of the operating principles of each of the devices.

8–9.3 Measurements

If any of the previous symptoms are observed, the technician will need to choose and use appropriate test equipment. For example, if a phase unbalance is suspected, a voltmeter measurement of the phase-to-phase voltage will be necessary. An over- or undervoltage, if suspected, can be measured with the same device.

8–10 MAINTENANCE

Motors and their control circuitry are used in industry as part of a manufacturing operation or a customer service. Failures of motors and their associated circuitry can be an expensive circumstance, not only in the cost of the parts and labor but also in the lost manufacturing and production time. It is important that equipment be kept in good operating condition. A good system of preventive and corrective maintenance will help keep equipment outages at a minimum. No program will eliminate the need for maintenance. It is imperative that maintenance personnel have a good understanding of motors, their associated circuitry, and proper maintenance principles.

8–10.1 Safety

Both industry and governmental agencies are committed to making the industrial workplace a safe one. No plant or workspace can, however, be made totally safe and accident free. All maintenance personnel should take appropriate safety precautions, such as the deenergizing of all equipment prior to starting work.

8–10.2 Manual Starters

Manual starters, due to their simplicity, need very little preventive maintenance. Tightness of connectors and overload relay heaters and contacts should be checked regularly. If the lever or pushbutton becomes sluggish, it may need to be cleaned or lubricated. Consult the manufacturer's specifications before applying any lubrication.

8–10.3 Magnetic Starters

As in the manual starter, the parts of the starter and associated electrical connection need to be checked regularly for tightness. The contactor part of the starter should snap into place quickly and freely. If sluggish operation occurs, you should check the alignment of the moving parts. Excessive amounts of dirt can also cause sluggish or weak operation.

The contactor or relay has one or more sets of contacts that make or break when the device is energized. These contacts are designed to carry electrical current flow. When contacts open, any voltage across the contacts will cause some arcing. Arcing will burn and pit the contacts, making them poor conductors. Electrical arcing erodes the contacts by vaporizing them and causing uneven wear. On larger devices, contacts that have been burned and cannot carry current efficiently should be replaced.

Contacts should be inspected regularly for signs of burning and pitting. Be sure to disconnect all power to the relay before inspection. New contacts, made of a silver alloy, have a uniform silvery color. As contacts wear, the silver color turns to blue, then to brown and black. The black color is a result of the formation of silver oxide when the contacts burn. Silver oxide is a conductor of electricity and, therefore, is not an impediment to current flow. All contacts, however, are not made of silver or silver alloys. In cases where other metals are used, blackened contacts can mean that significant contact resistance has developed. In such cases, it is best to read the manufacturer's data for the contacts and to measure the actual contact resistance.

When contacts wear, the contour changes, usually involving a slantwise type of wear. In Figure 8–25 note that the lower right edge of the contact has worn away more quickly than the other three. Contacts normally wear in this fashion, needing replacement only when the wear interferes with the electrical continuity.

Pitting or irregularities in the contact surface are caused by vaporizing

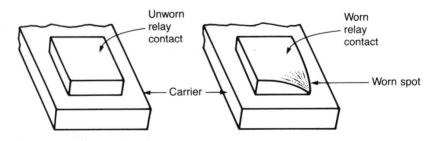

FIGURE 8–25 Relay contact wear

the contacts in some areas but not in others. The amount of pitting is directly proportional to the current carried by the contacts. The higher the current through the contact, the greater the pitting. Excessive pitting should prompt contact replacement.

In all cases, the contacts should be replaced when more than 75% of the surface is worn away, no matter what shape the contacts are in. Rather than a percentage of surface, some data sheets specify contact wear by the amount of contact left in inches, usually 0.015 in.

The decision to replace contacts should be based on a close inspection of the contact surface and measurement of wear. Extreme contact erosion and pitting are grounds for contact replacement. Silver alloy contacts should not be filed because filing merely increases contact wear. Filing silver contacts does not generally improve performance or extend contact life.

When the decision is made to replace the contacts, replace both the movable and stationary contacts as well as the contact spring. Kits are available from relay manufacturers and suppliers with all the parts necessary for contact replacement. On three-phase motor starters, replacing one pole's contacts necessitates replacement of the other two.

Contacts that do not require replacement but are burned and pitted can be resurfaced with a file or sandpaper. Never use emery paper. Contacts should be only lightly worked since they may have a certain defined shape necessary for proper operation. Heavy filing can destroy this shape, causing the contacts to seat improperly. It is a good practice to consult the manufacturer's data on the particular starter prior to any maintenance action. The starter contacts should be carefully examined. Infrequently, contacts may be welded together. This is a serious fault that can be detected by careful inspection. Contacts should NEVER be lubricated. Lubrication will limit current flow and collect dust and dirt, further impairing current flow.

Starter contacts are usually spring driven. The purpose of the springs is to hold the contacts together so that current can flow through them easily. The tension of these springs can decrease over time, causing the contact pressure to decrease. Decrease in contact pressure can result from excessive wear on the contacts and associated moving parts. Special tension measuring devices can be used to measure contact pressure.

Faults in coils can be classified as open, shorted, or partially shorted. The contacts of the starter will not move if the coil is open, even though power is applied. A shorted coil is a more dangerous situation. Excessive current will flow and can cause a fire if not interrupted by a fuse or circuit breaker. Partially-shorted windings are more difficult for the operator to detect. The device may energize in a slow or sluggish fashion. A simple resistance check will determine whether the coil is open, shorted, or partially shorted. Of course, all power should be disconnected from the circuit prior to measuring coil resistance.

All components associated with starters, such as resistors, capacitors, and inductors, should be checked regularly for proper parameters.

8–10.4 Environmental Conditions

Moisture Moisture in starting circuits and associated conduits can cause problems such as breakdown of insulation, ground faults, and other undesirable low resistance paths for current. Moisture problems can be dealt with by placing starting circuits in controlled atmospheres, such as in moisture-proof boxes or air-conditioned spaces. Slight amounts of heat, as in a low-wattage light bulb, may be sufficient to keep moisture at bay, especially in infrequently energized circuitry.

Dirt Some industrial environments produce large amounts of dirt and dust. These contaminants can cause improper operation of starters for several reasons. First, dirt buildup on contacts can cause high contact resistances. Second, dirt buildup can interfere with the free travel of moving parts in contactors, relays, and solenoids. Third, excessive dirt can cause low resistance current paths that can cause arcing, fires, improper equipment operation, and safety hazards. Equipment in dirty environments should be checked regularly for buildup and removed if excessive. Preventive measures are the same as for dealing with moisture.

Lubrication In general, motor starters should not be lubricated unless the manufacturer specifically recommends it. Lubricating where it is not necessary can cause dirt and dust to build up more quickly than it otherwise would, causing earlier equipment malfunction.

Oxidation Oxidation can be a problem in industrial circuits for two reasons. First, oxidation on contacts can limit current due to the high electrical resistance of oxide deposits. Second, oxidation can restrict the free travel of the moving parts in a relay, contactor, or solenoid. Oxidation is usually found in two types: ferric oxides and copper or cupric oxides. Ferric oxides, commonly known as rust, occur in metals that contain high percentages of iron, such as steel. Copper oxides are found especially on copper relay contacts, where they exhibit a very high resistance. Oxidation can be found in all industrial environments but is more of a problem where high temperatures, moisture, or chemicals speed up the oxidation process. Oxidation can be prevented by keeping starters away from moisture and chemicals. Removal of oxide deposits can be accomplished by filing lightly with sandpaper or a special file.

CHAPTER SUMMARY

- In electrical motor control circuit diagrams, the ladder diagram is the most popular because it simplifies complex circuit operation.
- Electrical motor control circuits are broken down into two-wire and three-wire circuits.
- Two-wire circuits provide undervoltage release circuit protection.

- Three-wire circuits will not reenergize a motor after power is restored.
- Full-voltage motor starting is popular because it is simple and low in cost.
- Full-voltage starting can cause a reduction in line voltage.
- Reduced voltage starting is used when full-voltage starting draws too much current from the line.
- Synchronous motor starters are similar to the starters used on induction motors.
- Large DC motors use a type of reduced voltage starting to reduce the risk of damaging the motor windings. Small DC motors will not usually be damaged by applying full voltage.
- Jogging is the momentary application of power to a motor to cause small shaft movements.
- Dynamic braking is used for both AC and DC motors. Dynamic braking slows a motor by the countertorque created when it acts as a generator.
- Plugging is a means of braking where the power is applied in a reverse direction for a short time.

QUESTIONS AND PROBLEMS

1. Explain the sequence of events that occurs when the start button is pressed in Figure 8–3. When the stop button is pressed.
2. Draw a three-wire and a two-wire control circuit and explain the difference between the two.
3. Define the following terms:
   a. undervoltage protection
   b. undervoltage release
   c. overcurrent protection
   d. full-voltage starting
   e. reduced-voltage starting.
4. Compare and contrast the different types of reduced voltage starting methods.
5. Which reduced voltage starting methods require a special type of motor?
6. Explain the differences in the starting circuits for a synchronous motor and an induction motor.
7. Explain how the induction and synchronous motors can be reversed. Draw a three-wire reversing control circuit for a three-phase induction motor with one start button, one stop button, and three overload relays.

8. Differentiate between CEMF and definite time DC motor starting circuits.

9. Define jogging. Draw a jogging control circuit for a three-phase induction motor to jog in one direction.

10. Differentiate between dynamic braking and plugging. Which method is more efficient at stopping a motor? Why?

11. Explain the effects of each of the following conditions on a motor and associated relay coils:
    a. overvoltage
    b. undervoltage
    c. unbalanced phases.

12. What causes burning and pitting in a relay contact? What is the nature of the black deposits on the relay contact? How do these deposits affect the operation of the relay?

13. List the environmental factors that could cause malfunctions in a motor system.

14. Suppose that a motor controller is able to disconnect the motor from the power supply, keep it disconnected, and then restart it automatically when conditions return to normal. What type of system is this?

15. The controller shown in Figure 8–26 uses what type of protection?

16. Suppose we push the button that starts the motor in Figure 8–26. Which contact closes to complete a holding circuit for energizing the contactor coil after the button is released?

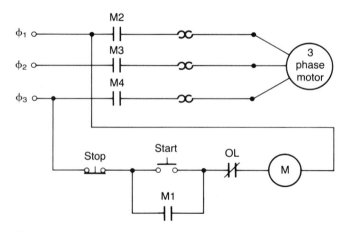

FIGURE 8–26 Three-phase motor controller

17. How can the circuit shown in Figure 8–26 be modified to permit reversal of the direction of the motor?

18. A reversing motor starter is used to control a three-phase induction motor and protect it from low voltages and overloads. If the line voltage fails, how is the motor started after the power is restored?

19. Describe how dynamic braking is accomplished in a series DC motor.

9

SEMICONDUCTORS IN MOTOR CONTROL

At the end of this chapter, you should be able to

- explain how the transistor and Darlington are used as power control devices.
- differentiate between the power transistor and Darlington in structure.
- explain how the MOSFET controls power in a circuit.
- define a thyristor.
- list the three thyristors that are used as power control devices and draw their schematic symbols.
- describe how the thyristor is used to control power.
- differentiate between the thyristors used as trigger devices in power control circuits.
- list three methods used to protect power semiconductors from damage.
- list three important areas in a preventive maintenance program for semiconductor equipment.

9–1 INTRODUCTION

In the previous chapter, motor control was performed by such electrical devices as the electromagnetic relay (EMR). The EMR has many disadvantages as a motor control device. In addition to being slow in reaction time and difficult to use in speed control applications, it is expensive and has a limited life due to contact wear. Designers of motor controls have turned to semiconductor technology to solve some of these problems. Semiconductors, if operated within their design parameters, are reliable, relatively inexpensive, and can be used to control a motor's speed in addition to starting it. Since the semiconductor is such an integral part of modern motor control, this chapter will review the principles and concepts behind the semiconductor.

A semiconductor is a material with an electrical resistance between that of conductors and insulators. The semiconductor controls the flow of electrons by changing the resistance of layers of semiconductor material. The resistance of this material can be changed by application of voltage, current, light, electric, or magnetic fields. Unlike electromechanical switches, semiconductor (or solid-state) switches control power without any moving parts. In motor control applications, we will divide semiconductors into two categories: power devices and drivers. Power devices control the motor directly, while the drivers control the power devices. Some semiconductor devices can be both power devices and drivers. Many drivers, however, do not have enough power to control motors directly.

9–2 TRANSISTOR

The transistor is a very common power control device. Transistors are three terminal devices capable of current gain. The normal input is the base current, and the normal output is the collector current. The current in the collector is greater than the base by a factor called beta (β). The ratio of collector current to base current is a dimensionless number called the current gain. Typical values for β are from 35 to 300 with 100 a nominal value. Although transistors can be used to amplify currents and voltages, their amplification capabilities are not used in such switching applications as motor controls. Transistors can be grouped into two categories: bipolar devices and field-effect transistors (FETs).

9–2.1 Bipolar Junction Transistors (BJTs)

Bipolar transistors have three leads—emitter, base, and collector—as shown in Figure 9–1. Application of a base voltage will control the current flowing between emitter and collector leads. There are two types of bipolar transistors: NPN and PNP types. NPN transistors have N-type semiconductors in emitter and collector leads and P-type material in the base leads. A positive voltage on the base lead (with respect to the emitter) will turn on the transistor current. Construction of the PNP transistor is opposite that of the NPN. The PNP transistors have P-type semiconductors in emitter and collector leads and N-type material in the base leads. A negative voltage on the base lead (with respect to the emitter) will turn on the transistor current. Since most high-voltage power transistors are NPN, our discussion will apply to this transistor construction.

FIGURE 9–1 Bipolar transistor

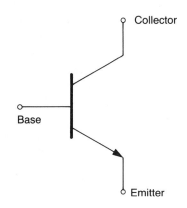

As stated previously, the BJT is used as a switch for most motor control applications. When no current flows in the base lead, the transistor will not conduct. This corresponds to the OFF-state (Figure 9–2a). When enough base current flows, the transistor conducts heavily, saturating the device. This corresponds to the ON-state of a switch (Figure 9–2b). Although the transistor has a linear region between cutoff and saturation, this area is not used in switching applications. Ideally, a switch should have zero resistance when closed. The transistor is not a perfect switch. A saturated transistor can drop between 0.2 and 0.4 V. A perfect switch would have a 0 V drop across it.

The transistor's major limitation in motor control applications is its peak current rating. In normal operation, the collector current will equal the base current times the current gain (β). If the collector current is forced above this value while holding base current constant, the voltage drop across the transistor rises considerably. When base current is above the base current saturation

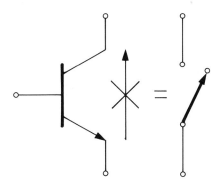

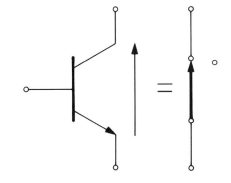

a. Transistor not conducting is like an "open" switch

b. Transistor conducting is like a "closed" switch

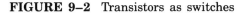

FIGURE 9–2 Transistors as switches

level, a larger base voltage will increase power dissipation drastically, decreasing efficiency and increasing temperature. Another problem with transistors is their sensitivity to damage from transient voltages and currents. When controlling inductive motor loads, a transistor switches on and conducts through the motor as shown in Figure 9–3a, developing the indicated voltage drop. When the transistor shuts off, the flux in the coils collapses, producing a high voltage of the opposite polarity, as shown in Figure 9–3b. This high reverse voltage can damage the transistor, an effect called secondary breakdown. Since motor controls are power applications, power transistors are used almost exclusively in motor controls.

The main design consideration in power transistors is power-handling capability. This capability is determined by the maximum junction temperature the transistor can stand and how quickly heat can be conducted away from the junction. Power transistors are designed to carry large amounts of current and to sustain large voltage fields. An important application of the power transistor is in switching. Large amounts of power can be switched with small losses in a motor control by using a base-driven (common-emitter) power transistor. In switching applications, the two most important considerations are: 1) the speed the transistor switches from cutoff to saturation and 2) the power dissipation. Power transistors are generally non-linear in operation over a wide signal range since they must operate in only two states. In power

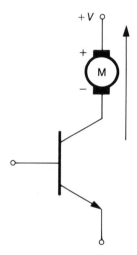

a. Transistor conducting through motor

b. Transistor not conducting - motor reverses polarity

FIGURE 9–3 Transistors driving motors

FIGURE 9–4   Darlington transistor configuration

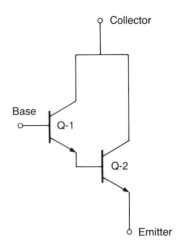

transistor switching circuits, the common-emitter configuration is by far the most widely used.

9–2.2   Power Darlington

Many modern motor controls use the Darlington configuration shown in Figure 9–4. The Darlington configuration can be made by connecting two individual discrete transistors. In this case, Q-1 drives the power transistor Q-2. The Darlington configuration increases the current gain of the circuit when used in place of a single transistor. A power Darlington is a Darlington configuration capable of handling high power levels, such as those needed to drive a motor. Darlington-connected transistors also come in a single package, as in the unit in Figure 9–5. This unit connects three transistors in Darlington configuration. It can handle up to 300 amps continuous current and, when off, up to 1000 V collector-to-emitter voltage. When saturated, this Darlington drops 2 V, collector-to-emitter, turning on in only 2.5 microseconds. At saturation,

FIGURE 9–5 Westinghouse power Darlington module schematic

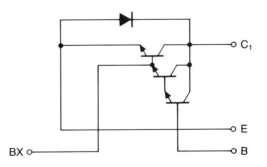

the device will dissipate 600 W. Heat sinks are normally used to dissipate the heat buildup caused by the power dissipation. Current gain at 300 amps is 100, which means that the unit requires 3 amps of driver current.

9–2.3 Power MOSFET

One of the disadvantages of the power transistor is the amount of drive required for high current operation. As we have seen, a Darlington with a current gain of 100 requires 3 amps of driving current to produce 300 amps of collector current. The situation is worse with single power transistors since they lower current gain more than the Darlington connection. Power transistors can have current gains as low as 10. A power transistor with a gain of 10 requires 30 amps of driving current to produce 300 amps of collector current.

The power MOSFET greatly simplifies the driving circuitry because it is a voltage-controlled device requiring little driving current, unlike the power transistor. Power MOSFETs have switching times of less than 100 nanoseconds and are more rugged than transistors. Power MOSFETs are not just a new type of BJT. MOSFETs use a totally different concept of operation.

The BJT is a current-controlled device. A current in the base-emitter junction causes a current flow in the collector terminal. The collector current is approximately equal to the base current multiplied by the transistor current gain. In the power MOSFET, the collector, base, and emitter terminals are replaced by the drain, gate, and source terminals. Figure 9–6 shows one of

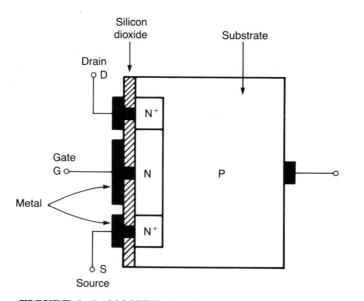

FIGURE 9–6 MOSFET structure

many power MOSFET structures. This device, called a double-diffusion MOS-FET (DMOS), has both P and N material. The P region has N material on either side. A depletion region forms where the N and P layers connect. Current must flow through the channel in the P material, since this is the only place majority carriers exist to carry current flow. The polysilicon gate, insulated by an oxide coating, creates a field over the channel when voltage is applied to the gate. The field reduces the size of the junction and effectively increases the channel width, increasing current flow. In this chip structure, the current flows from the drain terminal vertically through the chip and then horizontally through the channel into the source region.

The MOSFET is a voltage-controlled device. With no voltage applied between the gate and source terminals, the impedance between the drain and source is very high. When a potential is applied to the gate, current flows in the drain. The amount of current flow in the drain depends on the voltage applied to the gate. Because the gate draws only a small amount of current, the DC current gain is very high, typically around 10^9. In fact, since the device controls current by a gate voltage, current gain is not an appropriate parameter. A more useful parameter for the MOSFET is the transconductance, or change in drain current caused by a particular change in gate voltage.

The very low drive current requirement of the power MOSFET and its high power gain are major advantages over the conventional transistor or Darlington. Note the difference in the schematic symbols for a MOSFET (shown in Figure 9–7) and the transistor. Figure 9–7a shows an N-channel power MOSFET; Figure 9–7b shows a P-channel MOSFET. The N- and P-channel MOSFETs have identical functions. Note the difference in biasing between the NPN transistor (Figure 9–8a) and the N-channel MOSFET (Figure 9–8b). Both are biased so that they are conducting. Removal of the potential between gate and source will cause the device to turn off.

Power MOSFETs are not high-voltage devices. They can be used at volt-

a. N-channel MOSFET

b. P-channel MOSFET

FIGURE 9–7 Circuit schematic symbols for bipolar junction transistors and MOSFETs

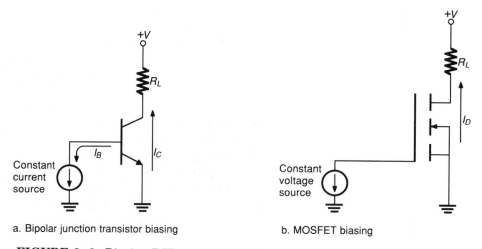

a. Bipolar junction transistor biasing   b. MOSFET biasing

FIGURE 9–8 Biasing BJTs and MOSFETs

ages up to 650 V and currents up to 100 amps. For example, the Siliconix VNE003 is a 60 A, 100 V device with an on-state resistance drain-to-source of 0.035 Ω. At rated current, the device will dissipate 350 W.

Power MOSFETs have very fast switching speeds. Switching speed depends on the time it takes to charge the gate input capacitance. If it were possible to charge the gate capacitance instantaneously, the switching time would depend on the time it takes the current carriers to go from the source to the drain. This time is typically in the picosecond range. Another reason the power MOSFET is faster than the BJT is that the power MOSFET is a majority carrier device. The transistor is a minority carrier device. The power MOSFET does not have minority carrier storage delays that slow down BJT switching. Switching speeds of power MOSFETs range up to 5 nanoseconds.

Power MOSFETs have an additional advantage over transistors. It is difficult to connect transistors in parallel to gain a greater current-handling capability. Unless the bipolar transistors are perfectly matched (an unlikely circumstance), one transistor will draw more current than the other. Because of the transistor's negative coefficient of resistance, the transistor drawing more current can also go into thermal runaway. Parallel operation of MOSFETs is relatively easy, due to their positive temperature coefficient of resistance. During parallel operation, devices that initially draw more current heat up faster. This increase in heat causes the drain-to-source resistance to increase, which, in turn, limits the current flow. The MOSFET behavior forces the two transistors to share current. Unlike BJTs, parallel operation of unmatched MOSFETs will generally work properly.

Because of their fast response times, the paralleled MOSFETs may break into oscillation. Oscillations cause excessive power dissipation and possible failure of the MOSFETs. Oscillations can be reduced or stopped by inserting ferrite beads or resistors in the gate lead.

Power MOSFETs offer great advantages over conventional bipolar transistors. As designers see these advantages, bipolar transistors are being replaced in many applications. Higher switching speed and lower driver requirements mean that circuitry can be greatly simplified, making a more inexpensive, reliable design. As we have seen, power MOSFETs have superior operating characteristics to those of power bipolar transistors. These better operating parameters occur because the MOSFET is a totally different type of device, not just an improved transistor.

9–3 THYRISTOR

The thyristor is a generic name for a type of semiconductor switch. Unlike the transistor, the thyristor has two stable states: on and off. In this regard it is like a mechanical toggle switch. No intermediate state exists between on and off. The thyristor is usually made up of alternating layers of semiconductor material. The most popular of the thyristor family is the silicon-controlled rectifier (SCR).

9–3.1 Silicon-Controlled Rectifier (SCR)

In 1957, General Electric Company introduced the first commercially available SCR. This small 35 A, 200 V device revolutionized power control and conversion. The popularity of the SCR is so great that it is difficult to imagine modern motor control equipment without it. With the advent of the integrated circuit and the microprocessor, we tend to view any device in electronics over five years old as obsolete. Work on the SCR continues, improving the performance of the SCR and inventing new types for specialized needs.

The name silicon-controlled rectifier, first given to the new device by General Electric, has assumed a generic quality. The true generic name for the SCR is *reverse-blocking triode thyristor,* which gives a clue to its structure and operation. The SCR schematic is shown in Figure 9–9a. As its name suggests, the device has three terminals: a cathode, an anode and a gate. The anode is attached to the P-layer (Figure 9–9a), with the cathode attached to the N-layer at the opposite end of the device. The gate lead is attached to the P region next to the cathode. We can also see that the device has three junctions, one between each of the four layers. If we connect a load and a power supply to the SCR so that the anode is positive with respect to the cathode, no current will flow. We can see the reason for this if we look at the diagram in Figure 9–9b. Note that, with a voltage applied so that the anode is positive with respect to the cathode, junctions *1* and *3* will be forward biased. Junction *2,* the middle junction, however, will be reverse biased. This reverse-biased junction will

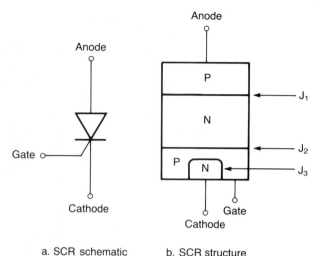

a. SCR schematic   b. SCR structure

FIGURE 9–9 Silicon-controlled rectifier (SCR)

block current flow through the device. Until a forward bias is applied to the gate, no conduction will occur, regardless of the polarity of the supply.

If a positive potential is applied to the gate (with respect to the cathode), junction *2* will be forward biased and the device will turn on and conduct current. After the current is established in the anode (a matter of a few microseconds), the gate potential can be taken away and the device will still conduct. This behavior is called *latching,* since it resembles the behavior of a latching relay.

SCR Characteristics We can learn a great deal about electronic devices by examining characteristic curves from the manufacturer's data sheets. The characteristic curve of an SCR is shown in Figure 9–10.

Note what happens when we apply an increasing forward bias potential to the anode, keeping the gate-to-cathode potential zero. As the anode-to-cathode potential is increased, only a small amount of current will flow. This small current is limited by the reverse-biased J_2 junction shown in Figure 9–9. Further increases in applied anode-to-cathode voltage cause a small increase in current, mainly due to avalanche breakdown. When the avalanche current becomes high enough, the J_2 junction will break over, causing the device to conduct and latch.

The critical voltage at which this conduction occurs is called the forward breakover voltage (V_{BO}). Note also that when the SCR breaks into conduction, the voltage drop across the device goes from a very large voltage to a small voltage. In most SCRs, the forward breakover voltage is several hundred volts. After the device breaks in conduction, the anode-to-cathode voltage falls to between 1 V and 3 V in most SCRs. This voltage is called the forward on-state voltage, usually symbolized by V_T or by $V_{F(ON)}$. Although it is possible to trig-

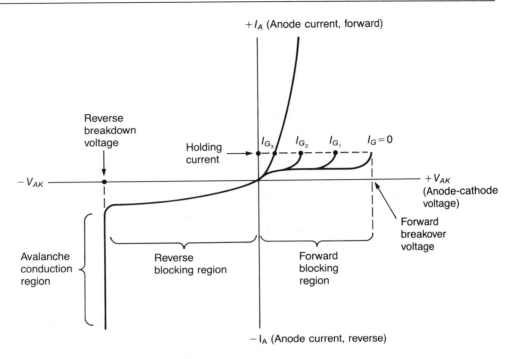

FIGURE 9–10 SCR characteristic curves

ger the SCR by breakover, the voltage required is generally too high. Most thyristors are triggered by a gate signal. We should note, at this point, that the voltage drop across most SCRs is at least twice that of transistors in the ON state. This means that the SCR has a greater power dissipation (wasted power) than the transistor. Thus, the transistor will deliver more power to the load, and is, therefore, more efficient than the SCR.

Gate triggering is the most common method used to turn on the SCR. When a gate-to-cathode potential is applied, the breakover voltage is decreased. The gate voltage at which the device breaks over into conduction depends on the amount of gate current drawn. The breakover voltage decreases with an increase in gate current. If gate current is high enough, the SCR junction J_2 will not block at all. The SCR then will switch into conduction under any forward anode-to-cathode voltage. When the SCR is to be triggered into conduction, a gate pulse is applied when the device is forward-biased. The gate voltage must be sufficient to draw more than the minimum gate current needed to fire the SCR. The voltage and current needed to turn on the SCR are called gate turn-on voltage and current, V_{GT} and I_{GT}, respectively. Once the SCR is triggered into conduction and a certain amount of anode current is flowing, the gate voltage can be removed and the SCR will remain conducting. This anode current is called *latching* current. Once the SCR is latched, anode current may be reduced to a level called the *holding* current. If the anode current goes below the holding current value, the SCR will return to a blocking state.

The SCR does not turn on instantaneously. Typically, SCRs take between 1 to 5 microseconds to turn fully on. Gate signals should have a width of greater than 10 microseconds to fire the SCR reliably. In addition, the gate voltage should be between 2 to 4 times greater than V_{GT} for reliable firing to take place.

9–3.2 Triac

As we have seen, the SCR controls current flow in only one direction. A triac is a bidirectional, three-terminal thyristor that controls current flow in both directions. It is therefore able to control AC in a load without rectifying it. Since the triac is a bidirectional device, it controls AC power well, up to the 20 kW range.

Figure 9–11a shows the basic triac structure. The vertical region directly beneath terminals MT_1 and MT_2 can be visualized as a PNPN switch in parallel with an NPNP switch. In other words, we can view this as a pair of SCRs connected back-to-back with a single gate. Figure 9–11b shows the triac schematic symbol. Note its orientation to the simplified diagram in Figure 9–11a. Since the triac is a bidirectional device, the terms *anode* and *cathode* have no meaning. The triac terminals are called *main terminals,* symbolized by the abbreviations MT_1 and MT_2 or, in some cases, simply T_1 and T_2.

The characteristic curve of a triac is shown in Figure 9–12, with MT_1 as the reference. Triac behavior relates to quadrants in the characteristic curve. Quadrant I is the region where MT_1 is positive with respect to MT_2. Quadrant III is the region where MT_1 is negative with respect to MT_2. The forward

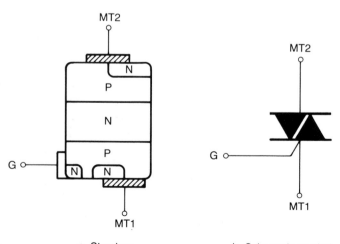

FIGURE 9–11 Triac   a. Structure   b. Schematic symbol

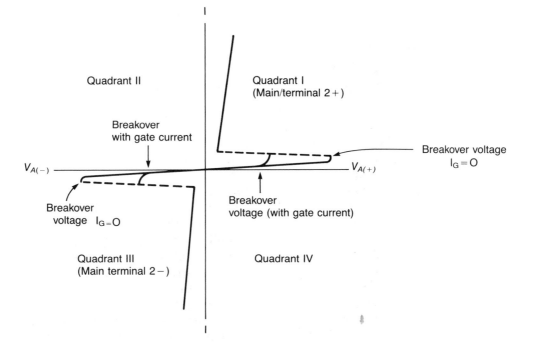

FIGURE 9–12 Triac characteristic curve

breakover voltage in either quadrant I or III must be larger than the applied AC or the gate will not control the firing of the SCR. A gate trigger voltage of either negative or positive polarity will fire the triac in either direction. The triac can be triggered with low values of gate current in quadrants I and III. The triggering modes and polarities for the triac are

MT_2+, Gate +; I+; First Quadrant, positive gate current and voltage

MT_2+, Gate −; I−; First Quadrant, negative gate current and voltage

MT_2-, Gate +; III+; Third Quadrant, negative gate current and voltage

MT_2-, Gate −; III−; Third Quadrant, negative gate current and voltage

The sensitivity of the triac is greatest in the I+ and III− modes. The sensitivity is slightly lower in the I− mode and lowest in the III+ mode. As a result, designers seldom use the III+ mode of triggering.

Some limitations exist in the use of the triac as a power control device. First, the triac is available in limited voltage and current ratings, about 1000 V peak-reverse voltage and 200 A average forward current. Second, internal construction characteristics make the triac useful only in frequencies near the 60-Hz line frequency. SCRs can be used at frequencies at least 10 times higher than the triac. If bidirectional control of AC power is necessary, two SCRs can be connected in inverse parallel to simulate the triac function. This connection is shown in Figure 9–13.

9–3.3 Gate-Turnoff Thyristor (GTO)

One major disadvantage of the SCR and triac relates to their latching behavior. Turning the SCR and triac on is relatively easy. Since the gate loses control after the SCR and triac are switched on, switching to the off condition cannot be done by the gate. Switching off can be accomplished only by decreasing current flow below the holding current value. The thyristor is then said to be turned off by low current dropout.

A device that allows the gate to switch the thyristor off is called the gate-turnoff switch (GTO), gate-turnoff thyristor, or the gate-controlled switch (GCS). The GTO turns an SCR off by applying a negative gate-to-cathode voltage. This characteristic is made possible by a special gate construction that allows a negative gate voltage to reverse-bias the J_2 junction. The GTO schematic symbol is shown in Figure 9–14.

The GTO needs a dual polarity gate drive circuit. As in other thyristors, the drive circuit must have a positive pulse to turn it on. The GTO also needs a negative pulse to turn it off. The positive pulse requirement is similar to the one used to turn on the regular SCR, although a slightly higher current is required. The gate current needed for GTO turn-on is in the tens of milliamperes. The negative turn-off pulse, however, is between five and ten times higher than the turn-on pulse. Where tens of milliamperes will turn on a GTO, hundreds of milliamperes are needed to turn it off. The larger turn-off requirement is not difficult to meet.

Small versions of the GTO are used in gasoline-engine ignition systems, TV horizontal deflection circuits, and computer print-hammer drivers. The larger versions are aimed at industrial power control markets, especially in AC and DC motor controls.

Besides the ability to turn off with a gate pulse, the GTO has another

FIGURE 9–13 Two SCRs simulate a triac

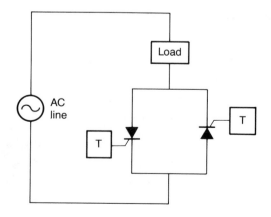

FIGURE 9–14 Schematic symbols for the GTO (GCS)

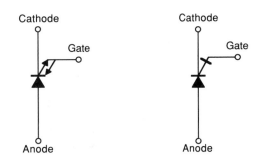

advantage over the SCR. The GTO has a faster turn-off time. This means that the GTO can run at faster switching speeds, especially at speeds greater than 10 kHz. As in all devices, the GTO is not perfect. It has a larger on-state voltage drop, typically 3 to 4 V, as compared to the 1 to 2 V on-state drop across the regular SCR. The larger on-state voltage drop makes the SCR a more power efficient device than the GTO, when controlling the same amount of current.

9–4 TRIGGER DEVICES

At this point, we have covered the basic semiconductor devices used for industrial control. In the transistor category, the bipolar transistor (including the Darlington) and the power MOSFET are most often used for power control. In the thyristor category, the SCR, triac, and GTO are the prominent power controllers. We have not, however, covered all the semiconductors you will see in power control circuits. Several other semiconductors are used to trigger the thyristor into conduction. The remainder of this section will be dedicated to discussing these devices.

9–4.1 Silicon-Controlled Switch (SCS)

A thyristor with a PNPN structure very similar to the SCR is called a silicon-controlled switch (SCS). Unlike the SCR, the SCS has terminals that make all four semiconductor regions accessible (Figure 9–15). Making the fourth region easily accessible greatly expands the circuit design possibilities. As we can see from the diagram, the SCS has two gates. The gate attached to the P-region and closest to the cathode is called the cathode gate. As in the SCR, the SCS can be triggered into conduction with an appropriate gate signal that is positive on the cathode-gate with respect to the cathode. The gate attached to the N-region and closest to the anode is called the anode gate. The SCS can also be triggered into conduction with the anode gate. If an appropriate gate signal is present that is negative on the anode-gate with respect to the anode, the device will conduct. The SCS is a relatively low-power device and is not used to control motors directly.

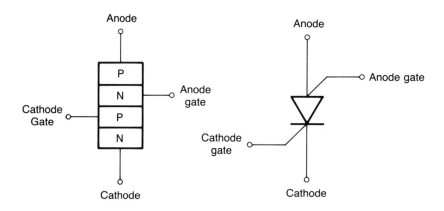

FIGURE 9–15 SCS (silicon-controlled switch)

9–4.2 Silicon Unilateral Switch (SUS)

A silicon unilateral switch (SUS) is a unidirectional, three-terminal thyristor. Its operation is most easily understood by examining the equivalent circuit diagram in Figure 9–16. The SUS is simply an NPN transistor, a PNP transistor (effectively a PUT), a 6.8 V zener diode and a 15 kΩ resistor. When a forward bias voltage of less than 7.5 V is applied anode-to-cathode, the device will remain in the blocking state. When the anode-to-cathode voltage rises above 7.5 V (the sum of V_{BE} and the zener voltage), the device will switch on. As in the SCR, the forward on-state voltage (anode-to-cathode) remains low, even if the current through it is greatly increased. The gate lead can be used to modify the turn-on characteristic of the SUS. For example, connecting a 3.9 V zener diode from gate to cathode will lower the turn-on voltage to approximately

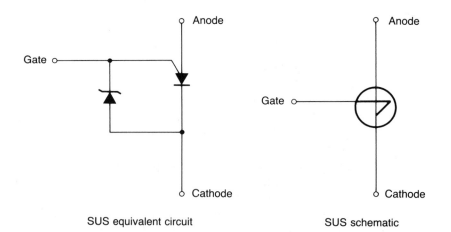

SUS equivalent circuit   SUS schematic

FIGURE 9–16 SUS (Silicon-unilateral switch)

4.6 V. A gate voltage must be used to turn the device on, as in the SCR. The SUS blocks reverse voltages in exactly the same way as the SCR.

Recall that the triac was considered to be functionally the same as two SCRs connected back-to-back with their gate leads connected together. This same technique is used to create a bilateral or bidirectional SUS, called the silicon bilateral switch (SBS). Operation is exactly the same as the SUS, except that the SBS conducts in both directions, either by breakover or by applying an appropriate gate signal.

9–4.3 Four-Layer Diode

Some thyristors are meant to be triggered by breakover only. This is the case for the four-layer diode, sometimes called the Shockley diode, a unidirectional, two-terminal thyristor. Note that the four-layer diode is similar in construction to an SCR without a gate lead (Figure 9–17). Since the device has no gate lead, the only possible way to trigger it into conduction is by breakover. It is turned off in the same way that the SCR is turned off, either by lowering anode current below the holding current level or by reverse-biasing it. The four-layer diode has reverse blocking characteristics similar to the SCRs. The four-layer diode is, therefore, a unilateral device.

As in the SCR and triac, two four-layer diodes connected back-to-back will make a new device that can break over and conduct in either direction. This new device is called a *diac*. Some diacs are nothing more than PNP or NPN devices similar to transistors without gate leads. Regardless of the different structures, the function is the same. Its schematic symbol is shown in Figure 9–18.

FIGURE 9–17 Shockley diode

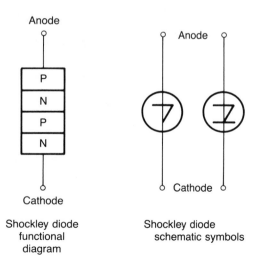

Shockley diode functional diagram

Shockley diode schematic symbols

FIGURE 9–18 Diac schematic symbol

9–5 MAINTAINING POWER SEMICONDUCTOR EQUIPMENT

Maintenance personnel often consider state-of-the-art uses of power semiconductors as something new, mysterious, and glamorous. To the contrary, industry has been using silicon power semiconductors since the 1960s. Since that time, knowledge of these devices has increased dramatically. Along with this knowledge has come a new state-of-the-art procedure for using and maintaining semiconductor equipment. The emphasis placed on selecting, purchasing, and installing devices is only part of the knowledge necessary to use semiconductors. The proper use and maintenance of equipment employing power semiconductors is of equal importance. In the first part of this chapter, we considered the types of semiconductors used in motor control equipment.

9–5.1 Suggestions for Maintenance

In this section we will consider some suggestions for maintaining motor controls containing semiconductors. Power semiconductors must be protected from excessive heat. Semiconductors are composed of materials with different rates of thermal expansion. This means that different parts of the semiconductor and case will expand at different rates and amounts. Both short and long term temperature changes not only change the electrical characteristics of the semiconductor, but also set up internal mechanical stresses at each of the material interfaces. These combined effects of temperature changes can ultimately lead to an equipment malfunction. In severe cases, these stresses can lead to the destruction of the device. The maximum temperature limits of various materials used in semiconductors vary from 150°C for soft solder to 1300°C for silicon. Once these materials are combined into a fabricated device, however, electrical deterioration begins at lower temperatures.

For example, the forward and reverse blocking capability of a semiconductor junction begins to decrease and leakage currents increase when allowable maximum junction temperatures are exceeded. (These temperatures are generally between 125°C and 200°C, far below the maximum temperature limit of silicon itself.)

Maintenance personnel using semiconductors should periodically monitor ambient temperatures, cooling fan operation, cooling water temperatures, and case temperatures of the semiconductors themselves. Close attention to temperature monitoring can pinpoint many malfunctions before they cause serious failures. Preventing high temperatures from developing in equipment containing semiconductors is, therefore, important to the continued safe and reliable operation of equipment. Monitoring must be emphasized in the operating procedures for the equipment operator and personnel involved in maintenance. Exercise care when measuring semiconductor case temperatures. Some semiconductors have high voltages between the case and ground. Such voltages, which can range into the thousands of volts, can be dangerous.

A second suggestion is to avoid overloading equipment and the semiconductors they contain. Avoiding semiconductor overloads can only be accomplished by protecting the equipment that the semiconductors are feeding and controlling. For example, overloading a motor operated with semiconductor controls puts a direct overload on the semiconductors. Even if this overload is present only for a short time, the semiconductor's useful life can be shortened. Motor controls are designed to include motor overloads, and the maximum values of overload should never be exceeded. Continued short term overloads above the control specifications can cause damage that is impossible to detect. Damage to a semiconductor accumulates over time, resulting in random, often unexplainable, semiconductor failures.

Operators sometimes produce overload conditions that can result in excessive semiconductor currents. These overload conditions include high on/off duty cycle or jogging for high production needs, rapid reversing, motor jam, and long acceleration time. Some mechanical problems that can cause motor overloads are: 1) high equipment temperature due to lack of cooling, 2) wiring or bearing failure due to improper installation or maintenance, 3) loss of a phase due to blown fuses or loose connections, 4) phase imbalance, 5) overvoltage, 6) transient overvoltage and 7) contaminants or dirty environment.

Table 9–1 shows various motor stress conditions and the common protection against those stresses. This table can be used to determine adverse load conditions and the type of protection that should be used. In properly designed systems, overload conditions are considered and provisions are made to accommodate the overloads. The user or operator has an obligation to operate the equipment within the design ratings. The table shows the equipment user and designer methods to prevent overloading a motor. Avoiding overload situations helps avoid early semiconductor failure and reductions of semiconductor life.

Power semiconductors should also be protected from transient overloads. Silicon semiconductors normally have long lives if they are adequately protected against surge currents and transient voltage spikes. The protection methods against failure due to these phenomena are relatively simple. This protection is usually designed into the circuit.

Table 9–1 Motor stress conditions

Problem	Cause	Motor Current Variance	Semiconductor Current or Voltage Variance	Common Semiconductor Protection	Comments
Overload	Operator's choice	Increases three-phase current approaching 200%	Increases semiconductor current; eventual motor burnout can cause surge on semiconductors	Thermal overload on semiconductors and motors. Design for semiconductor current overload.	Operation at thermal overload conditions can reduce semiconductor life.
Motor Jam	Load blockage	High operating time at locked rotor current to 600%	Increases semiconductor current; eventual motor burnout can cause surge on semiconductors	Thermal overload on semiconductors and motors. Design for semiconductor current overload.	Depending on the design up to 600°. Increases in semiconductor currents are possible, resulting in reduced life or shorted semiconductors.
High On/Off Duty Cycle	Jogging for high production needs	High operating time at locked rotor current to 600%	Increases semiconductor current; eventual motor burnout can cause surge on semiconductors	Thermal overload on semiconductors and motors. Design for semiconductor current overload.	
Rapid Reversing	Production needs	High operating time at locked rotor current to 600%	Increases semiconductor current; eventual motor burnout can cause surge on semiconductors	Thermal overload on semiconductors and motors. Design for semiconductor current overload.	Surge currents can also reduce semiconductor lifetime and cause semiconductor failures.
Long Acceleration Time	High inertia; slow starting loads	High operating time at locked rotor current to 600%	Increases semiconductor current; eventual motor burnout can cause surge on semiconductors	Thermal overload on semiconductors and motors. Design for semiconductor current overload.	High ambient temperatures produce the same results as high currents (i.e., reduced life and failed semiconductors).
High Equipment Temperature	High ambient temperatures; lack of cooling	No increase but can cause wiring and insulation failure	Increases semiconductor current; eventual motor burnout can cause surge on semiconductors	Thermal overload on semiconductors and motors. Design for semiconductor current overload.	
Motor Wiring or Bearing Failure.	Overcurrent; improper installation or maintenance	Locked rotor current to 600%	Increases semiconductor current; eventual motor burnout can cause surge on semiconductors	Thermal overload on semiconductors and motors. Design for semiconductor current overload.	Never operate semiconductors without proper cooling.

Problem	Cause	Motor Current Variance	Semiconductor Current or Voltage Variance	Common Semiconductor Protection	Comments
Phase Failure	Blow fuse; loose connection	Decrease in current until motor core reaches saturation and current increases	Increases semiconductor current; eventual motor burnout can cause surge on semiconductors	Thermal overload; phase failure relay	
Phase Unbalance	Unbalanced single phase loads on same line; poorly regulated service	Decrease in one phase and increase in the other two phases of current	Increases semiconductor current; eventual motor burnout can cause surge on semiconductors	Thermal overload; phase failure relay	
Overvoltage	High source	Slight voltage increase decreases current. Large increase may saturate core and increase current.	Decrease or increase semiconductor current	Overvoltage relay	Increase of currents can cause semiconductor failure and reduced life.
Transient Overvoltage	Lightning, opening inductive switches, etc.	Slight average current increases	High transient voltages across semiconductor when semiconductor is in the off position	Capacitor-resistor networks, Voltraps, MOV, etc.	Short duration high transients can cause semiconductor failure.
Underload	Operator's choice	Decrease in motor current	Decrease in semiconductor current	None	Light load increases semiconductor life.
Contaminants	Dirty environment	Causes corona	Causes corona	Clean room filters	Clean equipment periodically, even with filters. Corona carbonizes dirt and causes eventual semiconductor shorts.

Courtesy of Westinghouse Electric Co.

Causes of Voltage Transients In industrial plant environments, most voltage transients in semiconductor circuits arise from three major causes:

1. Switching is the most common source. Whenever current is switched on or off in an inductive circuit, a transient voltage is generated at the switch terminals. Transformers and motor windings are highly inductive components. The circuit wiring itself is inductive. Surges caused by lightning contribute to switching transients. These transients can originate in remote circuits and still feed back to the semiconductor circuit through the supply line. Protection of semiconductors from such transients is primarily a matter of attenuating the surge to a level the semiconductor can tolerate.

2. Commutation transients are associated with the reverse recovery characteristic of a rectifier junction. In normal use, a semiconductor is continually switching from a conducting state to a nonconducting state. This rapid switching causes changes in current. (You may see these rapid current changes, called high di/dt, in engineering literature.) For this reason, and especially where fast switching rectifiers are used, it is important to keep circuit inductance low. Here again, suitable suppression networks should be designed into the circuit.

3. Regenerative surges in inductive or dynamic loads are the third major source of reverse voltage transients. Such loads include motors, lifting magnets, solenoids, relays, and other devices that store inductive energy in the form of high induced voltage. Suppressing these transient voltages usually requires protective devices with high energy storage capacity. Some of the common devices used for transient protection are capacitors, zener diodes, and varistors (MOV/ZNR).

Protect power semiconductor circuits from current overloads. The power semiconductor circuit must be protected from heavy current loads caused by short circuits or other component breakdowns. Several methods are available for this type of protection:

1. Fuses—Semiconductor fuses protect against overloads. When properly applied, they remove the semiconductor from the power source when an overload occurs. The result, however, is downtime and higher operating costs. Therefore, fusing is generally limited to applications where the power source can damage components before slower breakers remove the power.

2. Mechanical breakers—Magnetically-operated breakers provide an inexpensive means of limiting current. This type of breaker operates best under short circuit conditions. It offers speedy, low-cost restarting, and relatively fast circuit interruption.

3. Thermal breakers—These employ heating elements to operate bime-tallic contact actuators. This type of breaker offers reasonable protection for wires and good protection for components with high thermal capacity. Their ability to protect semiconductors is, however, limited.

4. Overrated semiconductors—Semiconductors with current ratings high enough to accommodate anticipated current overloads are another method of protection. This approach allows the cost of fuses, wiring, and mounting to be put into the semi- conductor. Since it also allows minimal protection, safety considerations require a conventional circuit breaker to disable the circuit in extreme malfunctions.

5. Combinations—Combining circuit breakers with fuses, oversized semiconductors, or current feedback circuits is a frequently applied technique. Another approach is using special branch protection. For example, each leg of a single-phase bridge might be separately protected instead of using a single fuse or breaker, while the mains are fused. This arrangement prevents overstressing a motor while the semiconductor devices are operating at high current peaks that are within fuse limits.

Keep power semiconductors clean. Semiconductors and semiconductor equipment must be kept clean. On high power semiconductors, the glass or ceramic seal on the semiconductor package can be a source of trouble when dirt accumulates. Good engineering practices result in the proper positioning of components and support material. These practices reduce or eliminate excess voltage stress. In a clean system, high transient voltages can begin a process called *corona,* which is the ionization of the air around a high voltage potential. Corona often occurs before an arc is drawn. The corona will be extinguished when the system returns to normal voltage. With an accumulation of moisture or dirt on the insulating glass or ceramic surface, however, the corona will remain when the voltage returns to normal and can cause extensive damage to the circuit. Even semiconductors without a visible layer of contaminants can be covered with conductive particles that cause corona. Therefore, periodic cleaning is advisable. The time between cleanings can be lengthened by the proper use of filters, but filters will not eliminate the need for cleaning. A good maintenance program is essential to proper operation of semiconductors.

9–5.2 Preventive Maintenance

The key to successful and efficient operation of high power semiconductors is a good, well-planned maintenance program. Routine maintenance of semiconductor equipment involves putting into actual practice good operating procedures. Emphasis should be placed on avoiding: 1) temperature build-up, 2) dirt accumulation, and 3) loose mounting and/or connections. A regular schedule should be followed for checking and eliminating these conditions.

Temperature Build-Up   Temperature build-up can be caused by the following conditions:

1. Excess ambient temperature
2. Poor or blocked air circulation
3. Failure of cooling devices, such as fans or water circulating equipment
4. Dirty heat sinks and air filters
5. Equipment overloading

Since temperature build-up is usually gradual, it should be constantly monitored with a temperature sensor. A temperature sensor often used for this purpose is called a thermocouple. The thermocouple converts temperature to a voltage; the higher the temperature, the higher the voltage. A suggested method of attaching a thermocouple to a semiconductor flange or heat sink is to first drill a small shallow hole. (Exercise care when doing this. Holes should never be drilled in the semiconductor case.) Then, fit the thermocouple into it, and peen the surface around the hole to secure it. Equipment designed for forced air cooling or water cooling should never be energized without proper air or water flow.

Dirt Accumulation   Dirt accumulation on the glass or ceramic surfaces of semiconductor packages must be periodically and thoroughly removed. When cleaning semiconductors, the safest cleaning method that gives the desired results should be used. Solvents can be hazardous and should be used with care. Even a solvent that is considered non-toxic can kill or injure when used with improper ventilation. One of the safest methods of cleaning is by wiping with clean cloths. Often, due to space limitations, the areas that need to be cleaned cannot be properly reached. In these cases, a liquid cleaner must be used. Table 9–2 lists solvents that can be used as cleaners.

The safest method of cleaning semiconductors is to use a detergent wash and then flush with distilled water. Regular tap water contains many conductive impurities and should not be used. The system must be completely dry before reapplying power.

Methyl, ethyl, or isopropyl alcohol can be used for cleaning, but they can attack and dissolve sleevings and insulations. Before using alcohol or any solvent, it should be used on a small sample of sleeving and insulation to determine its effect. Since time of exposure to a solvent determines any adverse effects, the time of exposure during the test should be the same as the time expected during actual cleaning. Any elongation, softening, or actual decomposition should eliminate the use of the solvent. A temporary softening may not necessarily be a problem. Remember that alcohol is toxic and should be used with care.

Follow all safety precautions when using solvents. Always read the instructions before using any solvent. Do not mix solvents (or any chemicals)

Table 9–2 Solvents for cleaning semiconductors

Solvent	Fire Hazard	Explosion Hazard	Toxicity Hazard	Electrically Conductive	Attacks Rubber Insulation
Water, tap	None	None	None	Yes	None
Precautions: Rinse with distilled water and dry thoroughly.					
Water, distilled	None	None	None	No	None
Precautions: Dry thoroughly.					
Water and detergent	None	None	None	Yes	None
Precautions: Rinse with distilled water and dry thoroughly.					
Methyl alcohol	High	High	High	No	Slight
Precautions: Avoid skin contact and use ventilation.					
Ethyl alcohol	High	High	High	No	Slight
Precautions: Can cause internal damage; ingestion can cause blindness.					
Isopropyl alcohol	High	High	High	No	Slight
Precautions: Mix with distilled water to reduce hazards.					
Paint thinner (mineral spirits)	Low	Low	Low	No	Slight
Precautions: Use proper ventilation and limit exposure to rubber.					
Acetone	High	Moderate	Low	No	Slight
Precautions: Limit exposure to prevent rubber degeneration.					
Perchloroethylene (dry-cleaning solvent)	Low	Low	Moderate	No	Slight
Precautions: Use short exposure to prevent rubber damage; irritates eyes and causes headaches; heating causes fumes.					
Trichloroethylene	Low	Low	Moderate	No	Slight
Precautions: Use adequate ventilation and limit exposure time to prevent rubber damage.					
Freon	Low	Low	Low	No	Slight
Precautions: Use adequate ventilation and limit exposure time.					
Maltier XL-100	Low	Low	Low	No	Slight
Precautions: Use adequate ventilation and limit exposure time.					
Miller Stephenson MS-180	Low	Low	Low	No	Slight
Precautions: Use adequate ventilation and limit exposure time.					

Courtesy of Westinghouse Electric Co.

unless recommended. Information about solvents is given in Table 9–2 as a general recommendation for information only. Always check local maintenance procedures for proper use of solvents in particular situations.

Loose Mounting and/or Connections Along with scheduled periodic cleaning, a systematic check of mounting and terminal connection tightness should be specified. If a loosely-mounted device is discovered, it should be removed, both mounting surfaces thoroughly cleaned, and thermal compound reapplied. It should then be remounted with specified torque.

Safety Safety in operating and servicing solid state equipment is especially important. Unlike mechanical equipment with rotating parts, moving contactors, etc., there is nothing to warn you that the equipment is energized. There is also a tendency for novices to believe that voltage and current levels associated with semiconductor electronics are too low to be dangerous. Nothing could be further from the truth. Obvious safety measures must be carefully observed. Most high power equipment includes built-in safety *interlocks* to make certain that the equipment is turned off before anyone can gain access to the circuit areas.

Maintenance personnel, however, often defeat these interlocks to simplify their service work. In many cases, this practice is not necessary. It is certainly dangerous. In cases when circuits must be checked while energized, a few simple rules can help avoid both personal injury and equipment damage. First, if possible, always work with another person present, one who knows how to shut down equipment quickly. Second, when working around high voltages, be sure that the floor is covered with a rubber mat. Follow the practice of working with one hand in your pocket, if possible. This sounds simple, but it could prevent you from completing an electric circuit through the body. Finally, always use insulated tools like the ones shown in Figure 9–19 to avoid short circuits that could further damage electronic components. Some of the common shop practices for safety are too often overlooked or ignored:

1. Always wear safety glasses—hot metal from a short or a soldering iron can ruin an eye as quickly as a metal chip. Safety glasses can be either the type shown in Figure 9–20 or ordinary eyeglasses with special lenses.

2. Use a cap or net to cover long hair.

3. Lock out the disconnect or breaker for the circuit you are about to service. Tag it so that no one will turn it back on without your permission. A tag used for this purpose is shown in Figure 9–21.

4. Check circuits with a voltmeter to be sure the circuit is completely dead before touching them with your hands. NEVER check to see whether a circuit is live by touching it with your fingers.

5. Maintain "off limit" areas for high voltage equipment. Access should be allowed only to authorized personnel.

6. Use the following procedure whenever possible when measuring high voltages.
   a. Deenergize equipment.

FIGURE 9–19 Insulated tools can prevent short circuits

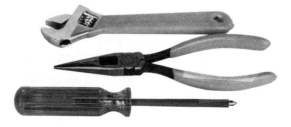

FIGURE 9–20 Safety glasses help prevent injury to the eyes

b. Discharge any capacitors that may still hold an electrical charge.
c. Connect probes.
d. Energize equipment and make the measurement.
e. Deenergize equipment and discharge any capacitors.
f. Disconnect probes.

Black lettering on red tag

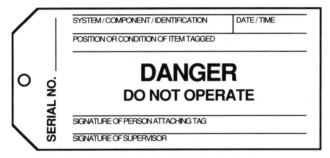

a. Front of tag

b. Back of tag

FIGURE 9–21 Safety tag

Too many accidents are caused by carelessness or laziness. Bear in mind that accidents can happen to anyone. Be careful.

9–5.3 Identifying and Replacing Semiconductors

Semiconductor Identification   During the troubleshooting process, you will find semiconductors that are not working correctly. Any device not working properly should be replaced by maintenance personnel. Proper identification of the semiconductor is the first step in replacing it. Many semiconductors are identified by a number known as a ***JEDEC*** number. JEDEC numbers for rectifier diodes begin with a "1N" prefix (i.e., 1N4001, 1N1206A, etc.). A power diode with its JEDEC number is shown in Figure 9–22. SCR and transistor JEDEC numbers begin with a "2N" prefix (i.e., 2N681, 2N3055, etc.). A JEDEC number is an attempt by the industry (Joint Electronic Device Engineering Council) to standardize electrical and mechanical parameters.

A benefit of standardizing is that products of one manufacturer will be interchangeable with those of another manufacturer. The user should, however, be careful when making a direct JEDEC substitution of another manufacturer because the JEDEC registered parameters do not always cover all of the critical parameters. As a result, manufacturers using completely different manufacturing processes often sell devices to meet the same JEDEC number. While this may not pose a problem in most general purpose motor control applications, you must recognize that various manufacturer's devices marked with the same JEDEC number can exhibit completely different secondary characteristics and safety margins. For example, an engineer may have un-

FIGURE 9–22   Power diode showing JEDEC number

knowingly designed his circuit around a particular JEDEC type number from a manufacturer with process A and a big safety factor (i.e., a 6-amp rated device that really has a 12-amp capability). Everything works fine until you choose a replacement JEDEC from another manufacturer. Then the new devices keep failing in the circuit. To be safe, all JEDEC substitutions should be evaluated thoroughly before use.

Most power semiconductors (especially those rated over the 40- to 100-amp range) are marketed under their respective manufacturer's part number. Even though each manufacturer uses its own device names and numbers, the mechanical packages and electrical ratings are reasonably well-standardized throughout the industry. Therefore, second sourcing and device substitution is not difficult, providing you have a good cross reference guide. Good technical data sheets for the devices being evaluated are also helpful. As with any cross reference, you must determine how acceptable a substitute is by reviewing the detailed electrical and mechanical characteristics of the devices being considered. This comparison will assure that the manufacturer's suggested replacement will perform properly in the given application.

In the event that the device number being sought is unknown, you should check with the manufacturer of the particular piece of equipment in question.

Semiconductor Replacement   Be sure to order the exact part number from the manufacturer. If you are in doubt, include the entire device marking on your order. The most likely place to get off-the-shelf delivery of an exact replacement is directly from the equipment manufacturer. The *original equipment manufacturer (OEM)* normally carries an ample supply of renewal parts to service its equipment market. When a replacement cannot be obtained from the OEM, you should contact a local industrial or electronic distributor that handles power semiconductors. Maintenance personnel should use replacement parts with the manufacturer's type number on it. These parts are generally available from the manufacturer's authorized distributor outlet or directly from the manufacturer's factory. If you use non-standard parts, you could receive inferior, counterfeit devices from an unknown source. It is best to deal only with a manufacturer or an authorized distributor whenever possible.

Maintenance personnel must be thoroughly familiar with equipment service manuals, with the power semiconductors and their technical data sheets, and with the various tools and instruments available for testing and replacing semiconductors. Carry spare semiconductor components in stock in case of equipment breakdown, especially for critical pieces of equipment.

Many motor control applications use rectifiers, SCRs, and transistors that have fairly broad parameters. As long as the replacement device you select has an adequate current and voltage rating and fits mechanically, it should operate properly. You must be extremely careful in selecting a replacement semiconductor when it is to be used in series and/or parallel combinations. Care must also be taken when replacing fast recovery rectifiers and fast switching SCRs, or when semiconductor devices are marked with special (non-catalog) part numbers. Specially selected, tested, or matched units may be required for the semiconductor to operate properly in the equipment. Failure to use the correct semiconductor device could result in device failures, equipment damage, and plant downtime.

When selecting a device for replacement, the user can always use one with a higher voltage rating, provided all other device ratings are equal or better. Likewise, a higher current rated unit can be selected so long as all of the other ratings are equal or better and the mechanical package is the same.

By following these guidelines, you can often locate a suitable replacement faster, reduce spare parts inventory by standardizing on a fewer number of replacement semiconductors, and gain increased reliability with greater voltage and/or current safety factors. Of course, the additional cost for the higher rated semiconductor must be weighed against the savings in inventory reduction, fewer failures, and reduced downtime.

In the absence of any other information, a good rule of thumb for specifying the proper device voltage rating (based upon the supply voltage to the semiconductor equipment) can be stated. If the device is used in a 110-VAC line, use a 300-V device; if used in a 220-VAC line, use a 600-V device; in a 440-VAC line, use a 1200-V device. If in doubt about what device to use, call the semiconductor manufacturer and ask for a recommendation. Most semiconductor manufacturers provide this service.

The user must realize that power semiconductors, like any other component, fail for a reason. Replace a suspect semiconductor if that is all that is wrong. Frequently, however, a suspect semiconductor fails as a result of a current or voltage overload elsewhere in the circuit. Simply replacing the suspect semiconductor in these instances will only result in destroying more semiconductors. Therefore, always look for the cause of failure before replacing any devices.

9–5.4 Installation, Mounting, and Cooling Considerations

Before installing power semiconductors, it is important to understand why so much emphasis is placed on using proper mounting techniques. Maximum efficiency and greater reliability is the goal when using power semiconductors. It is important, therefore, that the heat developed at the semiconductor junction be removed as fast as possible. There are usually three distinct obstacles in the path of heat transfer: (1) the interface and material between the semiconductor element and the semiconductor device mounting surface; (2) the interface between the device mounting surface and the heat sinks; and (3) the transfer of heat through the heat sink to the ambient, which might be natural convection air, forced air, oil, or water. These obstacles are referred to as thermal resistances:

- $R_{\theta_{JC}}$ (junction-to-case)
- $R_{\theta_{CS}}$ (case-to-sink)
- $R_{\theta_{SA}}$ (sink-to-ambient)

Thermal transfer is shown using an electrical analog in Figure 9–23.

The junction-to-case thermal resistance is generally a function of device construction and design. It is determined by the semiconductor manufacturer. Except for disk mount devices that require the user to apply the correct mounting force to assure the proper $R_{\theta_{JC}}$, the user has no control over $R_{\theta_{JC}}$ once a device has been selected. Your only option then is to select semiconductors that offer low $R_{\theta_{JC}}$. The other two thermal resistances, however, are quite variable, and you must carefully consider the various methods and techniques available to insure long life and efficiency.

The most important considerations in getting a low case-to-sink thermal resistance are (1) the degree of flatness and surface finish of the device and heat sink mating surfaces; (2) the use of a thermal joint compound; and (3) the proper amount of force (torque) applied. A couple of suggestions are worth mentioning. First, use a torque wrench to insure proper mounting force. If the mounting hardware is too loose, $R_{\theta_{CS}}$ will be high and the semiconductor may overheat. Second, use thermal joint compounds sparingly. It takes only a thin film between the device and heat sink mounting interface to be effective. Most people apply too much compound. Also remember, when you use insulating hardware, that the case-to-sink thermal resistances will increase about tenfold. Use insulating hardware only when necessary.

The final obstacle in the heat removal path is the sink-to-ambient thermal resistance ($R_{\theta_{SA}}$). You can minimize this value by choosing the proper heat sink and most effective cooling method. A semiconductor can be no better than its heat sink. A wide range of heat sink materials, configurations, sizes, and finishes are available that will meet almost any space, cost, and heat dissipation requirement. An example of a power semiconductor heat sink is shown in Figure 9–24.

The most cost-effective and popular air cooled heat sinks are made from aluminum extrusion. For natural convection air, aluminum heat sinks can be

FIGURE 9–23 Thermal resistance model

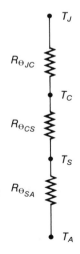

FIGURE 9–24 Power semiconductor heat sink

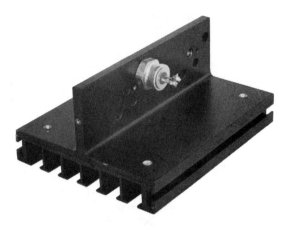

black anodized or painted black to optimize their ratings. For example, painting heat sinks black improves natural convection ratings by as much as 25% compared to unpainted heat sinks of equivalent size. For higher output current at less cost, forced convection is the answer. Forced air can frequently allow the user to more than double the output current of a given device/heat sink assembly over natural convection rating capability. A few fans and some baffling may provide a substantial savings in system design costs. Use forced air in lieu of natural convection air when possible.

Water cooling is the most efficient type of cooling in general use today. In order to optimize the thermal efficiency of a water-cooled heat sink, pure copper heat sinks should be used. Copper alloys, bronze, aluminum, etc., generally have much lower thermal conductivities and thus are not as efficient. Water cooled assemblies usually offer from 1.5 to 4 times more output current than the air cooled assemblies, weigh only one-third as much, occupy about 40% of the space, and cost about the same. Therefore, when heat sinks are compared, water cooling (if available) is usually the most economical choice; forced convection air is next; and natural convection air is last.

A final consideration when installing power semiconductors is to place them where the surrounding ambient temperature is as low as possible. Do not mount semiconductors near or above transformers or other heat generating components in a cabinet. Avoid "dead air" pockets in equipment enclosures. Mount semiconductors heat sinks vertically to take advantage of the natural chimney-effect or up-draft of air flow. When using forced air or water to cool various components in a cabinet, make sure the cooling medium reaches the power semiconductors. Good installation practices when mounting and cooling power semiconductors will insure good operating performance and long life.

CHAPTER SUMMARY

- The transistor is a three-terminal, semiconductor device that is used as a power control for motor control circuits.

- The Darlington transistor configuration has a higher current gain than the single transistor.
- Transistors require high drive currents while power MOSFETs do not.
- Power MOSFETs may be connected together to improve current-handling capabilities.
- The SCR is functionally equivalent to a latching relay, but has unidirectional current conduction.
- The triac is functionally equivalent to two SCRs connected back-to-back, which gives it bidirectional current conduction.
- The GTO is a unidirectional thyristor that can be turned on and off by a gate signal. SCRs and triacs can be turned on only by a gate signal.
- The key to a successful maintenance program is good preventive maintenance.
- The following factors should be avoided:
  a. excessive heat
  b. accumulated dirt
  c. loose electrical connections
  d. overload conditions
  e. voltage transients
- Always use approved safety practices in maintaining semiconductor equipment.
- When replacing semiconductors, always over-rate current and voltage specifications.

QUESTIONS AND PROBLEMS

1. Contrast, listing advantages and disadvantages, the following power control devices:
   a. transistor
   b. Darlington
   c. power MOSFET
   d. thyristors

2. Draw the schematic symbols and explain how the following devices are triggered:
   a. SCR
   b. triac
   c. GTO
   d. SCS
   e. SUS
   f. Shockley diode

3. Explain how the devices in problem 2 are commutated or turned off.

4. Describe the regenerative action that makes a thyristor latch.

5. List three factors likely to damage semiconductors and how to avoid them.

6. List the causes of transient overloads to semiconductors.

7. List and explain four methods of overload protection.

8. List the causes of temperature buildup in semiconductor equipment.

9. Explain the safest way to clean semiconductors and why.

10. List the steps involved in measuring a high voltage safely.

11. List at least three shop safety practices.

12. Explain the JEDEC classification for semiconductors.

13. Explain what OEM stands for.

14. Explain the steps you would go through to find a replacement semiconductor.

15. Explain the term *thermal resistance* and its relationship to semiconductor cooling.

16. List the considerations involved in installing semiconductors in power equipment.

10

ELECTRONIC DC MOTOR CONTROL

At the end of this chapter, you should be able to

- explain the operation of the following drives:
  - half-wave controller
  - full-wave half-controlled bridge
  - full-wave half-converter.
- describe the difference between a drive that uses open-loop control and closed-loop control.
- differentiate between the speed-torque curves of the following devices:
  - stepper motor
  - brushless DC motor
  - AC induction motor.
- list the requirements for a brushless DC motor controller.
- explain the operation of the following:
  - brushless DC motor controller
  - stepper motor controller
  - wound-field permanent-magnet DC motor.

10–1 INTRODUCTION

Until the advent of the thyristor (a semiconductor switch), AC and DC motor controls were performed by relays and other mechanical switches. The thyristor opened up the market for solid-state motor controls (commonly called drives today) about 30 years ago. Today solid-state devices control AC and DC motors up to 15,000 HP. Modern AC and DC drives account for 30 to 40% of the applications for thyristors. This chapter will discuss the DC drive and its applications in controlling DC motors.

A *DC drive* consists of a DC motor and a power converter that houses the control circuits. Dominating the consumer market today are the fractional HP DC drives used for portable tools, home appliances, and other low-cost applications. The large number of DC drives has been made possible because of the low cost and small size of the semiconductors.

Since most industrial power is supplied in the form of AC, DC drives require that the power supply be changed to DC, or rectified. In many cases, actual rectifier diodes are used. The diode is a PN junction that is rated high enough to stand the currents a motor will draw. DC drives fall into three general groups: single-phase input with motors up to 5 HP, three-phase input to motors between 5 and 500 HP, and three-phase input to motors above 500 HP.

We will consider DC drives in three different areas of application: the wound-field DC motor, the stepper motor, and the brushless DC motor.

10–2 WOUND-FIELD DC MOTOR SPEED CONTROL

Most industrial DC motor speed controls use a shunt or separately-excited field DC motor. The speed can be varied by changing the armature voltage while the voltage to the field is kept constant, as seen in Figure 10–1a. By using a separately-excited motor and keeping the field voltage constant, the motor will have a constant field flux. When the flux is held constant, the motor is said to operate in the *constant torque* area of its operating curve (Figure 10–1c). When applying rated armature voltage and rated field current, the motor will run at a speed called the base speed. Below the base speed, the maximum operating torque is a constant value. It is limited only by the maximum safe armature and field currents. This area of operation is known as the constant torque range of speed control. Since the power a motor produces is proportional to speed and torque, the power a DC motor produces is proportional to its speed. In the constant torque area of operation, we control the speed by varying the applied armature voltage from zero volts to its maximum rated armature voltage.

By varying the field strength, it is possible to operate the separately-excited DC motor above its base speed, as seen in Figure 10–1c. This region of operation is called the constant power region. It is achieved by applying the maximum rated voltage to the armature and decreasing the field voltage (Figure 10–1b). Decreasing the field voltage decreases the field current and the field flux, increasing the motor speed and decreasing the torque produced.

10–2.1 Single-Phase DC Motor Controllers

Since most industrial power systems are either single-phase or three-phase AC, the power converter can both rectify the incoming AC and control the amount of voltage applied to the armature. Fortunately, one device, the SCR,

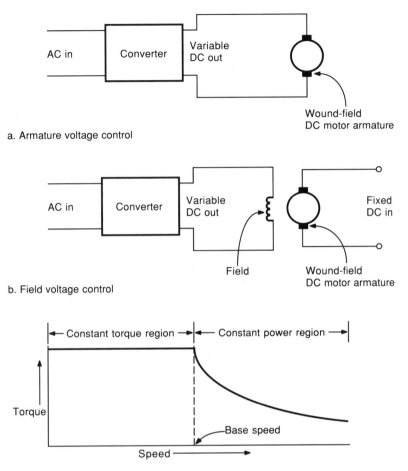

a. Armature voltage control

b. Field voltage control

c. Separately-excited motor speed-torque curve

FIGURE 10–1 Basic wound-field DC motor control

will do both jobs. In addition, when the SCR is used in an AC circuit, it can be self-commutated or line-commutated. *Line commutation* occurs in AC circuits when the SCR turns off (or commutates) as the line voltage forces the anode current to zero. As the SCR anode current approaches zero, the anode current will fall below the holding current level. The SCR then shuts off. No special circuitry is necessary to turn the SCR off. Another advantage of the SCR is that it does not drop much voltage across it, typically only 1 to 3 V. Use of the SCR as a power control device in DC motor controls is, therefore, an efficient use of energy.

Single-Thyristor Converters The simplest of the DC motor speed controls uses a single SCR, as shown in Figure 10–2. This controller, which controls

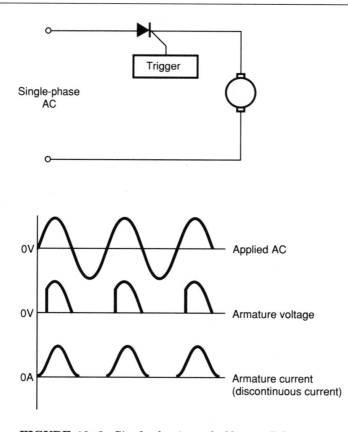

FIGURE 10–2   Single-thyristor, half-wave DC motor controller

one-half of the applied voltage, is called a half-wave controller. This controller is used in fractional HP drives for series DC motors, universal motors, and permanent-magnet DC motors, such as those used in hand tools and small domestic appliances. Although inexpensive, this drive is not very efficient, since it can use only a maximum of half of the available power. The half-wave drive does not have a high average DC compared to the other converters. If the motor is under a heavy load, the motor may slow down while the SCR is off. This causes the speed to fluctuate somewhat. When the SCR shuts off, no current flows through the motor. This type of situation is referred to as discontinuous current flow. ***Discontinuous armature current*** occurs when current stops flowing at some time during the motor's operation, as seen in Figure 10–2. Discontinuous armature current can cause severe speed and torque variations in the motor.

The amount of discontinuous current can be reduced by adding a ***freewheeling diode,*** shown in Figure 10–3 as diode D2. The freewheeling diode conducts when the SCR is off. When the SCR turns off, the flux built up around

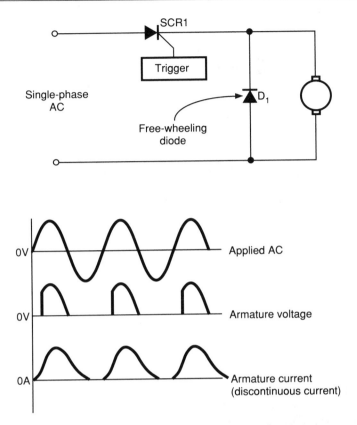

FIGURE 10–3 Single-thyristor, half-wave DC motor controller with freewheeling diode

the armature winding collapses, continuing current flow in the motor. The freewheeling diode also protects the SCR from damaging spikes when the SCR turns off and the field built up around the armature coil collapses. The effects of the freewheeling diode can be seen by comparing the wave forms in Figures 10–2 and 10–3.

Note also in Figure 10–2 that the half-wave converter gives one pulse of voltage to the motor for every cycle of line voltage. This type of converter is sometimes referred to as a single-pulse converter where q (the number of pulses) equals one. An example of a half-wave converter drive is shown in Figure 10–4. During the positive alternation, the capacitor $C1$ is charged up. A portion of the voltage across $C1$ appears at the wiper of $R2$, causing gate current to be drawn. When the gate current drawn exceeds the gate turn-on current, the SCR fires, providing current to the motor. The SCR remains on until the anode current falls below the holding current value. The SCR shuts off and remains off during the entire negative alternation. The diode $D2$ is reverse biased during the negative alternation. The diode $D2$ thus prevents current

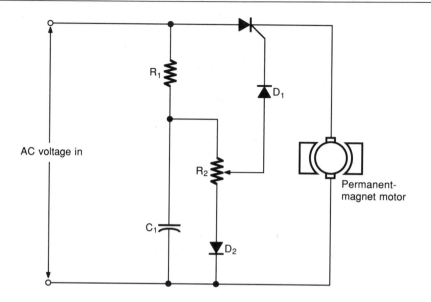

FIGURE 10–4 Example of a half-wave controller driving a permanent-magnet DC motor

flow and power loss in the resistors. Diode *D1* prevents damaging reverse current from flowing in *SCR1*.

Shunt and separately-excited DC machines can be driven by the circuit illustrated in Figure 10–5. Note that the SCR does not rectify the incoming AC. The diode bridge rectifies the input AC. You may see a half-wave rectifier driving this circuit, instead of the bridge. The field windings are not shown. Control circuits, although not shown, are usually added. Two popular additional control circuits are the braking circuit and the reversing circuit. The braking circuit can be mechanical or electrical. If electrical, the braking method is usually a dynamic braking resistor. The reversing switch reverses the armature leads, so that current can flow through the motor in the opposite direction.

Note that the wave form across the SCR rises until the SCR is fired. After firing, the voltage drops to the SCR's anode-to-cathode forward voltage for the remainder of the positive cycle. At the same time, there is no voltage across the motor armature until the SCR fires. After firing, the full supply potential is applied to the armature. The point at which the SCR fires serves as a reference point for two events. The point on the armature wave form where the supply just goes positive to the point where the SCR fires is called the firing angle and is measured in degrees. The point where the SCR fires to the end of the positive cycle is called the conduction angle, and it is also measured in degrees.

Note that the current wave form rises somewhat slowly and lags the voltage wave form slightly. We would expect this behavior from a highly in-

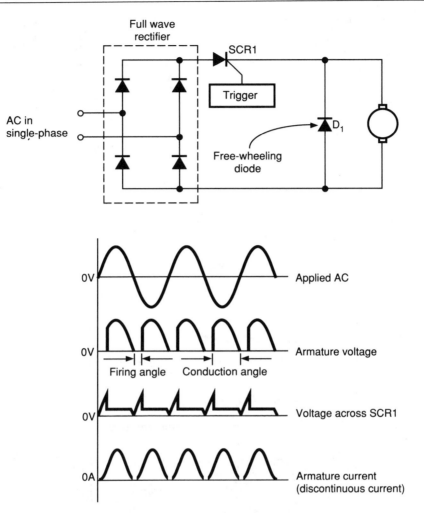

FIGURE 10–5 Single-thyristor control—full-wave diode bridge and a single thyristor with freewheeling diode

ductive circuit. The freewheeling diode prevents the SCR from conducting when the load voltage goes negative.

Multiple-Thyristor Converters One type of converter places SCRs in the bridge circuit as shown in Figure 10–6a. This configuration is called the full-wave, half-controlled bridge or the two-pulse, half-controlled bridge. This circuit gives better control than the half-wave converter, since both cycles are controlled by an SCR. When the AC supply goes positive, *SCR1* is forward-biased and can be triggered at any time in the positive cycle. When *SCR1* is fired, current flows from the negative supply potential, through the forward-biased diode D_2, up through the motor armature, through *SCR1*, and back to

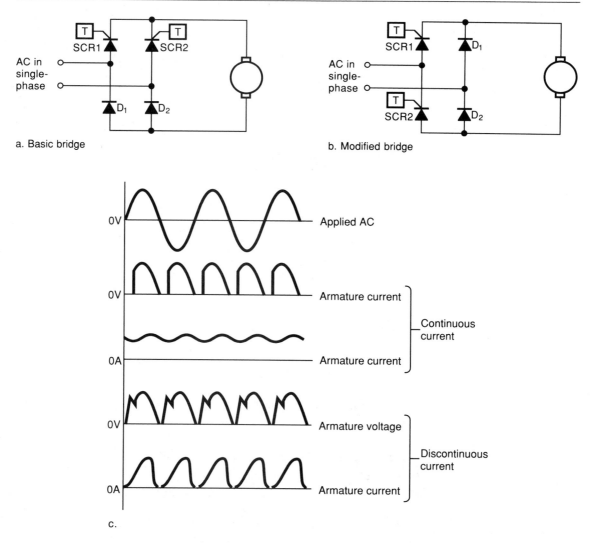

a. Basic bridge

b. Modified bridge

c.

FIGURE 10–6 Single-phase half-controlled bridge converters

the positive potential of the supply. When the supply goes negative and *SCR2* is triggered, current flows through forward-biased D_1, up through the motor armature, through *SCR2*, and back to the positive terminal of the supply. The freewheeling diode, connected across the motor armature, dissipates some of the energy stored in the highly inductive armature. A major problem of the half-controlled bridge in Figure 10–6 is half-waving. Half-waving occurs in this circuit when the firing angle is low and the inductance of the load is high. Let us assume that SCR1 has been turned on with a small firing angle, and the gate firing pulses have been removed. When SCR1 becomes reverse biased, the current from the collapsing field around the motor will conduct through SCR1 and D1. If the current has not reached zero before SCR1 becomes forward biased, SCR1 will start conducting at zero degrees, without the delay caused by the gate pulse. This situation is avoided by the modified half-controlled bridge in Figure 10–6b.

A modification of the full-wave half-converter circuit shown in Figure 10–6a is illustrated in Figure 10–6b. The diodes act as freewheeling diodes, conducting when the SCRs are off and energy is stored in the armature inductance. Voltage and current wave forms are shown in Figure 10–6c. Note that current and voltage wave forms are shown for both continuous and discontinuous current conditions. Whether current is continuous or discontinuous depends on the operating conditions, particularly the speed and torque. The half-wave, single-thyristor controller is the only single-phase controller in which the armature current is always discontinuous. The half-controlled bridge is used in DC motor drives up to 25 HP.

The full-wave full converter, shown in Figure 10–7, provides the ability to feed power back to the supply, unlike the previous converters. This process of feeding power back is called regeneration. It is one of the methods used to brake a motor and is called regenerative braking. The SCRs fire in pairs: *SCR1* with *SCR4* on the positive cycle, and SCR2 and SCR3 on the negative cycle. Note that no freewheeling diode is present or necessary. This converter is also called a two-pulse converter. The diagram in Figure 10–8 shows a single-phase AC input, DC motor controller, manufactured by Graham Inc. This controller is used in industry for conveyors, printing presses, packaging machinery, and welding positioners. The power supply provides a +15V IC regulated output, a zener-diode regulated -15 V supply for the IC $-$V, an unregulated 25 V supply for gating the SCRs, and a 25 V sync voltage. The primary of TP_1 can be connected for 120-VAC or 240-VAC operation. The speed regulation circuitry is in the upper left of the schematic diagram. A reference voltage is applied to the circuit through R_2. Feedback current, proportional to armature voltage, is applied through R_3, P_1, and R_4. The capacitor C_4 is a filter capacitor, reducing the ripple content of the armature voltage. Amplifier *1* is an op amp summing amplifier, which adds all frequencies that appear at its inputs. The output of IC_1 is applied to the top of the current limit potentiometer, P_2. Amplifier *2* is an integrator. It integrates the sum of two currents: one feedback current from R_C through R_9, and the other from the current limit potentiometer P_2 through R_8. With the current limit potentiometer P_2 set fully counterclockwise, no current is supplied to the motor. If the current limit potentiometer is set fully clockwise, the motor draws maximum current.

The sync voltage, S, is applied to the base of transistor Q_1, causing it to saturate early every half cycle. The diodes D_6 and D_7 limit the base voltage to about 1.2 V, preventing the transistor from being damaged by too high a base voltage. The voltage at the collector of Q_1 is a narrow positive pulse that coincides with the zero crossing of the AC lines. Amplifier *3* charges C_7 (as long as the collector of Q_1 is low) in a positive direction at a rate proportional to the

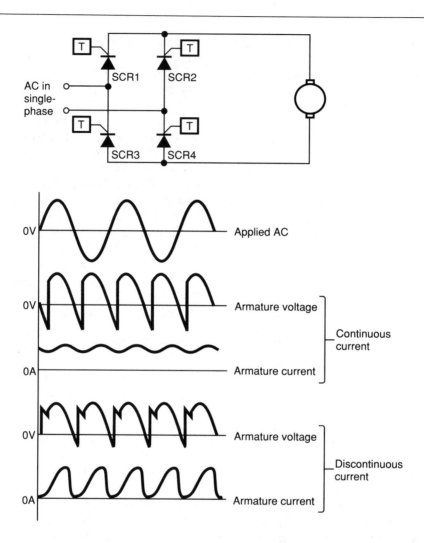

FIGURE 10–7 Single-phase fully-controlled bridge converter with voltage and current wave forms

voltage at the non-inverting terminal of amplifier *3*. Amplifier *4* is a comparator whose trip voltage is about +10 V. The output voltage of amplifier *4* is normally negative. When the output voltage of amplifier *3* exceeds the trip level, the output of amp *4* goes positive. The output remains positive until the end of the half cycle. The output of amplifier *4* is a positive pulse with a width corresponding to the conduction angle. The output of amplifier *4* is applied to the base of Q_2 through the voltage divider formed by R_{21} and R_{22}. The output of this divider sets the maximum gate voltage to about 7.5 V. Transistor Q_2 acts as a voltage follower. It transfers the charge on C_8 to C_8 and C_9. A charge on C_8 and C_9 produces a voltage across R_{24} and R_{25} and a gate current through

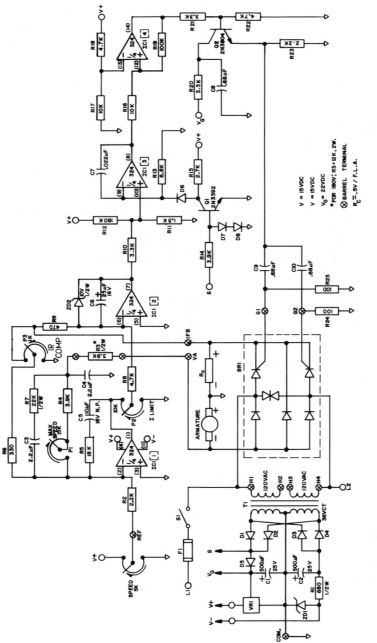

FIGURE 10–8 Example—single-phase motor controller—Magnapak Mark V (courtesy of Graham Co.)

327

the gate of the SCRs in BR_1. Both gates are fired every half cycle. The capacitor C_8 is charged to V_G through R_{20} during the interval before firing. A higher voltage on the speed potentiometer produces a higher voltage at the input to the non-inverting terminal of amplifier *3*. The higher voltage on amplifier *3* charges C_7 faster, causing the trip voltage to be reached earlier in the cycle, increasing the conduction angle.

10–2.2 Three-Phase DC Motor Controllers

Three-phase sources power large horsepower DC motors in industry. The frequency of the ripple voltage is higher in three-phase converters than in single-phase converters. Three-phase converters, therefore, have less need for filters to smooth out the ripple from the DC power source. The current flow is also more continuous in the three-phase converters than in the single-phase converters, producing smoother torque and less fluctuation in speed.

The diagram in Figure 10–9 shows a three-phase, half-controlled bridge, also called a semi-converter. The interval at which the thyristors are fired is equal to 120°. Recall that, in the single-phase converter, the interval between firing pulses is 180°. The current in the motor has less time to decrease, mak-

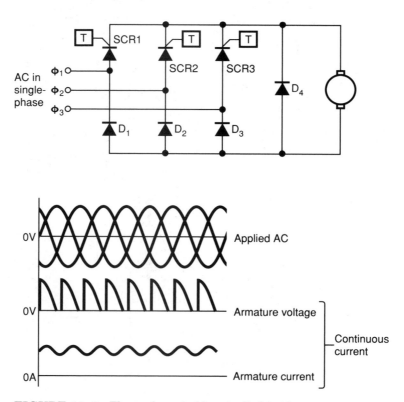

FIGURE 10–9 Three-phase half-controlled bridge converter

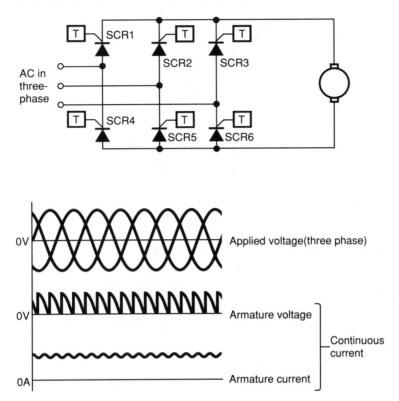

FIGURE 10-10   Three-phase fully-controlled bridge converter

ing the current less discontinuous. During the first interval, *SCR1* and diode *D3* will conduct. After the applied voltage passes through zero, the freewheeling diode, *D4*, conducts. In the next interval, *SCR2* and diode *D1* conduct, followed by *SCR3* and diode *D2*. The cycle then repeats. Current may be discontinuous at large firing angles if the torque demand is low and the speed high. This drive is used with DC motors between 10 and 150 HP.

A three-phase, fully controlled bridge converter (also called the full-converter) is shown in Figure 10–10. Note that the thyristors are fired at intervals of 60°, compared to the 120° interval in the semi-converter. This makes the current more continuous, producing smooth torque and speed operation. This converter is used on DC drives from 100 to 2500 HP.

All of the three-phase converters discussed up to this point have provided current in only one direction. Current could flow through the motor in the opposite direction only through a switching system, such as might be used in a plugging-type brake. The three-phase dual converter shown in Figure 10–11 allows reversal of both voltage and current at the motor terminals. This type of system is used to brake the motor by plugging or to reverse the direction of the shaft rotation. This type of drive can be seen on DC motors in the megawatt range, controlling speed in both directions.

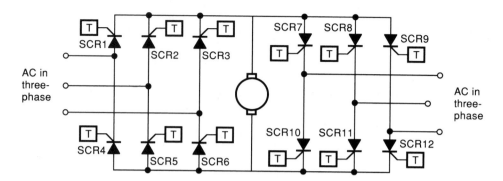

FIGURE 10–11 Three-phase dual converter

10–3 BRUSHLESS DC MOTOR CONTROL

Although the following section describes brushless DC motors (BDCMs) using transistors as power control devices, you should realize that thyristors (SCRs) are used for the same purpose in some controllers. Circuits using thyristors, however, are usually more complicated than a comparable transistor design. The more complicated design comes from the difficulty in turning off the SCR.

10–3.1 BDCMs versus Other Motor Types

The control circuits for a stepper motor or frequency-controlled AC motor at first glance may appear to be similar to some brushless DC motor controllers. The distinction between these types of controllers must be made at the outset. A brushless DC motor system should have the torque-speed characteristics of the conventional DC permanent-magnet motor. Figure 10–12 illustrates the torque-speed characteristics of a conventional DC motor. By our definition, the brushless DC motor should have the same basic characteristics.

If we examine a stepper motor system speed-torque characteristic (Figure 10–13), we see a non-linear relationship between the two variables. The stepper motor's poor low-speed performance limits its usefulness in velocity control applications. The stepper motor is a unique device that differs from a brushless DC motor in fundamental performance aspects.

The AC induction motor characteristic in Figure 10–14 shows an entirely different relationship between speed and torque. The usable part of the curve lies in the region between 90 and 100% of synchronous speed. The synchronous speed depends on the applied stator frequency. As we shall see in the next chapter, the AC motor speed-control system depends on a variable frequency, variable AC voltage. This voltage must be coordinated with the shaft velocity to produce a controlled slip frequency current in the rotor windings. Because the rotor and stator structure can be considered a transformer, it does not work well at low frequencies (low p701shaft speeds). This is a fundamental differ-

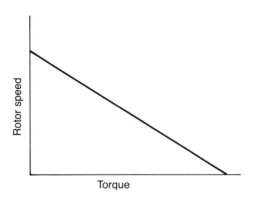

FIGURE 10–12 Speed-torque curve—permanent-magnet DC motor

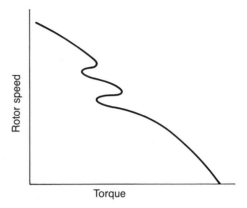

FIGURE 10–13 Speed-torque curve—stepper motor

ence between the AC motor and the brushless DC motor. In the BDCM, torque is produced by the interaction of a magnetic field produced by a permanent-magnet rotor and a magnetic field due to a DC current in the stator.

When engineers design BDCM controllers, they generally have two things in mind. First, they want a design that requires the least number of power semiconductor switches to adequately meet the performance requirements. The second guideline is that permanent-magnet materials should be used wherever possible. These materials eliminate the need for slip rings in the rotor assembly. In a brushless DC motor, it is usually most practical to provide a stator structure as shown in Figure 10–15. Note that the windings are placed in an external, slotted stator. The rotor consists of the shaft and a

FIGURE 10–14 Speed-torque curve—AC induction motor

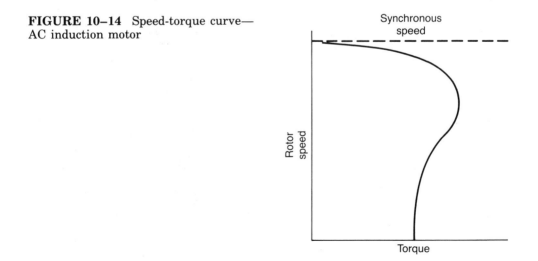

FIGURE 10–15 Cutaway view of brushless
DC motor (courtesy of Electro-craft Corp.)

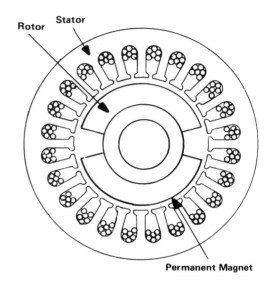

hub assembly with a magnetic structure. The picture shows a two-pole magnet. For contrast, Figure 10–16 shows the equivalent cross-sectional view of a
conventional DC motor. Note that the permanent magnets are situated in the
stator structure and the rotor carries the various winding coils. There are
significant differences in winding and magnet locations. The conventional DC
motor has the active conductors in the slots of the rotor structure. In contrast,
the brushless DC motor has the active conductors in slots in the outside stator.
Removing heat produced in the active windings is easier in the brushless DC
motor, since the thermal path to the outside of the motor is shorter. Since the

FIGURE 10–16 Cutaway view of conventional
PM motor (courtesy of Electro-craft Corp.)

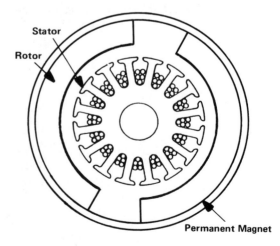

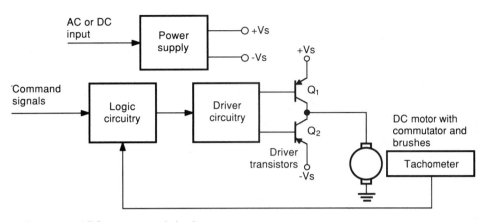

a. Conventional DC motor control circuit

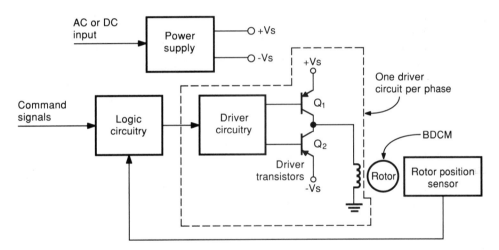

b. BDCM controller

FIGURE 10-17 Comparison of conventional and brushless DC motor controllers

permanent-magnet rotor does not create any heat, the brushless DC motor is a more stable mechanical device from a thermal point of view.

A comparison of conventional and brushless DC motor systems is presented in Figure 10-17. Figure 10-17a shows the parts of a conventional DC motor and control system. A bidirectional controller and driver stage are shown together with a power supply and driving circuitry. The equivalent brushless DC motor system is shown in Figure 10-17b. The main difference is in the shaft position encoder. Note that the conventional DC motor uses a

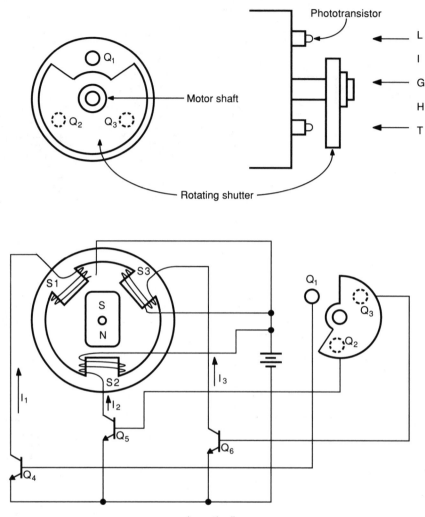

a. schematic diagram

FIGURE 10–18 Three-phase BDCM schematic diagram

tachometer to indicate the speed of the motor. The BDCM uses sensors to determine the rotor's position. The shaft encoder generates logic signals that control the commutation or switching of the windings. We will examine the most common commutation configurations.

10–3.2 Brushless DC Motor Controllers

Three-Phase Controllers The three-phase brushless DC motor, shown in Figure 10–18a, is one of the simplest and most practical brushless DC motor circuits.

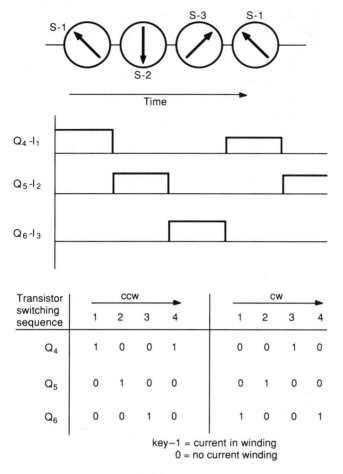

Transistor switching sequence	CCW				CW			
	1	2	3	4	1	2	3	4
Q_4	1	0	0	1	0	0	1	0
Q_5	0	1	0	0	0	1	0	0
Q_6	0	0	1	0	1	0	0	1

key—1 = current in winding
0 = no current winding

b. firing sequence

FIGURE 10–18 *(Continued)*

It is a half-wave control circuit with a conduction angle of 120°. Each winding is used one-third of the time. The phototransistor Q_1 turns on as it is illuminated by the light through the shutter. Transistor Q_1 turns on Q_4, which drives current through stator winding S_1. The current I_1 creates a north pole on the edge of the pole face, attracting the south pole of the rotor and pulling it CCW. In the next part of the sequence, Q_1 is turned off by the shutter, which turns off Q_4. Then Q_2 turns on Q_5, producing a north pole in S_2. The rotor is then attracted to the bottom pole S_2. The entire firing sequence is shown in Figure 10–18b. The logic control of the system is simple. Varying the power supply voltage controls the speed and torque output of the motor. Torque reversal is achieved not by reversing the power supply voltage, as in a conventional

DC motor, but by shifting all logic functions 180°. The rotor then turns in a clockwise direction. This example shows one of the basic differences between brush and brushless DC motors. Although only one winding at a time was energized in this example, two windings are sometimes energized together. Energizing two poles (phases) at a time uses the windings more efficiently and produces more torque.

The bipolar controller, shown in Figure 10–19a, drives the three-phase BDCM more efficiently than the unipolar drive shown in Figure 10–18 because current flows in both directions in the stator winding. The winding configuration is based on a delta connection stator arrangement. In this configuration, each winding is oriented 120° from the other. The six transistors are connected to the end points of each stator leg. They form a three-phase full-wave motor control. This system differs from the previous one in that current flows continuously in one leg when current in the other is being turned off. We

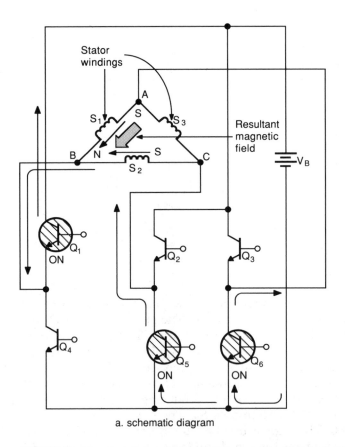

a. schematic diagram

FIGURE 10–19   Three-phase, full-wave (bipolar) brushless DC motor controller

can see that when Q_1 is energized, Q_5 and Q_6 are also conducting. Current flows from point A to point B through stator S_1 and from C to B through stator S_2. This produces the resultant field direction seen in Figure 10–19b.

In the next sequence, Q_1 remains energized, Q_5 is shut off, and Q_6 remains on. Current will then flow from point A to point C and from point A to point B. The current in these windings produces the resultant field shown. A reversal of the sequence produces a clockwise rotation.

Four-Phase Controllers A four-phase BDCM controller is shown in Figure 10–20. In this example, two Hall-effect devices are used to detect the rotor position. The Hall-effect device produces a voltage potential output in the presence of a magnetic field. With the rotor in the position shown, HE_2 will detect the rotor's presence, turning on Q_2. The north pole created at S_2 will pull the rotor through 90°. As the pole leaves S_1, the flux decreases in HE_2, decreasing its output voltage and turning off Q_1. At the same time that HE_2 turns off Q_2, HE_1 turns on Q_3, pulling the rotor south pole to the bottom of the motor. In this way, the rotor is pulled around the motor, alternately being attracted to the next pole.

Transistor \ Step	1	2	3	4	5	6	7
Q_1	1	1	0	0	0	1	1
Q_2	0	1	1	1	0	0	0
Q_3	0	0	0	1	1	1	0
Q_4	0	0	1	1	1	0	0
Q_5	1	0	0	0	1	1	1
Q_6	1	1	1	0	0	0	1

key—1 = transistor ON
0 = transistor OFF

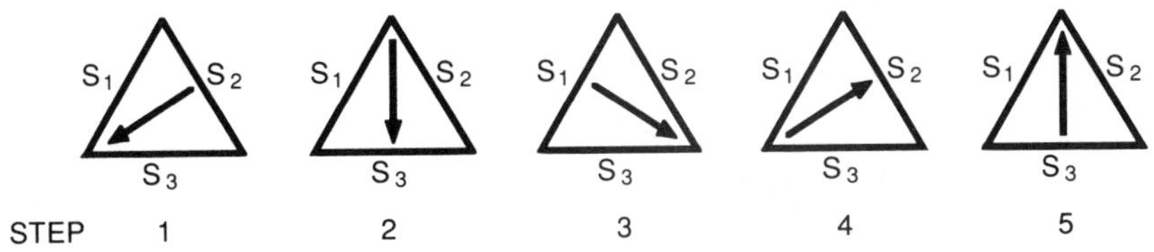

b. firing sequence, ccw rotation

FIGURE 10–19 *(Continued)*

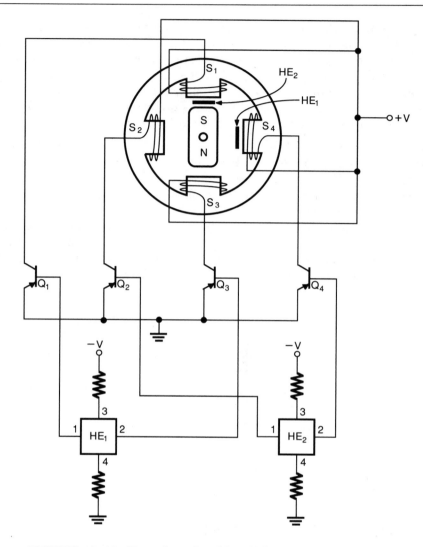

FIGURE 10–20 Four-phase brushless DC motor controller using Hall-effect sensors

10–3.3 Power Control Methods

We have discussed commutation control without reference to power control of the brushless DC motor. One method of power control is to vary the supply voltage to the commutation system. An example of this method is shown in Figure 10–21. The six switching transistors control commutation at the proper angular intervals. The series-connected power transistor handles velocity and current control of the brushless motor by linear (Class A) control or by pulse-

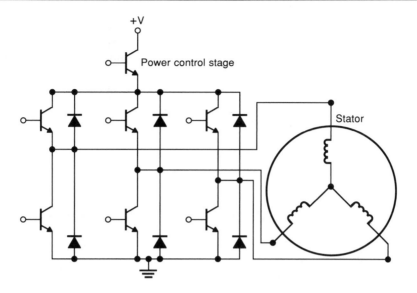

FIGURE 10–21 Series regulator power control—three-phase BDCM

width or pulse-frequency modulation. If directional control is needed, the commutation sequence must be adjustable 0 to 180°. In effect, then, a series regulator is controlling the power supply voltage for a switching stage commutation controller.

Another way of controlling voltage and current in the brushless motor is to let the commutation transistors control the motor current either by linear control means or by pulse-width or pulse-frequency modulation. Such a control method uses available semiconductor devices efficiently.

The pulse-width or pulse-frequency control scheme is well suited for control of voltage and current to a brushless motor. Since logic circuitry capable of switching the appropriate transistors on and off is already in place, the implementation of such control is possible.

Any high performance brushless DC motor will require some form of current limit control. Limiting current protects the controller stage and the magnetic circuit. With either of the controller schemes discussed previously, it is easy to apply such current limit control. The switching controller can sustain current limit conditions without much circuit power dissipation. The linear control system, however, dissipates more excess power in the power transistors. These reasons point clearly toward pulse-width or pulse-frequency modulation as a superior control scheme for brushless DC motors, especially where the controller deals with significant amounts of DC power.

The brushless DC motor can be controlled with very efficient amplifier configurations. In cases of severe environmental conditions, the controller can be located away from the motor. The control system can easily interface with digital and analog inputs. BDCM controllers are, therefore, well suited for incremental motion (as in the stepper motor) and for phase-locked loop speed control systems. The BDCM and controller have a lower level of radio frequency-emission than the conventional DC motors and controls.

The brushless DC motor controller is, in general, more complex than the controller for an equivalent conventional DC motor, but it may be similar in size and complexity to a closed-loop stepper motor controller. The commutation sensor system can, in some cases, provide some incremental motion applications. An encoder system is usually added to the commutation system to suit the application.

10–3.4 Integrated Circuit (IC) BDCM Controllers

The BDCM controller has been integrated onto a single chip, as we have seen with other types of DC motor controllers. An IC BDCM circuit is shown in Figure 10–22. The heart of the controller is the LS7263 IC, manufactured by

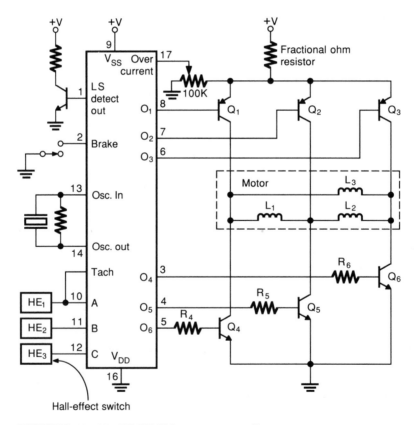

FIGURE 10–22 IC BDCM motor controller

LSI Computer Systems Inc. This chip controls the speed of a three-phase BDCM to speeds around 3600 rpm. This circuit uses a 3.58-MHz crystal to provide its speed regulation time base. Pin *1* incorporates a low speed detection circuit. When shaft speed goes lower than 1100 rpm, the output at pin *1* goes high, turning on the transistor attached to that pin. A high voltage on the transistor's input will cause the collector to go low. This circuit can be used to sound an alarm or enable another circuit function.

In normal operation, the outputs O_1 through O_3 and O_4 through O_6 drive transistors that, in turn, drive the motor coils in proper sequence. The outputs O_1 through O_3 drive PNP transistors, while outputs O_4 through O_6 drive NPN transistors. The outputs turn on in pairs. For example, when Q_1 and Q_5 are on, current flows up through Q_5, L_1, Q_1, and the fractional ohm resistor. Note that current cannot flow in any other coil because there is no complete path for current flow.

Pin *2* provides a static positive braking system. When pin *2* receives a high voltage, outputs Q_1 through Q_3 turn off and Q_4 through Q_6 turn on. This action effectively shorts the windings together, creating an electrical load on the motor. The motor will quickly come to a stop. Pins *13* and *14* allow for connecting a clock crystal for timing. The frequency of the clock can be tested at pin *15* without loading the oscillator down. Another useful feature is the overcurrent protection offered at pin *17*. A fractional ohm resistor is placed between the positive supply and the common emitters of the PNP drivers. Pin *17* is then connected to the wiper of a potentiometer, in this case 100 Kohm, with one end tied to ground and the other to the common emitters. The wiper is adjusted so that the outputs O_1 through O_6 are off for currents greater than the limit. Note also the tachometers attached to pins *10* through *12*. The information from the tachometers is important in the commutation process. This chip permits highly accurate speed regulation, to within +/– 0.1%.

10–4 STEPPER MOTOR CONTROL

A basic stepper motor system is shown in Figure 10–23. The system is composed of a DC power supply, a DC input signal usually provided by a pulse generator, a stepper motor, and a driver.

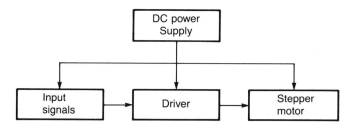

FIGURE 10–23 Parts of a stepper motor system

Driving the stepper motor requires a power supply for both the motor and the driver. The power supply for the motor should apply a rated voltage to the motor winding and cause a rated current to flow. Power supplies generally use a DC voltage regulator to accomplish this action. The driver also requires that the power supply furnish a constant voltage under changing load conditions. Power supplies may be required to supply up to 100 V at 20 A.

Driving a stepper motor requires that current be switched from one stator winding to another. A circuit that provides this switching function is called a driver. The driver arranges, distributes, and amplifies pulses from the pulse generator. The basic stepper motor driver accepts a digital (or pulse) signal and converts it into the form required by the stator windings. Stepper motor drivers do more than merely drive the stepper motor. Drivers determine the stepper motor performance in an application. By properly designing drivers, stepper motor performance can be enhanced considerably.

For example, stepping rates can be increased from hundreds of steps per second to thousands of steps per second by using a different driver design. Simple stepper motor drivers use a source with a low output impedance to drive the stepper. The source produces pulses that cause the rotor to come to a complete stop between steps. Using this type of drive, most permanent-magnet steppers are limited to 200 to 300 steps per second, while variable reluctance steppers can reach 700 to 800 steps per second. At very high stepping rates, the impedance of the winding limits the rotor current and, therefore, the rotor torque. Since the rotor current affects the torque produced, stepper motors should be driven by a ***constant-current source.***

Before discussing stepper motor drivers, we need to elaborate on the stepper motor stator. Stepper motor stator windings can be classified into two types: the ***monofilar*** winding and the ***bifilar*** winding. An example of a monofilar, four-phase stepper is shown in Figure 10–24. Note that only one

FIGURE 10–24 Four-phase, monofilar winding configuration for a PM stepper motor

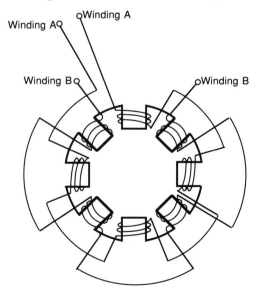

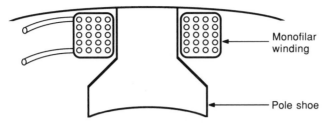

a. Cutaway diagram of a monofilar winding and pole shoe

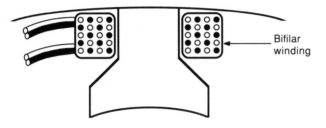

b. Cutaway diagram of a bifilar winding and pole shoe

FIGURE 10–25 Monofilar and bifilar winding construction

winding exists on each pole. The cutaway diagram in Figure 10–25a shows the pole structure of a monofilar winding. Instead of the three coils shown on each pole in Figure 10–24, the monofilar winding has many turns of wire. The bifilar winding, shown in cross section in Figure 10–25b, is made up of two overlapping wires wound on the same pole. The two wires are separated from each other.

10–4.1 Stepper Motor Excitation Modes

Stepper motor drivers can be divided into three types of excitation modes: *one-phase, two-phase,* and *one-two phase modes.* In the one-phase drive, also called the *wave drive,* only one phase is energized at a time. The popular two-phase excitation mode is shown in Figure 10–26.

Two-Phase Excitation The electrical input to this circuit, shown with manual switches, is a four-step switching sequence. Clockwise rotation is started by placing both switches up, in the positive position. For the next step, the right switch is moved down, to the negative position. For the third step, the left switch is moved down into the negative position. Continuing this sequence causes the rotor to move in the clockwise direction. Reversing the sequence reverses the direction of rotation. The stepper motor can easily be controlled by a pulse generator. If the motor is operated at a fixed frequency, the resulting wave form applied to the stator windings is as shown in Figure 10–27. The electrical input is a two-phase, 90° shifted square wave. Although the input

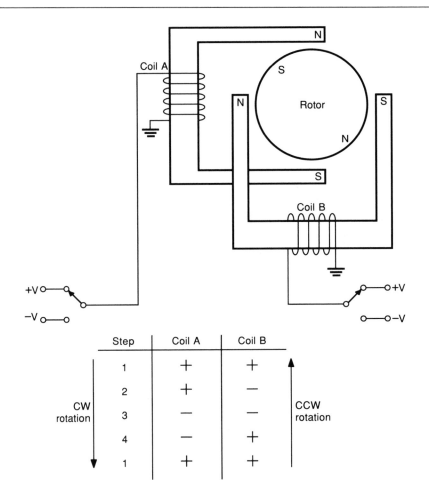

FIGURE 10–26 Two-phase, four-step excitation

current is twice that of the one-phase excitation mode, the two-phase mode does not oscillate as much as the one-phase mode. The two-phase mode can respond to a wide range of input frequencies. The two-phase drive offers a 20% gain in the available torque per watt over the one-phase drive. The one-phase logic for the driver is more complicated than the two-phase.

Wave Excitation   Energizing only one winding at a time is called wave excitation. An example of wave excitation is shown in Figure 10–28. Note that it produces the same type of four-step sequence as in the two-phase excitation mode. Starting at step *1*, switch *A* is positioned in the up position, applying a positive voltage to the *A* stator coil. At the same time, switch *B* is in the off position. No voltage is, therefore, applied to the *B* stator coil. For the next step (step *2*), switch *A* is turned off, while switch *B* is moved to the up position.

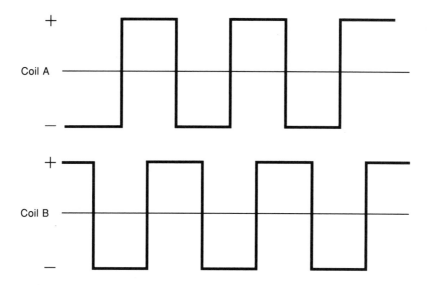

FIGURE 10–27 Voltage wave form of a fixed frequency, four-step drive sequence

These switches apply a positive voltage to the B coil, while the A coil has no applied voltage. In the next step, a negative voltage is applied to the A coil, while the B coil receives no power. In the wave drive, no more than one coil receives power at one time.

Since only one winding is on, with rated voltage applied, the holding and running torque in the wave drive will be reduced, compared to the two-phase drive. Reduction in holding and running torque is approximately 30% less than the two-phase drive. Within limits, the stator voltage can be increased, which will bring the output power close to the rated torque value. The one-phase wave drive has the advantage of low power consumption and good step angle accuracy. This mode usually requires damping because the rotor tends to overshoot and oscillate.

Half-Step Excitation It is also possible to half-step a stepper motor in an eight-step sequence, using the switching mode found in the wave drive. This mode of excitation, shown in Figure 10–29, is called the half-step or one-two phase excitation mode because it can be used to get a stepper motor to step in between the normal step angle, such as a 3.75° step from a 7° motor. The half-step drive combines both the wave and two-phase methods. The half-step method alternates between the two basic excitation modes. During one step, two stator windings will be energized. In the next step, only one will be energized. When the two windings are energized, holding and running torques are equal to those of the two-phase excitation mode. When only one winding is energized, torque output is equal to that of the wave drive. The motor may exhibit a strong step, followed by a weak step. Simply put, the holding and

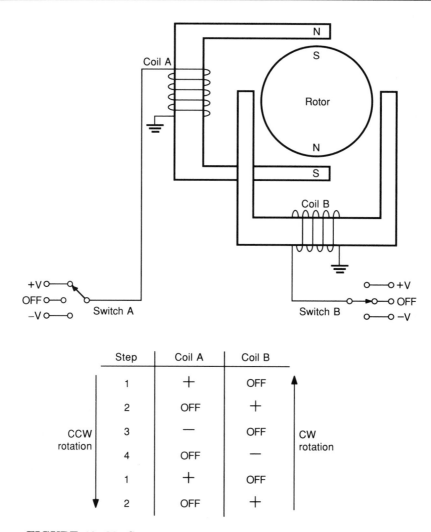

FIGURE 10–28 Stepper motor wave excitation

running torque vary every other step. Since the winding and flux conditions are not similar for each step when half stepping, the step accuracy will not be as good. If the load torque is less than 30 to 40% of the maximum, the difference in torque output is not important. At higher loads and at low frequency applications, however, the weak step shows a significant loss in torque. At high frequencies, less torque is needed. The torque difference between strong and weak steps can be reduced by designing the drive so that more torque is produced for the weak step. The motor will then produce close to its rated torque at every step.

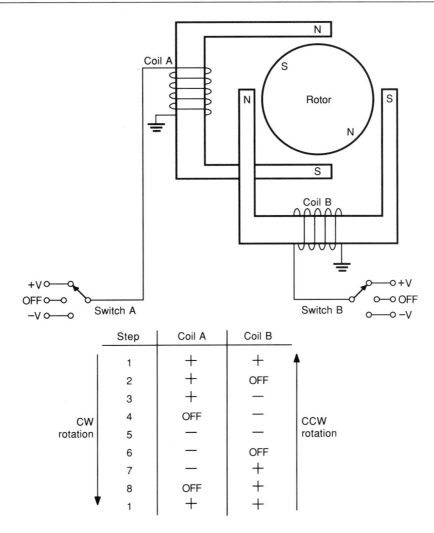

Step	Coil A	Coil B
1	+	+
2	+	OFF
3	+	−
4	OFF	−
5	−	−
6	−	OFF
7	−	+
8	OFF	+
1	+	+

FIGURE 10–29 Half-step wave excitation

10–4.2 Driver Circuits

With PM stepper motors, drives are classified as either ***unipolar*** or ***bipolar***. With a unipolar drive, current flows through the stator coil in only one direction (Figure 10–30a). In the bipolar drive, current flows in both directions (Figure 10–30b,c). The unipolar type of drive is used because of its simplicity, which is achieved by having the current flow in only one direction. The stepper designed to be driven by a unipolar drive usually has a total of four coils. One transistor drives each coil, so that the circuit diagram in Figure 10–30a would be duplicated four times. Bipolar circuits, where current flows in both direc-

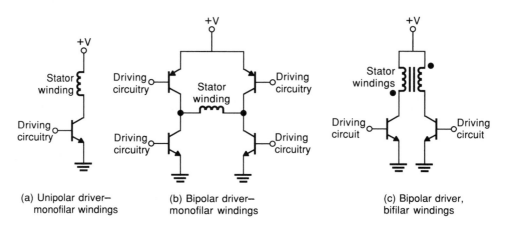

(a) Unipolar driver—
    monofilar windings

(b) Bipolar driver—
    monofilar windings

(c) Bipolar driver,
    bifilar windings

FIGURE 10–30 Stepper motor drivers

tions in the stator windings, are used with both monofilar and bifilar construction.

The unipolar drive is the method most widely used to drive stepper motors. As shown in Figure 10–31a, the unipolar drive uses a high voltage, applied through a large series resistance. In this arrangement, the source behaves as if it were a constant-current source. This driving method suffers the disadvantage of high power dissipation and is, therefore, inefficient. Another disadvantage is the slow current rise time. Current rise time can be increased by adding a capacitance in parallel with the series resistance, as in Figure 10–31b.

Chopper Driver A more efficient method of unipolar drive is found in the chopper drive shown in Figure 10–32. As we have seen, adding a series resistor

FIGURE 10–31 Unipolar supply

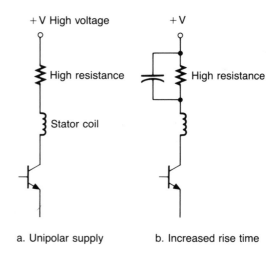

a. Unipolar supply

b. Increased rise time

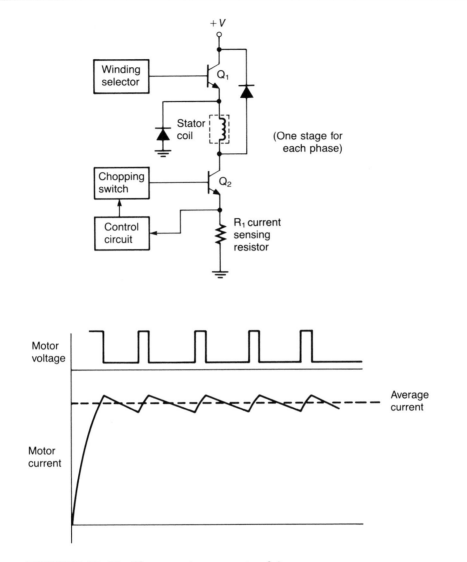

FIGURE 10–32 Chopper stepper motor driver

to decrease the time constant in small stepper motors is inefficient. In larger motors, the larger power losses in the series resistors make this method almost impossible to use. In the chopper system shown, a high voltage is applied to the stator initially. The applied voltage is typically 10 to 20 times the stator voltage rating. This high voltage is applied until the correct stator current is reached. The circuit is then opened by the chopper. The voltage is then reapplied when the current reaches a predetermined value and turned off again when the high current level is reached. This voltage cycling continues as long as the power is supposed to be applied to the stator winding. When the driving pulse is ended, the transistor turns off and the energy in the stator coil will bleed to ground through the diode to the left of the coil. Figure 10–32 shows the relationship between the voltage and current of the chopper drive, plotted with respect to time. Choppers operate well in start-stop applications. The motor accelerates quickly due to the high voltage applied to the stator.

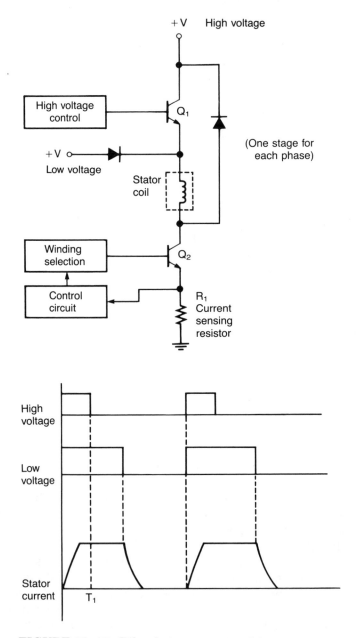

FIGURE 10–33 Bilevel stepper motor drive

Bilevel Driver Another unipolar drive sometimes used in place of the resistance limited drive is the bilevel or dual-voltage drive. The bilevel drive, shown in Figure 10–33, requires two supplies: a high voltage supply and a low voltage supply. The driver uses the high voltage supply for starting and stopping operations. Like the chopper, the high voltage is applied until the current reaches a predetermined level. Unlike the chopper, which turns the stator voltage off at this point, the bilevel drive switches over to the low voltage supply at time *T1* in Figure 10–33. This part of the figure shows the relationship between the voltage and current of the bilevel drive, plotted with respect to time. At the end of the pulse, the low voltage is turned off and the pulse energy is returned to the high voltage supply. This action begins and ends the pulse quickly with small losses.

The resistance-limited unipolar drive is the most popular drive because it is the simplest and the least expensive. However, it is not very efficient, especially with higher power stepper motors. The higher efficiency drives (chopper and bilevel) improve motor performance, but with added cost and complexity.

10–4.3 Transistor Protection

We have seen transistors used as driving devices in the previous circuits. The basic unipolar transistor driver is shown in Figure 10–34. The transistor energizes the stator coil. Each stator coil in the stepper motor has a similar driver circuit. We have not shown the circuits used to drive the transistor. In practice, the transistor can be driven by TTL logic or an optoisolator for isolation. The stator coil, since it is made up of coils of wire, is a very inductive load. When the transistor is deenergized, current stops flowing in the stator coil. The flux built up in the stator coil collapses, producing a transient voltage spike in a process called inductive kick. This transient voltage spike can destroy the switching transistor.

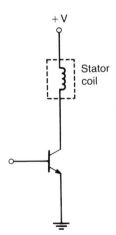

FIGURE 10–34 Transistor driving stator coil

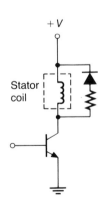

FIGURE 10–35 Diode protects driver transistor

A common method used to protect transistors is shown in Figure 10–35. The diode is connected so that when the transistor is conducting and current is flowing through the stator coil, the diode is reverse-biased. When the switching transistor turns off, the flux built up in the coil collapses, reversing its polarity and forward-biasing the diode. The transient pulse is then shorted across the low impedance of the forward-biased diode. The resistor in series with the diode is used to lower the RL time constant. Unlike the RC time constant, the RL time constant is inversely proportional to the resistance. If the resistance is low, the time constant will be long and the diode will not quickly discharge the flux built up in the coil. Adding the resistor lowers the time constant and allows the transient pulse to be shorted more effectively.

10–4.4 Power MOSFET Drivers

Recent developments in field effect transistor (FET) technology have made it possible to replace the transistor driver with the power MOSFET. Power MOSFET differences in comparison with the bipolar transistor are presented in Table 10–1.

Table 10–1 Comparison of the Power MOSFET and the BJT

Parameter	MOS	Bipolar
Zin	High (10^{10} Ω)	Intermediate (10^4 Ω)
Current gain	High (10^6)	Intermediate (10^2)
Switching frequency	High (100–500 kHz)	Intermediate (20–80 kHz)
On resistance	High	Low
Off resistance	High	High
Voltage capability	Intermediate (500 V)	High (1200 V)
Ruggedness	Excellent	Good
Maximum operating temperature	High (200°C)	Intermediate (150°)

One of the most significant differences between the power MOSFET and the power transistor relates to the fundamental nature of how the devices are driven. Since the transistor is basically a current-driven device, the load and the supply voltage determine the collector current. The amount of collector current then determines the amount of base current needed for saturation. Since the current gain of power transistors can be as low as 10, the transistor drive requirements are sometimes quite high. High drive requirements reduce circuit efficiency and increase power dissipation requirements and make drive circuitry more complicated and expensive.

The power MOSFET is a voltage-driven device, similar in function to the vacuum tube. The power MOSFET can be driven by a voltage potential with little current drain. The reason for this behavior lies in the structure of the power MOSFET. An oxide layer separates the gate from the channel region, giving the power MOSFET its high input impedance and low gate current drive.

The diagram in Figure 10–36 shows a full-step center-tapped stepper motor controller. The controlling logic is provided by a CMOS 4-bit presettable shift register. The stator coils are driven by four N-channel power MOSFETs. Note the four freewheeling diodes that prevent the voltage from the coil from damaging the MOSFETs when the MOSFETs are turned off. The full-step sequence shown in the figure is two *on* periods followed by two *off* periods. This sequence can be preset into the 4-bit shift register MC14194. The motor steps clockwise by right-shifting the MC14194, while left-shifting yields counter-clockwise stepping. The control signals on *S0* and *S1* control the stepping,

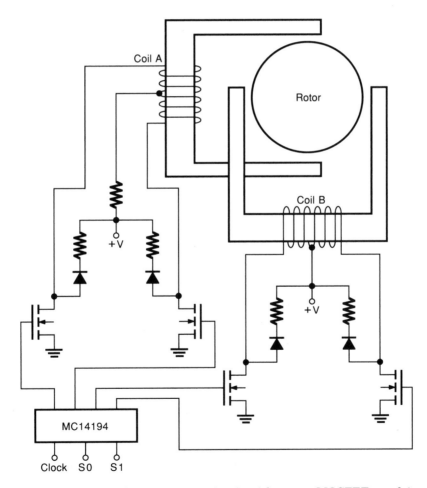

FIGURE 10–36 Stepper motor circuit with power MOSFETs as drivers

along with the clock line input. On power-up, the MC14194 requires a leading-edge clock pulse, with *S0* and *S1* each set to a *1* for presetting. When this preset value is established, the logic is put into a known state. After this preset value has been established, a value of *0* for *S0* and *1* at *S1* causes the motor to step clockwise. When the bits are reversed, a *1* at *S0* and a *0* at *S1* cause the motor to turn counterclockwise. The motor will turn at a stepping rate determined by the clock frequency.

Integrated Circuit Stepper Driver Several manufacturers have introduced stepper drivers completely integrated on a single chip. Stepper motor manufacturers have responded by introducing stepper motors designed to operate efficiently with integrated circuit stepper drivers. The Airpax Company, for example, has introduced a four-phase stepper that can be directly driven by an SAA1027 integrated circuit. Both the IC and the motor can operate from a single +12 V supply. Because the IC has been designed with a high level of noise immunity, a stepper driver circuit using this chip can be used successfully in an electrically noisy environment.

The SAA1027 IC is produced in a 16-pin DIP plastic package. Table 10–2 shows a brief summary of the electrical characteristics of this device. Figure 10–37 shows a system using the SAA1027. The system is powered by a 12 V-supply with a current output compatible with the motor being driven. The current applied to pin *4* of the IC must be adjusted by proper selection of the resistor R_B. Resistor R_B is chosen to correspond with the current-handling capabilities of the motor used.

Table 10–2 Characteristic of the SAA1027 IC stepper driver

Characteristic	Abbreviation	Value
Supply voltage	Vp	9.5 V to 18 V
Load current (each output)	Iq	350 mA max
High logic level	Vth	7.5 V to Vp
Low logic level	Vtl	0 V to 4.5 V max

The stepping rate of the motor is determined by the frequency of the pulses applied to the *T* input terminal at pin *15*. Steps are begun by the positive-going edges of the pulses in the sequence shown in Figure 10–37. The output switching sequence can be set to a predetermined logic state by applying a low voltage level to pin *2* (the *S* input). If the trigger input is high at the time *S* is driven low, Q_1 and Q_3 outputs will be low and Q_2 and Q_4 will be high. Maximum noise immunity is achieved by connecting the *S* input high constantly, if it is not needed. Direction of rotation is controlled by the voltage at pin *3*, the *R* input. If the voltage level at pin *3* is high, the drive shaft will rotate counterclockwise. A low voltage level at pin *3* causes the shaft to rotate

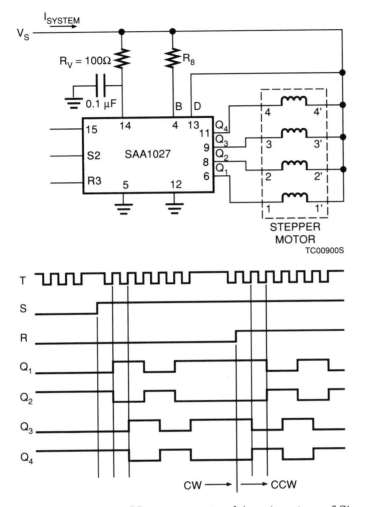

FIGURE 10–37 IC stepper motor driver (courtesy of Signetics Corporation)

in a clockwise direction. The level of the voltage applied to pin *3* can be changed at any time, regardless of the logic state of the other two inputs. The best noise immunity is achieved by connecting the *R* input either high or low, not allowing the input to float.

As is sometimes the case, the power supply will supply both the motor and the drive circuitry. When this is done, special precautions must be observed to prevent transient voltage spikes, which are caused by switching off the motor windings, from interfering with the switching sequence. The RC network attached to the Vp supply prevents this interference. The 0.1 μF capacitor must be located as close as possible to pins *14* and *5*.

10–5 DC-DC MOTOR CONTROL

Most of the controllers discussed up to this point have converted AC into DC for DC motor control. Some controllers use a device called a chopper to change constant DC to variable DC. The concept of the chopper is illustrated in Figure 10–38. The basic parts of the chopper are the supply, the switch, the load, and the freewheeling diode. The switch turns on and off, decreasing the average voltage to the load. During the time that the switch is on, current flows from the supply to the motor and back to the supply. When the switch is off, current flows from the motor coils through the forward-biased freewheeling diode.

The average voltage in the chopped wave form is given by

$$V_{avg} = V_{app}\frac{t_{on}}{t_{on} + t_{off}}$$

where t_{on} is the time that the switch is on and t_{off} is the time that the switch is off.

You will recognize the following expression as the duty cycle of the chopper.

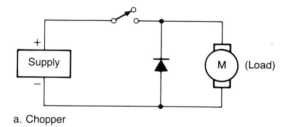

a. Chopper

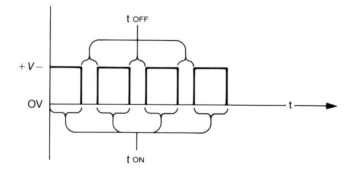

b. Chopper output

FIGURE 10–38 Basic chopper theory

$$\frac{t_{on}}{t_{on} + t_{off}}$$

If the time on and time off are equal, the duty cycle is 50%. The average voltage is then one-half of the applied voltage, and the load voltage is directly proportional to the duty cycle. With a constant period, as the time on increases, the duty cycle will increase, increasing the average voltage across the load. An increase in the duty cycle will increase the speed of a DC motor. Average voltage control that keeps the period constant and changes either the time on or the time off is called pulse width modulation. Pulse width modulation is the most popular method of voltage control. The second method, sometimes called frequency modulation, keeps the time on constant and varies the period (and, therefore, the frequency). Chopper frequencies range from 500 Hz to 2500 Hz. Choppers are widely used in industry in transportation applications such as forklifts and other battery-operated vehicles. Other applications involve traction motors such as those used in electric trams, subways, trolleys, and streetcars. Choppers are also widely used in switching power supplies.

The chopper referred to previously is called a step-down chopper. The average voltage is always less than the applied voltage. In this case, we refer to the chopper as a step-down device. Another type of chopper, called a step-up chopper, is shown in Figure 10–39. When the switch is on, current flows from the supply to the inductor. The inductor stores the energy transferred from the supply. When the switch opens, energy from the inductor adds to the supply, resulting in a higher voltage at the load.

Chopper circuits have used various semiconductors as switches. One of the first to be used was the SCR. Using the SCR as a chopper presented an engineering challenge in commutation. You will recall from our discussion on SCRs that the SCR is easily turned on in DC circuits. The problem in using an SCR in DC circuits is turning it off. Recall that SCR turn-off is not a problem in circuits supplied by an AC line source. This method of commutation is called line commutation.

10–5.1 Thyristor Choppers

An example of a chopper using an SCR as a switch is shown in Figure 10–40. Sometimes known as the Jones circuit, it is more often used with a series motor

FIGURE 10–39 Step-up chopper

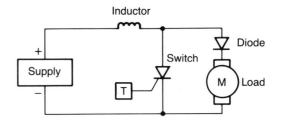

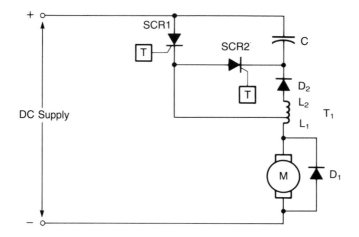

FIGURE 10–40 Thyristor chopper

than with a shunt motor. The high inductance of the series motor armature smooths out the current pulses from the chopper better than the shunt motor. A cycle of operation will start with the capacitor charged positive at the top and both SCRs off. When SCR_1 is triggered into conduction, current flows from the supply, through the motor, L_1, and SCR_1. Current through L_1 causes current to be induced in L_2, since T is an autotransformer. The current in L_2 discharges the capacitor C and charges it in the opposite direction, negative on top and positive on the bottom. This voltage is held on the capacitor by the reverse-biased D_2. The capacitor will remain charged until SCR_2 is triggered. When SCR_2 is triggered, the voltage on the capacitor reverse-biases SCR_1, turning it off. The capacitor will discharge until the current through SCR_2 falls below the holding current value. SCR_2 commutates naturally with the discharge of the capacitor. The cycle starts again when SCR_1 is triggered into conduction. Note the importance of the autotransformer T. The autotransformer insures that whenever current comes from the source to the load (the motor), the correct voltage is induced in L_2 to charge the capacitor.

10–5.2 Transistor Choppers

A simplified transistor chopper is shown in Figure 10–41. The driver, in this case, is a pulse-width modulator. A freewheeling diode, $CR1$, is provided to maintain continuous current flow in the motor during the time the transistor is off. Note the wave forms in Figure 10–41. When the transistor is turned on, current increases through the motor.

When the transistor turns off, current in the motor continues to flow through the freewheeling diode, $CR1$. Note that I_M decreases during this time, but not to zero. A Darlington transistor is often used in place of the transistor. As you will recall, the Darlington configuration has a higher gain when com-

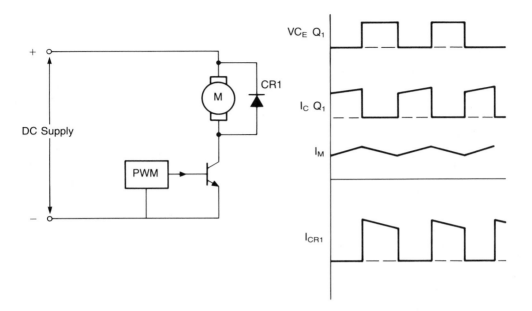

FIGURE 10–41 Transistor chopper

pared to the regular transistor. A higher transistor gain keeps driver current requirements low.

The system illustrated in Figure 10–41 operates only in a single quadrant. That is, it cannot produce a negative torque, nor can it reverse direction of rotation without a reversing switch. A reversing switch reverses the armature terminals and, therefore, the direction of rotation. The motor can then operate in the third quadrant. Motor operation in four quadrants is shown in Figure 10–42.

Door openers, wiper controls, and similar applications operate in first and third quadrants. Applications such as electric vehicle controls, numerically

FIGURE 10–42 Four quadrants of motor operation

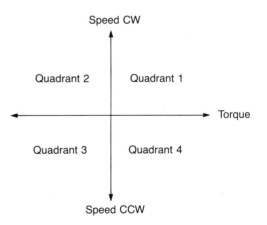

controlled machine servo drives, and hoist controls operate in all four quadrants, as illustrated in Figure 10–42. This figure shows the four quadrants of motor operation. In the first quadrant, both speed and motor output torque at the shaft are positive. An example of a positive speed and torque is a hoist raising a load against the action of gravity. In the second quadrant, which is entered only momentarily when a load is on the hook, the motor torque is reversed to decelerate. In the third quadrant, both speed and torque are negative. This quadrant is entered momentarily when a load is on the hook. The motor, assisted by the load, is accelerating the rotating parts of the mechanism at the beginning of a lowering operation. With a light load, the motor operates in the third quadrant throughout the greater part of the downward movement of the load. In the fourth quadrant, the motor torque tends to raise the load, which is moving downward. This means that the motor resists the tendency of the load to accelerate downward and run away with the system.

In quadrants one and three, energy must be supplied by the electrical system to the mechanical system. In the chopper, the power is supplied by a DC power supply. In quadrants two and four, energy is supplied, either momentarily or continuously (depending on load), by the mechanical system to the electrical system. This process is called regenerative braking. Two types of configurations allow operation in all four quadrants. Figure 10–43a shows a T (complementary) type output stage, and Figure 10–43b shows the H, or bridge,

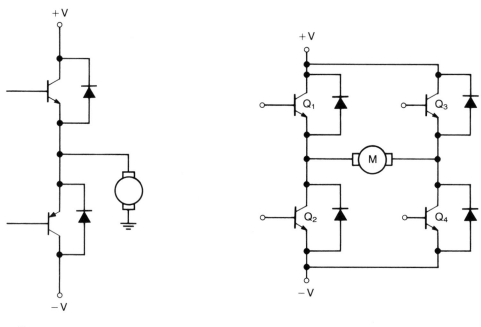

a. T output stage   b. H output stage

FIGURE 10–43 Transistor chopper output stages

configuration. The T (complementary) output stage requires two power supplies. The H bridge requires only one power supply. The H bridge is, therefore, more commonly used for a variety of applications.

The H Bridge Output Stage Chopper controls often use pulse width modulation at a fixed frequency to modulate or change the average motor voltage. The resulting motor current has a ripple content. The size of the ripple voltage depends on the duty cycle and the motor time constant. In the bridge circuit shown in Figure 10–43b, the motor current ripple is used to regulate the switching of transistors by comparing a set point with the motor current. The motor current wave form, indicating when the devices conduct, is shown in Figure 10–44.

The motor can be a permanent-magnet type or one with a separately-excited field. To run the motor in the forward (CW) direction, transistors Q_1 and Q_4 are turned on (Figure 10–43b). Once the motor current (I_M) reaches a peak value, Q_1 is turned off to let the current free-wheel through D_2 and Q_4. By

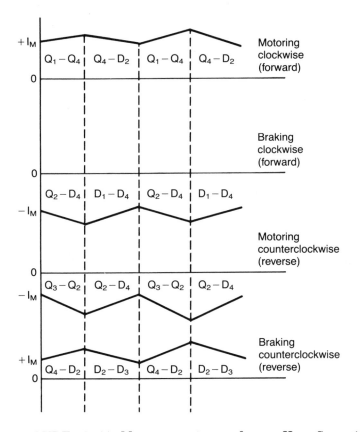

FIGURE 10–44 Motor current wave forms—H configuration

monitoring I_M, Q_1 is turned on again at a set value of I_M. A constant ripple current results and is maintained at all speeds. During forward (CW) braking, both Q_1 and Q_4 are turned off, and Q_2 is turned on. This action forces the motor current to reverse. Now the current flows from the motor through Q_2 and D_4. The energy stored in the motor (which is proportional to the moment of inertia of the motor) can be fed back to the supply by turning off Q_2. In this case, the current flows from the motor through D_1 to V_{BB}. The regenerative current can be controlled by monitoring the voltage build-up on the input capacitor and turning on Q_2 again. The cycle of operation is just the opposite when the motor rotates in the reverse (CCW) direction.

Where the bridge circuit is powered by a battery, a complementary bridge circuit is used, as shown in Figure 10–45. Fractional HP motors use the complementary bridge circuit in control circuitry.

By turning on Q_1 and Q_4, the motor can be run in the forward direction. To reverse the direction of rotation, Q_1 and Q_4 are turned off, and Q_2 and Q_3 are turned on after a fixed time. By driving the transistors into hard saturation and operating at higher frequency, the circuit can be operated most efficiently. This circuit has a limited application, as PNP transistors are not available in high current and high voltage ranges. For DC voltage applications of 12 V or less, General Electric's D44VH and D45VH have the lowest saturation voltages in the industry and are often used for this type of application. A modified version of this circuit is shown in Figure 10–46.

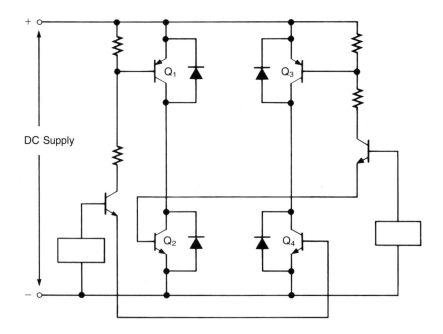

FIGURE 10–45 Complementary bridge circuit

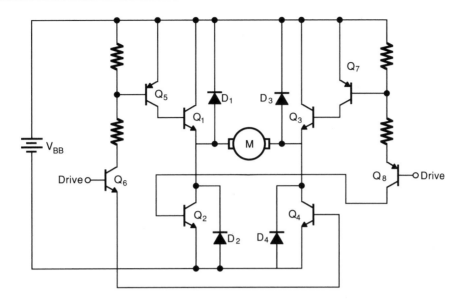

FIGURE 10–46 Modified bridge with NPN transistors (courtesy of General Electric Company)

Q_1 and Q_3 are NPN transistors, and Q_2 and Q_4 are Darlington transistors. When Q_6 is turned on, Q_5 turns on, supplying base drive to Q_1, and Q_4 turns on, enabling the DC motor to run in the forward direction. To run the motor in the reverse direction, Q_6 is turned off, and after a fixed time Q_8 is turned on, which turns on Q_3 and Q_2. In this circuit, the PNP transistors Q_5 and Q_7 must provide base current for the NPN power transistors Q_1 and Q_3.

10–5.3 Power MOSFET Choppers

Since they can be turned off easily, power MOSFETs have become popular in choppers. The power MOSFET has other advantages. First, it is a voltage-driven device. The gate is electrically isolated from the source by a layer of silicon dioxide. Since the power MOSFET is a voltage-driven device, only minute amounts of current are necessary to drive it. The small amount of driving current simplifies the driving circuitry. Since the gate-source does have some capacitance, the only important factor is having enough current available to charge and discharge the capacitance. This can be accomplished by having a source that has a low output impedance.

Another advantage of the power MOSFET is its ability to be paralleled to gain greater current-handling capability. Such paralleling is difficult to achieve with SCRs or bipolar transistors. Power MOSFETs are easy to parallel because they have a positive temperature coefficient. As temperature in one paralleled device increases, the resistance increases, cutting down current and forcing a current sharing between devices. Power MOSFETs can be paralleled to build choppers to handle currents in the hundreds of amperes.

A further advantage of the power MOSFET over the thyristor and transistor choppers is the power MOSFET's ability to operate at high frequency. Choppers using bipolar transistors and thyristors are operated at low frequencies to keep them simple. Transistor and thyristor choppers also need a series-connected field winding when operating at low frequencies to smooth out the ripple. With the higher operating frequencies made possible by the power

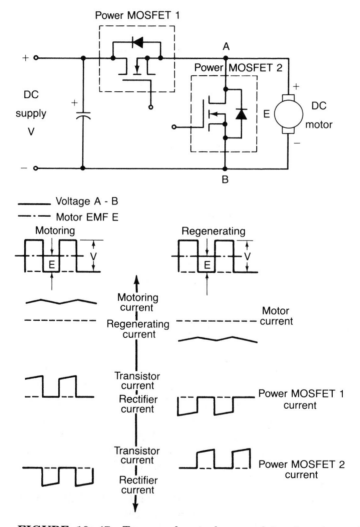

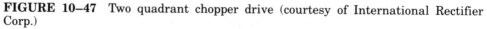

FIGURE 10–47 Two quadrant chopper drive (courtesy of International Rectifier Corp.)

MOSFET, the motor armature inductance itself is sufficient to smooth out any ripple, even at low speeds.

A basic two quadrant chopper using a power MOSFET is shown in Figure 10–47. This circuit gives continuous control of motor speed in quadrant *1*. Operation in quadrant *1* is called motoring. While motoring, the DC motor receives power from the DC source. In the motoring mode of operation, MOSFET *#1* is on, while *#2* is off. Although power MOSFET *#2* is off while motoring, the body-drain diode acts as a freewheeling diode when power MOSFET *#1* is off. When the motor is to act like a generator and return energy to the DC source, the power MOSFET *#2* is on and power MOSFET *#1* is off. This mode of operation is called the regenerating mode and is related to quadrant *#2*. To regenerate, the motor must be either a permanent-magnet field or a shunt or separately-excited wound-field motor.

A practical 48 V, 200 A chopper schematic is shown in Figure 10–48. This circuit uses ten IRF150 power MOSFETs connected in parallel for the

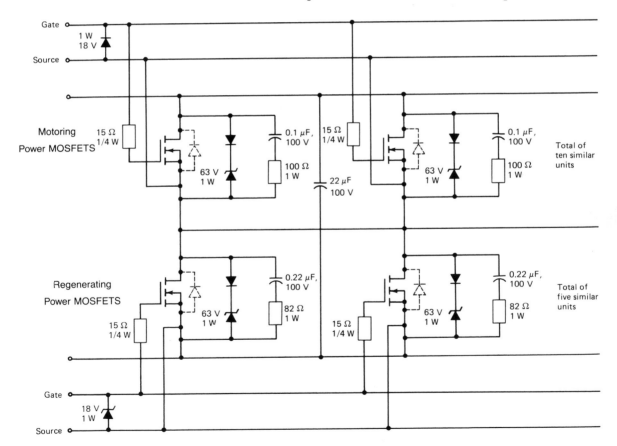

FIGURE 10–48 Practical two-phase chopper (courtesy of International Rectifier Corp.)

motoring switch. Five IRF150 power MOSFETs are connected in parallel for the regenerating switch. This system is capable of delivering a total of 200 A while motoring and 140 A of regenerating current.

Although we have illustrated three choppers—the SCR, transistor, and power MOSFET versions—other semiconductors are used in industry as choppers. Another semiconductor used in choppers is the the GTO (gate turnoff thyristor).

CHAPTER SUMMARY

- A DC drive consists of a DC motor and a power converter.
- The main disadvantage of the half-wave converter is its highly discontinuous current.
- The full-wave, half-controlled bridge uses a diode bridge to rectify the DC before it is applied to the SCR.
- The flywheel diode decreases the amount of discontinuous current and helps prevent damage to the thyristor.
- Fully controlled bridges have smoother torque and speed characteristics when compared to half-controlled bridges.
- The brushless DC motor uses a sensor to detect the position of the rotor. The rotor position information is used to decide the best time to apply DC power to the stator winding.
- The brushless DC motor driver requires a circuit that applies DC to the stator windings at the correct point needed to rotate the rotor.
- The brushless DC motor controller is usually more complex than a comparable DC motor system with brush commutation.
- The four parts of a stepper motor drive system are the DC power supply, a DC input stepping signal, a stepper motor, and a semiconductor driver.
- Driving a stepper motor requires that current be switched from one stator winding to another. This is usually accomplished without knowing the position of the rotor.
- Three stepper motor excitation modes are: 1) one-phase, 2) two-phase, and 3) one-two-phase
- Stepper motor drivers are either unipolar or bipolar. In the unipolar drive, current flows in one direction through the stator winding. In the bipolar driver, current flows in both directions.
- Stepper motor and BDCM drivers now use power MOSFETs and other types of semiconductor drivers to replace the transistor. IC drivers, which cut circuit cost and complexity, are now common in low power drive circuits.

- The chopper DC drive takes a constant DC input and changes it to variable DC by varying the duty cycle of the chopper.

QUESTIONS AND PROBLEMS

1. Explain the operation of the following drives by drawing synchrograms:
   a. half-wave controller
   b. full-wave half-controlled bridge
   c. full-wave half-converter

2. Explain the purpose of the flywheel diode and how it operates.

3. Draw the the speed-torque curves of the following devices, discussing the ways in which they differ:
   a. stepper motor
   b. brushless DC motor
   c. AC induction motor

4. Draw and explain at the block diagram level the following systems:
   a. a brushless DC motor controller
   b. a permanent-magnet DC motor controller
   c. stepper motor controller

5. Define the term phase control and explain how it relates to DC motor control.

6. Describe the difference between the following modes of stepper excitation.
   a. one-phase
   b. two-phase
   c. one-two phase

7. Draw a two-phase stepper motor and describe a normal, full-step motor step sequence.

8. Explain the differences between and the advantages of unipolar and bipolar stepper drives.

9. Explain the operation of the following types of stepper drives:
   a. series resistance
   b. chopper
   c. bilevel

10. Compare the transistor and the power MOSFET as a stepper driver.

11. With a +10V supply, calculate the output of a chopper with the following duty cycles:
    a. 50%
    b. 25%
    c. 78%

12. What is the difference between a normal chopper and a step-up chopper? Describe the operation of each.

13. What is meant by the term line commutation?

14. Describe the operation of the Jones chopper.

15. Explain what is meant by quadrants and the operation of a DC motor in each of the four quadrants.

11

ELECTRONIC AC MOTOR CONTROL

At the end of this chapter, you should be able to

- describe three basic methods for changing the speed of an AC induction motor.
- differentiate between a converter and an inverter.
- describe the following types of adjustable-frequency speed control:
  - variable-voltage inverter
  - current-source inverter
  - pulse-width modulated inverter.
- define the volts/Hz ratio.
- describe the operation of the cycloconverter.

11-1   INTRODUCTION

For many years the choice of motors in industry was simple. Virtually every motor was a DC motor. The only choice to be made was in the type of drive to be used. Since the introduction of the SCR in 1957, solid-state electronics has reduced the cost of AC motor drives to the point where they are cost-competitive with DC motor drives. When environmental conditions, efficiency, higher operating speeds, and closer control are considered, AC drives are often chosen. AC motors have much to commend them. They are 80 to 85% more efficient than a comparably-sized DC motor. AC motors are also smaller, lighter, have less inertia, and can be run at much higher speeds than DC motors. Maximum speed for most wound-field DC motors is about 2500 rpm, while the AC motor can run at twice that speed without special construction.

This chapter will consider control of the AC induction motor, synchronous motor, and universal motor. We will concentrate, however, on induction motor speed control techniques, since the induction motor is the most popular AC motor in industry.

In the last 30 years, many methods have been used to control the speed of AC motors. In one speed control scheme used in the 1950s, adjustable speed alternators produced a variable frequency that was applied to the stator of the AC motor. The variable frequency power applied to the stator changed the speed of the motor. Since then, variation in AC motor speed control has been done by changing the frequency, although the method of frequency change has been different.

11–2 INDUCTION MOTOR SPEED CONTROL

The polyphase induction motor has been popular in industrial machine tools, conveyors, pumps, and fans because of its ruggedness, reliability, and low cost. The DC motor has been more popular in variable speed applications since DC motor speed is more easily changed than induction motor speed. In general, the speed of the induction motor is determined by four factors: the number of poles, the frequency and amplitude of the stator voltage, and the size of the load. When a fixed frequency source, such as a 60-Hz power system, is used, the speed can be varied by changing the applied stator voltage. In this method, however, the motor torque is proportional to the square of the applied voltage. At low speeds, the capacity to drive a load is greatly diminished. This variable voltage approach to speed controls is used to control shaded pole, permanent-split capacitor-wound rotor, and universal AC motors.

If the stator voltage is held constant and the stator frequency varied, the maximum torque produced by the motor will not decrease with the speed. Changing the frequency of the stator voltage produces the best induction motor speed control for polyphase induction motors. Varying stator frequency in the induction motor gives a speed range of 30:1 while producing rated torque.

The basic adjustable frequency speed control system has three parts: an ordinary three-phase motor, a power conversion unit, and an operator's control station. Normally, the motor used is a standard NEMA design B squirrel cage polyphase induction motor. The power conversion unit typically receives 220 V or 440 V, 3-phase, 60-Hz input power. The output power provided to the motor has a frequency that is variable from 0 to 60 Hz without steps. The output voltage is normally proportional to output frequency. This provides a constant ratio of voltage to frequency, as required by the characteristics of the induction motor. The operator's station has pushbuttons for starting and stopping the motor and an adjustment for setting the speed.

The power conversion unit in an adjustable frequency AC drive unit is often called a **converter.** The converter uses two separate stages of power conversion, as shown in Figure 11–1. The first stage is either a controlled or uncontrolled rectifier section. The purpose of the rectifier is to convert fixed-frequency AC power to DC power. The second stage is called an **inverter,** which converts DC to adjustable frequency AC power. By applying a rectifier and an

FIGURE 11–1 Converter block diagram

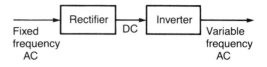

inverter in various configurations, the motor can be supplied with an adjustable frequency stator voltage over a wide range of frequencies.

In the inverter section, three approaches are used to convert DC to adjustable-frequency AC. First, the *variable-voltage inverter (VVI)* takes input power in the form of an adjustable DC source. This source presets the input DC voltage to provide the required output voltage amplitude from the inverter. In one type of VVI, a phase-controlled bridge rectifies the incoming AC voltage. The volts/Hz ratio is kept constant by changing the amplitude of the rectified DC as the frequency is changed. A second type of VVI replaces the phase-controlled rectifier bridge with a diode bridge and a DC regulator or chopper. This system, therefore, has a rectifier that is divided into two parts: the diode bridge, which converts fixed frequency AC to a constant voltage DC, and the regulator or chopper, which changes the constant DC voltage to a variable DC voltage. Normally, the VVI drives lack the ability to apply regenerative braking. Of the types of inverter drives, the VVI drives are the simplest in construction, used in industry for applications up to 400 HP.

The *current-source inverter (CSI)* takes input power for an adjustable current source, not a voltage source, as in the VVI. Except for the current source, the CSI drive is similar in construction to the VVI. The CSI drive, however, can apply regenerative braking to a motor.

The *pulse-width modulated (PWM) inverter* takes voltage from a fixed voltage source. The peak output voltage applied to the motor is, therefore, constant. The average value of the output voltage wave form is controlled by changing the width of the zero voltage interval in the output wave form.

Several techniques are used to control the speed of the induction motor: (1) variable voltage constant frequency or stator voltage control, (2) variable voltage, variable frequency control, (3) variable current, variable frequency control, and (4) regulation of slip power.

11–2.1 Stator Voltage Control

This form of AC induction motor control varies speed by controlling the stator voltage. This method should be familiar since it is virtually identical to the phase-control method used to control DC motor speed. Figure 11–2 shows two possibilities for accomplishing speed control by varying voltage. In Figure 11–2a, a triac is controlling a single-phase AC motor. Virtually the same type of control can be accomplished by connecting SCRs back-to-back, as shown in Figure 11–2b. In either case, varying the firing angle will change the rms value of the voltage applied to the stator. Increasing the firing angle will decrease the conduction angle, causing the rms voltage to decrease. Since a some-

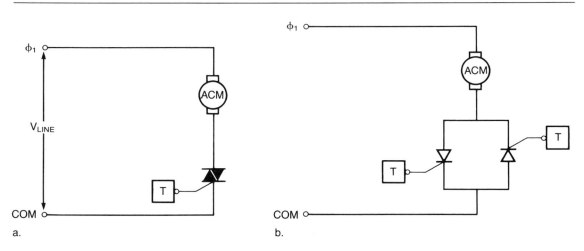

FIGURE 11–2   Single-phase AC motor control

what choppy sine wave is applied to the stator, the harmonic content of the applied voltage is high. Nevertheless, this type of drive has been successfully applied to motors larger than 100 HP.

Controlling an induction motor by varying applied voltage is different from controlling the DC motor. Changing the stator voltage changes the slip of the induction motor. Recall that slip is the difference between rotor and stator speed in an induction motor. In a normal induction motor, variable stator voltage speed control gives a maximum 10 to 15% change in speed. Use of a special high slip motor can give speed control over a 2 to 1 range. Figure 11–3 shows the speed torque curve when varying the stator voltage, controlling a fan load. Note that as the SCRs reduce the voltage from 100% (when the SCRs are on all the time), the speed-torque curve settles downward, producing a decrease in speed. This type of control works best with an induction motor that has a high rotor resistance. Motors with low rotor resistance have a speed torque curve that is more peaked, giving poorer control characteristics.

Figure 11–3 shows that the motor's torque production decreases with speed. In fact, the torque decreases as the square of the speed. This type of response is especially useful in those applications where the load torque decreases with a decrease in speed, as in fan and pump applications. It is not useful in applications where starting torque is high.

Compared to inverter circuits that control speed by changing the frequency of the voltage applied to the stator, the stator voltage control is inexpensive because it is simple. Although this circuit is simple in design, it has definite disadvantages and limitations. Although it seems that in theory at least, we could decrease the motor speed to zero, in practice, this is not the case. At the lower end of the speed range, the slip is high. At a 2 to 1 speed range, the slip at the lowest speed is about 0.5. At these low speeds, the rotor must dissipate a large fraction of the input power. These higher rotor losses at lower

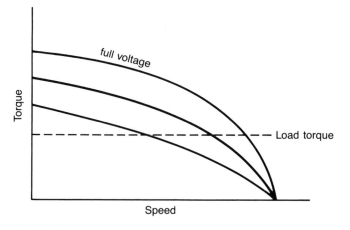

FIGURE 11–3 Speed-torque curves—varying stator voltage

speeds tend to overheat the motor. The reason for the tendency for the motor to overheat is twofold. First, the fan used to cool the motor is attached to the rotor. When the rotor turns slower, the fan is not as efficient at moving the air into the motor to cool it. Second, the heat generated by the rotor is difficult to dissipate. The heat is generated in the interior of the motor where it is difficult to remove efficiently. How much heat the rotor and stator can stand determine how low in speed a drive can take a motor. Another problem with the stator voltage speed control concerns vibration. As speed is reduced, motor vibration tends to increase. The abruptness and separation of the power pulses at low speeds cause the vibration. Power factor is also poor, which makes the drive less efficient.

Three-Phase Induction Motor Control The circuit shown in Figure 11–2 is a single-phase AC motor control, shown without the gate-triggering circuit. The concepts, applications, and problems discussed apply equally well to the three-phase AC induction motor. The arrangement shown in Figure 11–4 uses a triac in each of the three stator windings. Once again, although triacs are shown, back-to-back SCRs are often used, especially when the triac cannot provide enough current.

The circuit adjusts the trigger angles of three line-commutated triacs. Each triac controls the voltage applied to one phase. Since each phase in three-phase AC is displaced from each other by 120°, the same number of degrees displacement is maintained between the gate pulses delivered to the triacs. When the triacs are gated on at the beginning of each alternation, the full amount of the voltage is applied to each stator winding. This would correspond to a firing angle of about 0° and a conduction angle of 180°. As the conduction angle is decreased and the firing angle increased, the rms value of the stator voltage decreases. When the conduction angle is 0° and the firing angle is 180°,

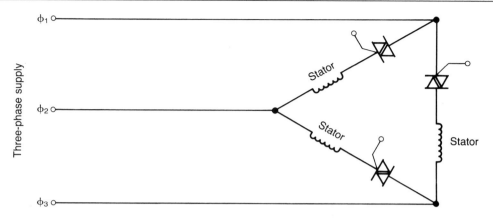

FIGURE 11–4 Triacs controlling three-phase motor

the rms voltage applied to the stator is zero. In this way, the voltage applied to the stator can be varied in a linear fashion.

Alternate arrangements of the thyristors are shown in Figure 11–5a,b. Figure 11–5b shows a different type of delta connection, while 11–5a shows a wye connection. Some variable-voltage speed controls use a feedback arrangement to keep speed constant. The speed of the motor is converted to a voltage, either through a sensor/**voltage-controlled oscillator** or a generator attached to the rotor shaft. In either case, the output voltage will be proportional to the speed of the rotor. The tachometer output voltage is then compared to a DC reference voltage, which represents the correct speed. (In process control terminology, the reference voltage would be called the **set point.**) The difference between the reference voltage and the tachometer output voltage is called the **error voltage.** This error voltage controls the firing angle of the thyristors. An example at this point will help our understanding of this circuit. Let us say that the load torque increases, causing the motor to slow down. The output voltage from the tachometer will then be less than the reference voltage. This difference in voltage makes the firing angle decrease, increasing the conduction angle. The increase in conduction angle increases the rms value of the voltage applied to the stator. The increase in stator voltage will increase the speed to a value close to what it was before the increased load.

The disadvantage of this scheme of control is poor efficiency at low speeds and poor power factor, as in the single-phase motors. Generally, it is used for pump or blower type loads. In these applications, the torque is low at starting but increases, typically as the square of the speed.

11–2.2 Regulation of Slip Power

This type of drive requires a wound-rotor induction motor. The slip power of the rotor is first rectified by a diode rectifier (Figure 11–6). Next, it is applied back to the AC line through a line-commutated inverter. Lowering the speed of

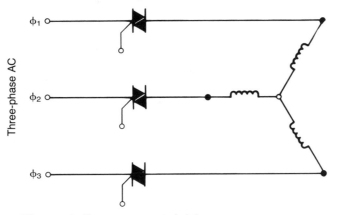

a. Triacs controlling wye-connected stators

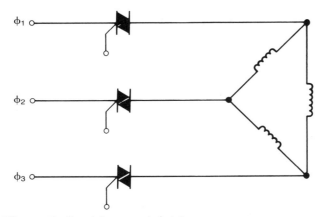

b. Triacs controlling data-connected stators

FIGURE 11–5 Triacs controlling three-phase motors

the motor will increase the slip, which will increase the current caused by the slip. The slip-current energy is put back onto the line by the inverter. The power factor of the drive system is poor at low speeds because lagging reactive power is drawn by the machine. The power is used by the motor to establish the magnetic flux in the air gap and by the thyristor inverter for line commutation. At full load and rated speed, the power factor is approximately 0.7, but it decreases as torque and speed decrease. This scheme is used for large horsepower pump and blower type applications where a limited range of speed control is required. The previous two schemes make use of SCRs.

11–2.3 Variable-Voltage, Variable-Frequency Control

This type of induction motor speed control uses a circuit called an inverter. The purpose of the inverter is to control the speed of the motor by adjusting the

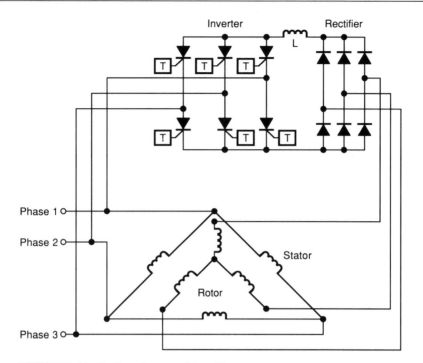

FIGURE 11–6 Speed control by slip power recovery

frequency. To produce a constant torque for the motor, the inverter drive must keep a constant V/Hz ratio. The way in which the inverter adjusts the frequency and voltage is determined by the particular type of inverter used. The variable-voltage, variable-frequency drive will be discussed first.

The variable-voltage, variable-frequency inverter is also known as a voltage-fed inverter or, simply, as a variable-voltage inverter (VVI). The VVI can be further broken down into two types: six-step (quasi-square wave inverter) and pulse-width modulated inverters.

Six-Step Inverter Figure 11–7a shows the power circuit of a three-phase inverter. A three-phase bridge rectifier converts AC to DC. The output voltage of the rectifier section is varied by a DC chopper. A thyristor chopper is preferable to a transistor chopper, which must use several transistors connected in parallel. Regardless of the type of chopper used, the chopper varies the constant DC voltage from the rectifier, which is then applied to the inverter. This type of inverter is called voltage fed because a large filter capacitor provides a stiff voltage supply to the inverter.

The inverter output voltage wave forms are not affected by the nature of the load. Figure 11–7b shows another way to vary the input voltage. In this method, the uncontrolled diode rectifier and the chopper regulator are replaced

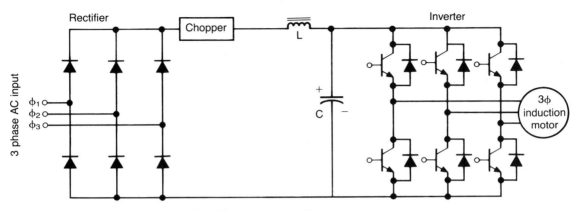

a. Six-step inverter with a conventional rectifier, chopper, and inverter

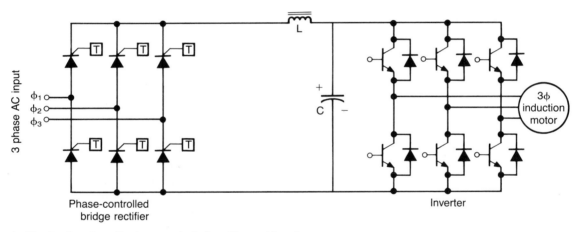

b. Six-step inverter with phase-controlled rectifier and inverter

FIGURE 11–7   Three-phase inverter

by a phase-controlled bridge rectifier. The principle of the variable-voltage, variable-frequency speed control method is shown in Figures 11–8 and 11–9.

The motor used in this drive has a low slip characteristic that improves efficiency. The speed of the motor can be changed by simply varying its synchronous speed. Varying the inverter frequency changes the synchronous speed. As the frequency is increased, however, the machine air gap *flux* falls, causing low developed torque capability. The air gap flux can be maintained constant, as in a DC shunt motor, if the voltage is varied with frequency so that the ratio remains constant.

Figure 11–8 shows the desired voltage-frequency relationship of the motor. Below the base frequency, the air gap flux is kept constant by the

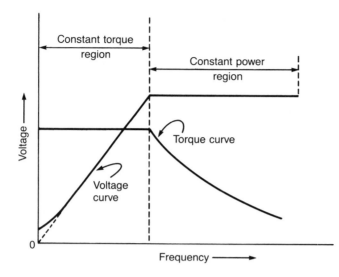

FIGURE 11–8 Voltage-frequency curve of induction motor

constant V/Hz ratio, which keeps the torque constant. At a very low frequency, the stator resistance is greater than the leakage inductance. To counter this effect, additional voltage is applied. At the base frequency, the input voltage regulator establishes full-motor voltage. Beyond this point, as frequency increases, the torque decreases because of loss of air gap flux. From this point on, the machine operates in a constant horsepower mode, as shown in Figure 11–9. In the constant horsepower mode, each torque-speed curve corresponds to a particular voltage and frequency combination at the machine terminal.

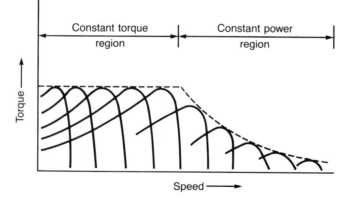

FIGURE 11–9 Torque-speed curves of induction motor with variable-voltage, variable-frequency power supply

Torque-Speed Relations Assuming constant air gap flux, the rotor EMF is proportional to the amount of slip. In a purely resistive rotor circuit, the current and torque are also proportional to slip. The induction motor, therefore, has a shunt torque-speed characteristic, which means that the speed decreases linearly with increasing load torque. This analysis is valid for low slip operation when the rotor or slip frequency is only a few Hz. At these low frequencies, the rotor leakage reactance is negligible, but it becomes more significant as the speed falls. The rotor current is reduced as a result of the increased rotor impedance. In addition, the current lags the EMF by the phase angle θ of the rotor circuit. Thus, if I_2 is the rms current in a rotor conductor and θ is the rotor phase angle, the torque is determined by the rotor current $I_2 \cos \theta$. The rotor torque (T) is also proportional to the air gap flux per pole ϕ and,

$$T = K_T \, \phi \, I_2 \, cos \, \theta$$

where

K_T is a constant.

We have seen in a previous chapter that the air gap flux is proportional to the V/Hz ratio and that the torque is proportional to the V/Hz ratio squared and the slip speed. Recall that the slip speed is the difference between rotor and stator speed.

Controlling the Motor A block diagram of an AC motor speed controller is shown in Figure 11–10. The system receives a nominal AC input voltage that is converted to a variable DC output voltage. The output voltage is applied to a voltage-controlled oscillator that, in turn, produces a frequency proportional to the DC power supply output voltage. The output of the voltage-controlled oscillator is then used to drive the six-phase logic that will provide properly-timed pulsed outputs to the optical coupler, buffer drivers, and power inverters.

Figure 11–11 shows a simplified six-step inverter diagram that will be used to show the proper switching sequence. Each of the switches shown in Figure 11–11 is actually a transistor or thyristor. The output voltage and current for a resistive load (connected in place of a motor) is shown in Figure 11–12. The current wave forms consist of six distinct steps when the switches are properly sequenced—hence, the name six-step.

Each of the voltages applied to the three phases is displaced 120° from each other, as shown in Figure 11–12. This figure shows the line-to-line voltages, V_{AB}, V_{BC} and V_{CA}. These voltages were found by adding the voltages algebraically. During periods *1* and *2*, the voltage from A to B = +V, since B is at the −V potential. During period *3*, the voltage A to B is 0 V, since both A and B are at +V. In this way, a six-step wave form is achieved.

The output AC voltage can be changed by varying the input DC voltage. The output frequency can be varied by varying the switching frequency of the

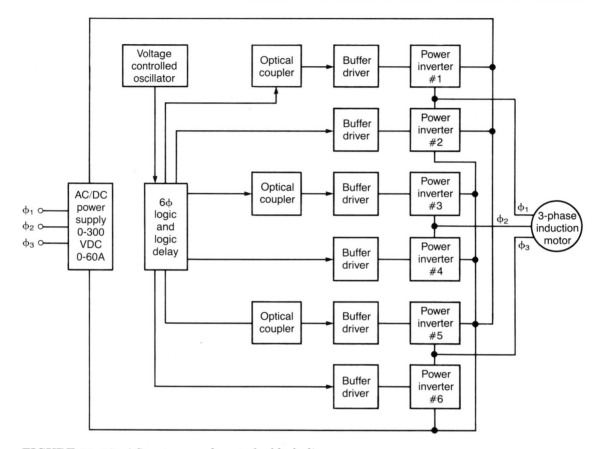

FIGURE 11–10 AC motor speed control—block diagram

transistors (*S1* through *S6*). Typically, the maximum frequency for motor speed control using a six-step inverter is 200 Hz.

Figure 11–12 shows that at any time three switches (transistors) are conducting, one is conducting in each leg of the bridge, and the successive legs are switched with delays of 210°. As shown in Figures 11–13 and 11–14, transistors *Q1* through *Q6* theoretically conduct for 180°.

However, in a practical situation, it is necessary to provide some time delay (typically 10° to 15°) between the positive-to-negative transition period in the phase current. This time delay enables the complementary transistor (Q_1 is complement to Q_2, etc.) to turn off before its opposite member turns on. This action prevents cross-conduction and eventual destruction of the power transistors. Therefore, the maximum conduction time will be 165° out of a 360° period. The diodes connected in parallel with each transistor conduct current when the transistor is turned off, represented by $-I_c$ in Figure 11–15.

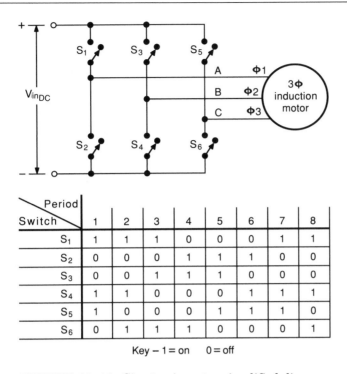

Period Switch	1	2	3	4	5	6	7	8
S_1	1	1	1	0	0	0	1	1
S_2	0	0	0	1	1	1	0	0
S_3	0	0	1	1	1	0	0	0
S_4	1	1	0	0	0	1	1	1
S_5	1	0	0	0	1	1	1	0
S_6	0	1	1	1	0	0	0	1

Key – 1 = on   0 = off

FIGURE 11–11 Six-step inverter simplified diagram and switching sequence

Another type of VVI is illustrated in Figure 11–16. This diagram is a simplified diagram. Single-phase power, 220 VAC, is applied to the bridge rectifier (D_9—D_{12}) through fuses F_1 and F_2 and line reactor L_1. The DC voltage produced by the bridge is filtered by filter capacitor C_4. A fixed value, filtered DC voltage is then found on lines *1* and *5*. Note that no SCRs appear across the AC line. This design is, therefore, more immune to line noise, hash, and spikes that might affect the firing of the SCRs in the DC section.

Six SCRs (*SCR1*—*SCR6*) and six diodes (*D1*—*D6*) carry the current in the adjustable voltage bridge. A pair of SCRs is switched on and turned off for each phase. This action causes each phase to become alternately positive-negative-positive-negative, etc. *SCR1* and *SCR2* are used in phase *A, SCR3* and *SCR4* in phase *B,* and *SCR4* and *SCR5* in phase *C.* The energy for the adjustable voltage bridge comes from capacitor C_3. In other words, the SCRs are drawing power from C_3 to run the motor, and they are continually trying to discharge C_3. The capacitor C_3 is charged by the fixed voltage bridge.

Each pair of SCRs controlling each phase is turned on and commutated off in the proper timing sequence to supply three-phase power to the motor by the main control circuit board (not shown in the simplified diagram). The faster the switching on and commutating off, the higher the frequency.

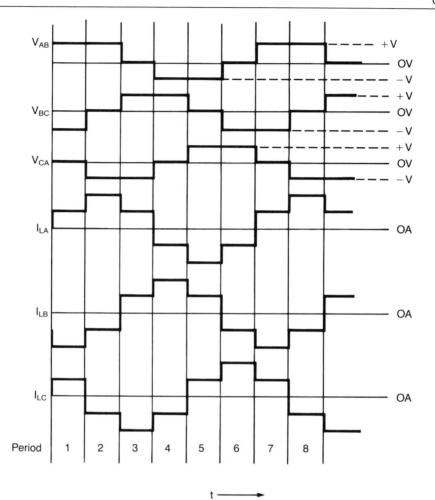

FIGURE 11–12 Switching sequence synchrogram

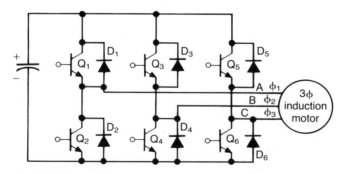

FIGURE 11–13 Six-step inverter using transistors as switches

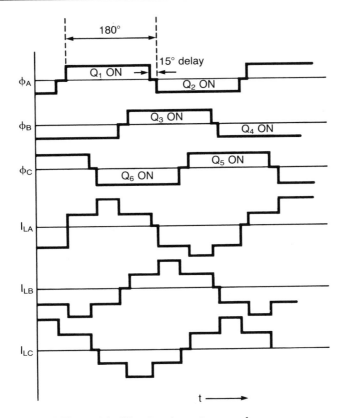

FIGURE 11-14 Six-step inverter synchrogram

FIGURE 11-15 Typical collector current with motor load

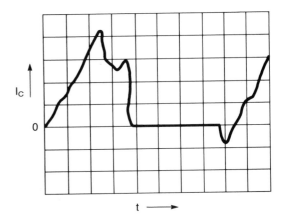

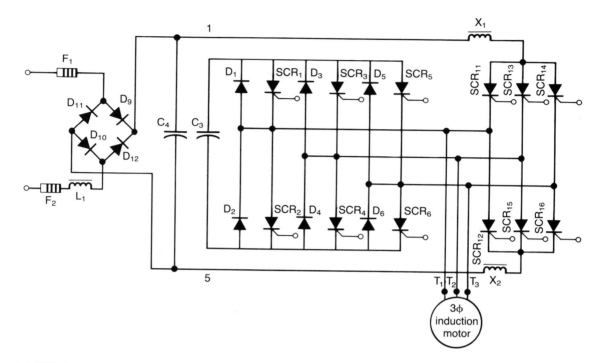

FIGURE 11–16 VVI AC motor drive using SCRs as switches

The six SCRs in the fixed voltage bridge (*SCR11—SCR16*) operate in parallel with their equivalents in the adjustable voltage bridge (*SCR1—SCR6*). The purpose of the fixed voltage bridge is to furnish energy to C_3 as the motor load uses it and to commute (turn off) the SCRs in the adjustable voltage bridge.

When an SCR in the adjustable voltage bridge, for example *SCR1*, is turned on, its equivalent in the fixed voltage bridge (*SCR11*) can be turned on at 10 kHz rate. It will be turned on for one pulse any time C_3 voltage is too low. When it is turned on for a pulse, energy comes from the DC line (lines *1* and *5*) through X_1 or X_2. Some of this energy goes to the motor, and the excess goes through the back diode (*D1—D6*) and helps recharge C_3. If C_3 voltage continues to be too low, another pulse of energy will be called for and once again the SCR in the fixed voltage bridge will be turned on. Up to 10,000 pulses of energy per second can be obtained from each SCR in the fixed voltage bridge in this manner.

To commute off an SCR in the adjustable voltage bridge, for example *SCR1*, its gate is turned off. Then its mate in the fixed voltage bridge (*SCR11*) is turned on for a pulse, causing current to flow. Some current flows to the motor; the excess flows through the back diode *D1*. This action shunts *SCR1*, causing it to stop conducting.

The SCRs in the fixed voltage bridge turn themselves off naturally after each pulse because they conduct into a tuned circuit. The voltage rises at the end of the pulse to block any further current flow.

Troubleshooting VVI Inverter Drives   Troubleshooting inverter-based AC drives is not as difficult as it may seem at first. Many manufacturers provide schematic diagrams, troubleshooting hints, and technical advice and assistance to plant maintenance personnel. Most problems after the installation of a new drive are caused by improper wiring. Before proceeding with any troubleshooting, be sure to check and double-check to see that the wiring is correct. Also, check the incoming AC voltage. If rechecking the wiring does not solve the problem, return all adjustments to their original positions and proceed. A safety note should be mentioned here. NEVER touch the inside of a drive unless you disconnect the main power from the drive and discharge all capacitors. Although the drive may not function correctly, dangerous voltages may still be present on internal components.

Through proper fault isolation procedures, you may determine that the SCRs or diodes need to be checked. They can be checked with an ohmmeter on its RX100 scale. First, disconnect AC power. Next, discharge the large capacitors (like C_3) by applying a 5-W, 200-Ω resistor across their terminals or else waiting five minutes for them to discharge themselves. Disconnect the motor at terminals T_1, T_2 and T_3. Check the resistance between the large terminal on each SCR (thyristor) and the heat sink on which it is mounted. There are 12 SCRs. A low resistance reading indicates a possible problem. Note that *SCR1* through *SCR6* have diodes in parallel with them. This gives a low resistance reading in one direction and a high resistance reading in the other direction. Be sure to check all 12 SCRs. Note those with possible problems. Remember to disconnect the motor load by removing wires connected to *T1*, *T2*, and *T3*.

On those still showing low resistance, unbolt SCR terminal connections and again check resistance from the large terminal to the heat sink. Any SCRs still showing low resistance are shorted and should be replaced. At this time, also check the resistance of all diodes mounted on the same heat sink with faulty SCRs. Replace diodes that have a low forward to reverse resistance ratio (less than 10:1). A short circuit from anode to cathode is a common semiconductor fault. A guide to checking semiconductors with an ohmmeter is included in the Appendix of this text.

When replacing SCRs and diodes, it is very important that proper torque and silicone heat transfer compounds be applied. Many manufacturers recommend using a specific heat transfer compound. For example, a maintenance manual may suggest using General Electric Silicone G624 or Dow Corning Silicone Compound #4 on both the bottom of the semiconductor and on the heat sink to assure good heat conductivity.

When replacing and checking SCRs and diodes, remember that it is important that the bolts attaching the case to the sink have the proper tightening torque. A manufacturer may, for example, recommend the following tighten-

ing torques. The stud-mounting nuts should be tightened with a torque wrench
to the torques specified:

SCRs and diodes with a 1/4 inch stud	25 lb · in
Diodes with 10 to 32 studs	25 lb · in
SCRs with 1/2 inch studs	137 lb · in

Pulse-Width Modulated VVI   In the six-step VVI control of induction motors, there are four main disadvantages:

a. The power must flow through two or three sets of semiconductor
switches (depending on the configuration). This increases the cost of the drive
and decreases the efficiency.

b. The separate inverter increases the physical size of the drive package.
The size benefits of the AC induction motor are offset by the larger inverter
section.

c. The shape of the six-step output wave form from the inverter leads to
decreased performance in the motor. The additional losses result from the high
harmonic content of the resulting motor current, particularly at low speed
operation.

d. The six-step wave form produces appreciable motor *cogging* (pulsating
torque) at low speeds. This causes the motor to run in a jerky fashion, rather
than smoothly.

These disadvantages have led to a better form of power control, called
pulse width modulation (PWM).

In the variable-voltage, variable-frequency inverter drive previously described, the DC link voltage (from the capacitor) can be kept constant by a
diode rectifier, as shown in Figure 11–17. The fundamental frequency output
voltage can be controlled electronically within the inverter by using a pulse
width modulation technique. In this method, the transistors are switched on

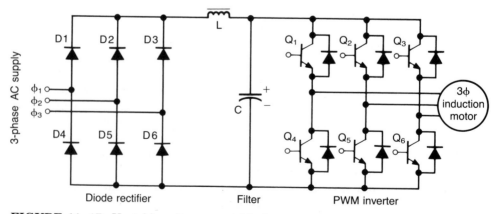

FIGURE 11–17   Variable-voltage, variable-frequency inverter

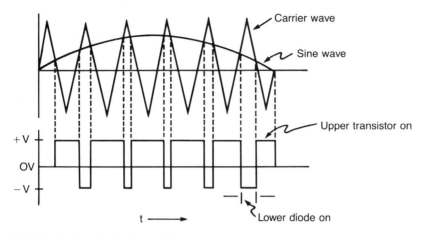

FIGURE 11–18 Pulse-width modulation by sine wave

and off many times within a half-cycle to generate a variable-voltage output, which is normally low in harmonic content. There are several PWM techniques. Sinusoidal PWM, shown in Figure 11–18, is common. A triangle wave is compared with the sine wave signal, and the crossover points determine the turn-off points for the transistor. Except at low-frequency range, the triangle wave is synchronized with the signal. An even integral (multiple of three) ratio is maintained to improve the harmonic content. The fundamental output voltage can be varied by variation of the modulation. PWM voltage control is applicable in the constant torque region (see Figure 11–18). In the constant power region, the operation is identical to that of a square wave drive.

A typical triangle wave carrier frequency lies between 1.5 and 2.5 kHz. Low harmonics and small torque pulses in the low frequency region permit a wide range of speed control, from almost standstill to the full torque capability of the machine. This characteristic of the PWM inverter makes it attractive for a variety of applications, such as machine tools, textiles, printing, and paper mills.

11–2.4 Current Source Inverter (CSI)

The current source inverter (CSI), sometimes called the current-fed inverter, is very similar to the VVI circuit just discussed. Its name suggests that current, not voltage, is varied in this inverter. A large inductor is used in place of the large capacitor in the VVI.

The CSI requires a *stiff* DC current source as opposed to a voltage-fed inverter. In this case, stiff refers to the capability to provide a large amount of current without loading down the circuit. Figure 11–19 shows the power circuit of a current-fed inverter using power Darlingtons as switches. A phase-controlled rectifier generates variable DC, which is converted to a current source by connecting a large inductor in series. A diode rectifier, followed by a

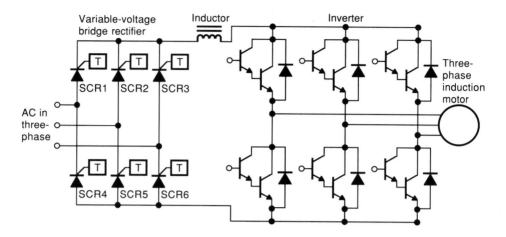

FIGURE 11–19 Variable-current, variable-frequency motor drive

DC chopper, can also make a variable DC source. The mode of control of the inverter could be either six-stepped or pulse-width modulated, similar to that of a voltage-fed inverter.

The diagram in Figure 11–20 shows a block diagram of a Model 1580 Graham adjustable frequency AC drive. This drive uses the CSI. The drive is made up of two main sections: the control section and the power section. The first part of the power section is the input bridge, which converts the incoming three-phase AC power to a fixed DC. The fixed DC is applied to the current source chopper. The chopper converts the fixed DC to a pulsating DC through a large inductor. The inductor becomes a source of current for the load, which is usually an induction motor. The last part of the power section is the inverter bridge. The inverter bridge directs the output of the chopper to the correct phase of the three-phase motor for the right amount of time. The control section has the regulating circuitry for voltage and current and the SCR gating logic circuitry.

The input bridge, shown in more detail in Figure 11–21, is a conventional three-phase diode bridge rectifier. It consists of diodes $D13—D18$. The purpose of the rectifier is to change the incoming three-phase power to a constant DC voltage bus. Capacitor C_1 represents an electrolytic capacitor bank. This capacitor is charged to bus potential. With a 460 VAC line voltage, C_1 will be charged to about 620 V.

The second section of the power circuit is the chopper. The chopper is an electronic switch that alternatively connects and disconnects the coils of a large inductor from the DC bus. A simplified diagram of the chopper is shown in Figure 11–22.

When the coils are connected to the constant potential DC bus, the current will increase. This position of the electronic switch is called increase or

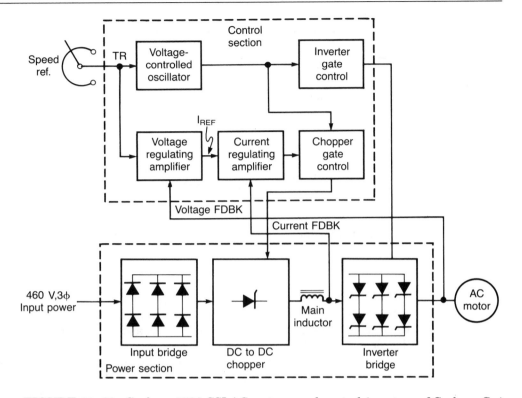

FIGURE 11–20 Graham 1580 CSI AC motor speed control (courtesy of Graham Co.)

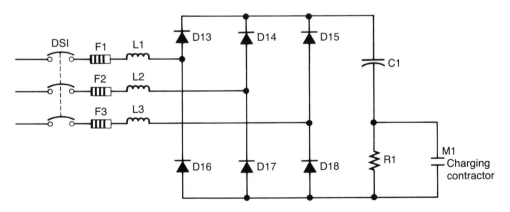

FIGURE 11–21 Input bridge, Graham CSI control (courtesy of Graham Co.)

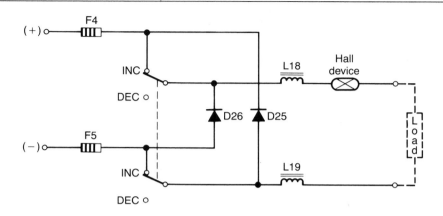

FIGURE 11–22 Chopper circuit, Graham CSI controller (courtesy of Graham Co.)

INC. When the switches are in the open position, the inductors *L18* and *L19* maintain the current flow without help from the DC bus. This takes energy out of the inductor and causes the current to decrease. This position of the switch is called decrease or DEC.

The load current is sensed by a Hall-effect device, which gives an output voltage proportional to the flux created by the load current. This proportional signal is then fed back to the regulator circuit where it is compared to a value that the system requires to make the load (motor) perform. The regulator circuits then vary the time spent in the decrease state versus the time spent in the increase state. This action controls the load current at the desired value.

The current-regulated section responds to changes in load impedance. If the load changes its impedance or becomes a short circuit, the current will not exceed the value the system demands. Any tendency for the current to rise too high will be sent back to the comparator and will cause less time to be spent in the increase state. The physical presence of the inductance of *L18* and *L19* prevents the current from changing more rapidly than the rate the regulator can cope with. Although mechanical switches are shown in Figure 11–22, actual switches are semiconductor devices.

The schematic in Figure 11–23 is a simplified schematic of the inverter section. It is a six-step inverter with *SCR1—6* switching the load current at the proper rate as determined by the control circuitry. The switching rate of the SCRs establishes the output frequency. Commutation capacitors *C28* through *C33* store energy necessary for turning off the SCRs by reversing their terminal voltage. Series diodes *D1* through *D6* isolate the capacitors from the load. Only two SCRs are on at any one time, with each one conducting the 120°. They are commutated when the adjacent SCRs in the next phase are fired in the order numbered.

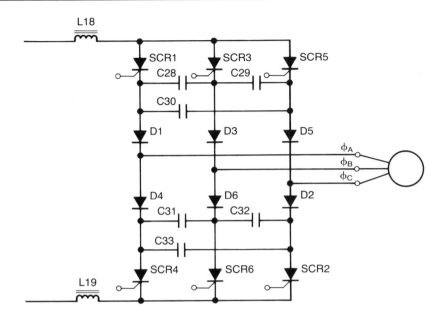

FIGURE 11–23 Inverter section, Graham CSI controller (courtesy of Graham Co.)

The inductance of the motor is a significant factor in the commutating scheme. The inductance stores energy necessary for commutation, and this energy charges the capacitors for the next cycle. There is also a precharge circuit consisting of a diode and resistance in series with each capacitor. The precharge circuit is present to insure that enough capacitor voltage is present to commutate the SCRs during starting and low frequency operation.

The chopper controls the current in the inverter coming from *L18*. Even when there are two parallel paths for current, the sum of the two can never be greater than the output current from the chopper. The inverter controls only the time that the current flows through each motor phase.

The DC voltage at the input terminals of the inverter will vary with the demand of the load. At no load, the voltage will be near zero. At rated load, it will be maximum.

The response of the motor and the load and the applied current and frequency determine the counter EMF (CEMF) of the motor. It is approximately sinusoidal with a spike occurring each time an SCR is commutated. Figure 11–24 shows a typical phase current and voltage in the inverter section.

We can see that current flows for 120° in the positive direction, ceases for 60° and repeats, but in the negative direction. The positive current for phase *A* is when *SCR1* conducts with either *SCR6* or *SCR2*. Negative current is drawn when *SCR4* conducts with either *SCR3* or *SCR5*. The sawtooth on top of the

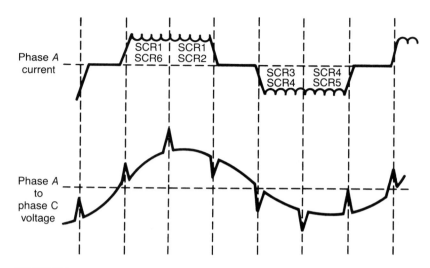

FIGURE 11–24 Synchrogram showing currents and voltages in inverter (courtesy of Graham Co.)

current coincides with the chopper changing between the increase and decrease states.

Some of the advantages of current-fed inverters are:

a. Rugged and reliable power circuit.
b. Lower peak current through the transistor.
c. Fault on the inverter side causes slow rise of fault current, which can be cleared easily.
d. Operates well at low speeds.

Some of the limitations of current-fed inverters are:

a. Limited frequency range of the inverter, and the starting torque developed is very low compared to that of voltage-fed machines.
b. Large size of the DC link inductance makes the inverter somewhat bulky.
c. Drive response is sluggish and tends to be unstable at light loads and high speeds. The current-fed inverters are used in medium to high horsepower ranges.

To realize best performance of the induction motor, a hybrid inverter can be used. This method involves a transition for operation of the inverter as a current source to operation as a voltage source. This has the good low speed performance of the current inverter and the good high speed performance of the voltage inverter.

11–3 SYNCHRONOUS MOTOR SPEED CONTROL

In a previous chapter, we learned that the synchronous motor runs at synchronous speed. Classification of the synchronous motor depends on how the rotor gets its flux. Small synchronous motors are usually either reluctance, hysteresis, or permanent-magnet types. Generally, induction motors are chosen over synchronous motors in small sizes, except where the application requires that motors run in synchronization. This is difficult to accomplish with induction motors. Applications that use synchronous motors, therefore, either run at a constant speed or use an inverter for speed control similar to the ones discussed for the induction motor. In applications under 50 HP, synchronous motors are rarely used unless the application absolutely demands it, due to the greater cost of the synchronous motor. The synchronous motor is also used where applications demand high power output, low speeds, or high efficiency. The synchronous motor has one of the highest motor efficiencies, between 92 and 96%. Synchronous motors are almost always used in applications demanding more than 1000-HP output power. In one recent application, a mill installed an 8-MW synchronous motor (over 10,000 HP). In addition, the synchronous motor is almost always used when speeds are lower than 500 rpm. In low speed applications, typically in milling applications, some form of speed control is sometimes necessary. Where low-speed synchronous motor speed control is necessary, the cycloconverter is used.

11–3.1 Cycloconverters

The *cycloconverter* is a circuit that converts fixed AC from the mains directly to a low frequency AC. Unlike the rectifier inverter, in the cycloconverter no other form of energy besides AC is used. While both the inverter and cycloconverter produce adjustable AC voltage and frequency as their final product, they do it differently.

The schematic diagram in Figure 11–25 shows a single-phase cycloconverter. This circuit is similar to a single-phase, center-tapped rectifier. A close look will reveal that two sets of controlled rectifiers (SCRs) are used instead of one pair of diodes. The two sets of SCRs allow current reversal through the load. Current reversal is necessary in an AC load.

SCR1 and *SCR2* provide positive current to the load, while *SCR3* and *SCR4* provide negative current. If *SCR1* and *SCR2* are triggered, the DC output voltage shown in Figure 11–26 will have the positive polarity shown on the left side of the output wave form figure. If *SCR3* and *SCR4* are triggered instead, the output polarity will be reversed, as shown in the right side of the output wave form. Although not shown, assume that a triggering circuit is provided that can change the firing points for the SCRs by applying a control voltage. If we use a sine wave control voltage that has a frequency lower than the line frequency, the firing angle will change with the lower frequency mod-

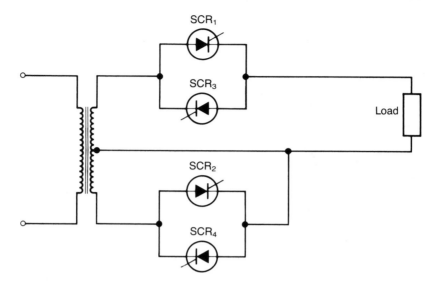

FIGURE 11–25 Single-phase cycloconverter

ulating frequency. You can see this occurring in the output wave form in Figure 11–26. Note that the firing angle starts out large, decreases to a minimum at the peak of the modulating wave, and then decreases. If we look at the average value of the output wave form, we can see that it reproduces a load voltage similar to the low, modulating frequency. A filter is usually used to eliminate the ripple in the wave form. It should be obvious from this circuit

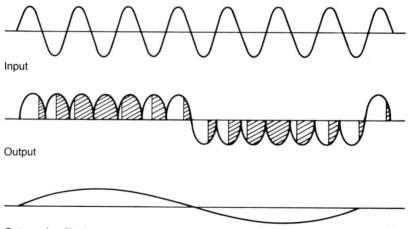

Input

Output

Output after filtering

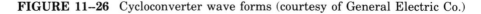

FIGURE 11–26 Cycloconverter wave forms (courtesy of General Electric Co.)

that the control of the final output wave form frequency is achieved by varying only one parameter, the firing points of the SCRs.

From this discussion, we can see that the low frequency output voltage is made from higher frequency line voltage. The basic circuit can be extended to three-phase motor applications. The basic power circuit for a three-phase motor is illustrated in Figure 11–27.

The cycloconverter has several characteristics that bear mentioning. First, it is generally used as a step-down frequency changer. We can conclude this from the fact that the low frequency AC is made up of sections of the higher frequency input voltages. Second, the direction of rotation can easily be changed by reversing the phase firing sequence. Third, we can see a clear economic disadvantage of this type of drive over the inverter when cost is a consideration. The cycloconverter needs more power semiconductors than the inverter. We can see from Figure 11–27 that a fixed three-phase input to variable three-phase output requires 18 semiconductors. A comparable rectifier-inverter requires only 12 thyristors. The cycloconverter also generates a lot of noise on the line because many SCRs are turning off and on. Nevertheless, as semiconductors become more inexpensive and higher in power, the cycloconverter is likely to become more popular.

Although the cycloconverter is used in synchronous motors, it is also used for induction motors. As we have seen, the induction motor does not run well at

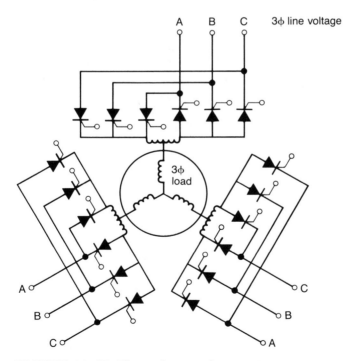

FIGURE 11–27   Three-phase cycloconverter

low speeds with the basic inverter drive. With a cycloconverter, however, the induction motor can run from 0 rpm to one-third of its full-load speed without any ill effects. In addition to running the induction motor efficiently and without clogging, cycloconverter control provides maximum torque from the induction motor at slow speeds. Because of its high torque/low speed operation, the induction motor/cycloconverter combination can be used in high starting torque applications that were previously in the exclusive domain of the series DC motor.

11−4 NASA POWER FACTOR CONTROLLER

Of great concern today is the shrinking supply of energy and the search for ways to conserve our energy assets. With the rapidly rising costs of petroleum products and the increasing problems with nuclear power, all electrical power control systems should be energy efficient. One area in which technology can make significant contributions is improving the efficiency of electric motors.

Billions of electric motors are in use today in the world, with over 50 million new motors made every year. Large numbers of electric motors use large amounts of electrical power. Even a small improvement in electric motor efficiency can give significant gains in energy conservation. NASA estimates that a 4% reduction in the amount of power consumed by electric motors could save 250,000 barrels of oil in the USA alone.

A circuit developed by Frank Nola of NASA has been used to increase energy efficiency. This circuit improves the power factor of induction motors operating at less than full load. When operated on light loads, induction motors become very inefficient. The no-load current of an induction motor can be 50 to 90% of full-load current. This means that lightly loaded induction motors waste large amounts of energy since no productive work is done. These high no-load currents cause heat losses in the motor and in the power distribution system.

Since the current remains high in an unloaded induction motor, the phase angle between voltage and current shifts with the motor load. The NASA circuit detects this shift and adjusts the phase angle shift to improve the power factor. The graph shown in Figure 11−28 shows the kind of efficiency improvement made possible with the NASA power factor control circuit (PFC). The PFC circuit senses the phase shift between voltage and current and converts this information into a voltage signal. Another reference voltage, proportional to the desired phase angle, is compared to the first voltage signal. The difference between these two voltages is used to generate a trigger signal to a phase control semiconductor device, usually a triac or an SCR. The ON time of the power semiconductor forces the phase angle of the current and voltage to remain at the desired value.

The original PFC circuit was designed for use with small, fractional HP induction motors. This type of motor is used on home appliances, air condition-

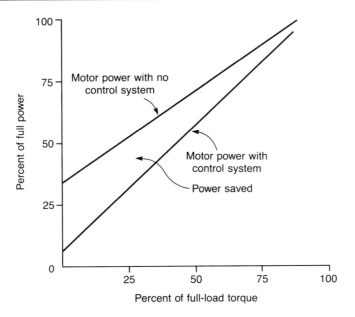

FIGURE 11-28 Efficiency improvement with a power factor controller

ers, and small machine tools. Recently, more work has been done on larger, three-phase industrial motors. A recent study has shown that at light loads, the NASA PFC can save more than 50% of the power used in three-phase motors in the 1 to 5-HP range.

The PFC is now being used in more industrial applications, such as large air conditioning systems, pumps, fans, and large machine tools. The higher power applications have forced changes in the original NASA design, although the basic principle remains the same. The original circuit used a triac as the power control device. As we have seen, triacs have voltage ratings under 800 V and current ratings under 50 A. This makes it difficult to use triacs in 440-V, three-phase applications, which are common in industry. Fortunately, SCRs are available in a wide variety of ratings, which are suited to almost any industrial application. Since triacs are bidirectional and SCRs are unidirectional, two SCRs must replace one triac in PFC circuits.

The NASA PFC seems to be at least one answer to the continuing problem of energy conservation.

11-5 SMARTPOWER

Most modern industrial motor controls use a combination of ICs for the control circuits and high-power semiconductors to drive the motors. Power semiconductors switch at high voltage and current levels while the IC control circuits use low voltages and currents. Until recently, the low-power control circuits have been unable to exist in the same high voltage and current AC environment. Cumbersome control circuitry has been required to connect the two classes of devices. This circuitry tends to be expensive, bulky and noisy.

Today, smartpower gives motor controls the combination of low-power control circuitry and high-power semiconductors in the same package. This combination greatly reduces the size and complexity of motor controls. Figure 11–29 shows a smartpower device made by General Electric, called a GESmart power module. This module combines an insulated gate power transistor (IGT), a fast-recovery flyback diode, and a high-voltage IC. Although intended for PWM motor drives, the unit can provide up to 5 A continuous current at 100° C. The high voltage IC performs the functions of control, level shifting, and the IGT/driver protection. The internal protection circuitry prevents both IGTs from conducting at the same time, which could short the power supply to ground.

With a bias voltage connected to pin *1*, a high voltage on pin *3* (TOP) turns on the upper IGT. A high voltage on pin *4* (BOT) turns on the bottom IGT. When the output of pin *2* goes low, a fault condition has occurred. This output pin may be used to turn on an alarm or remove power from the motor.

A simple full-bridge, bidirectional DC motor control is shown in Figure 11–30. Let us assume that current flow from right to left indicates a CW rotation. The motor will turn when the top IGT of module *1* and the bottom IGT of module *2* both turn on. The speed may be controlled by varying the time on with respect to the time off. The higher the time on, the higher the average voltage and the faster the motor will turn. Reversal of rotation will occur when the bottom IGT of module *1* and the top of IGT of module *2* turn on. Current will then flow from left to right through the motor.

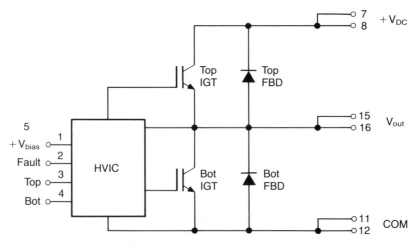

FIGURE 11–29 GESmart™ smartpower module

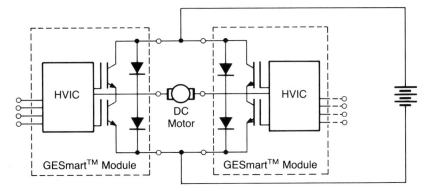

FIGURE 11–30 Full-bridge bidirectional DC motor drive using GESmart™ smartpower modules

CHAPTER SUMMARY

- The speed-torque characteristics of an induction motor can be controlled by adjusting the stator voltage, the frequency of the AC source, the resistance of the rotor, and the slip power in the rotor.
- Stator voltage control is limited in the amount speed can be changed. This type of control also reduces the torque produced by the motor when voltage is lowered.
- The variable-voltage inverter changes a DC input into a variable-frequency AC voltage that is then applied to the stator of the induction motor.
- The current-source inverter changes a DC input into a variable-frequency AC current that is then applied to the stator of the induction motor.
- Both CSI and VVI require DC inputs.
- The cycloconverter, often used to drive the AC synchronous motor, changes fixed-frequency AC into variable-frequency AC of a lower frequency.

QUESTIONS AND PROBLEMS

1. Describe the three basic methods for changing the speed of an AC induction motor listed below:
   a. variable-voltage, constant frequency
   b. variable-frequency, variable voltage
   c. variable number of poles.

2. List the advantages of the three types of induction motor speed control methods listed in problem number 1.

3. Describe the function of the inverter and what part it plays in an AC motor speed control system.

4. Describe the following types of adjustable-frequency speed control by drawing and explaining at a functional block diagram level:
   a. variable-voltage inverter
   b. current-source inverter
   c. pulse-width modulated inverter.

5. Define the volts/Hz ratio and explain how this ratio is important in AC induction motor speed control.

6. List several disadvantages of variable-voltage, constant-frequency induction motor speed control. In what application is this type of control used? Are there any advantages to this type of control? If so, what are they?

7. List the four main disadvantages of VVI speed control.

8. When discussing the CSI, we say we need a stiff current source. Define what is meant by the term stiff and explain why supply is necessary for a CSI.

9. List at least three advantages of a CSI. List at least two disadvantages.

10. How does the cycloconverter change the speed of a synchronous motor? Can a cycloconverter be used to control the speed of an induction motor?

12

DIGITAL MOTOR CONTROL

At the end of this chapter, you should be able to

- Draw and explain the meanings of the relay logic circuits for the following logical functions:
  - AND
  - OR
  - NOT
  - NOR.
- Explain the difference between TTL digital logic and hard-wired relay logic in motor control.
- Compare and contrast the following methods of digital motor control:
  - hard-wired relay logic
  - TTL digital logic
  - microcomputer
  - programmable controller.
- Draw a basic functional lock diagram of a DC motor speed control using feedback, explaining the function of each block.
- List and explain the function of each of the five parts of a programmable controller.
- List and describe three methods of programming a programmable controller.

12-1 INTRODUCTION

Since the advent of the microprocessor in the mid-1970s, many motor controls have used digital technology. This should not be surprising, since the relay circuits use an on-off type of control well suited to digital logic circuitry. The ladder diagrams used to represent industrial control systems are said to use *relay logic.* This chapter will examine how relay logic corresponds to the digital logic in such semiconductor logic circuits as TTL and CMOS. We will then take a simple relay logic circuit and note its similarity to a functionally equivalent semiconductor logic circuit. In principle, any relay logic circuit can be converted to a functionally equivalent semiconductor digital logic circuit. Both of these types of circuits suffer one major disadvantage in that they cannot be easily changed.

The logic of a semiconductor or relay logic circuit can be performed, in turn, by a microprocessor circuit. Unlike the relay and semiconductor digital logic circuitry, the logic of the microprocessor can be easily changed. The logic in a microprocessor circuit is changed by modifying the computer program. The computer program, usually called software, can be written in a high-level language like BASIC or PASCAL, or it can be written in machine code. Machine codes are program commands in binary format. Machine codes are called low-level languages and are different for each type of microprocessor. Programming a microprocessor, whether in a high- or low-level language, requires a high degree of skill. It is not practical for users of a microprocessor to be required to learn how to program it every time the program needs to be modified. This factor and others led to the development of an easily programmed, microprocessor-based device that could perform relay logic operations. This device was called the ***programmable controller (PC).*** In this chapter, we will also discuss and show examples of microprocessor-based motor control programs as well as the popular programmable controller.

12–2 RELAY LOGIC

12–2.1 Basic Logic Gates

We stated previously that relay and semiconductor digital logic are functionally equivalent. This is demonstrated in Figure 12–1. In Figure 12–1a, current will flow through the relay coil (C) if and only if switches A and B are closed. The switches are electrically in series. Current can flow through the relay only if there is a complete path for current through the series circuit. This corresponds to the *AND* digital logic function. Both A and B are required for C. The corresponding semiconductor digital logic gate is shown in Figure 12–1b, with its truth table in Figure 12–1c. Highs at both A and B will give a high out at C. Any other combination of inputs results in a low out. Both IEEE (Institute of Electrical and Electronics Engineers) and NEMA have attempted to standardize logic symbols. The ANSI symbol for the *AND* function is shown in Figure 12–1d, while the NEMA symbol is found in Figure 12–1e. Since some of these new logic functions are appearing in industrial diagrams, you should be familiar with them.

A	B	C
1	1	1
1	0	0
0	1	0
0	0	0

a. Relay logic circuit diagram
   AND function

b. Old IEEE AND logic gate
   schematic diagram

c. AND logic function truth table

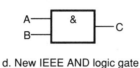

d. New IEEE AND logic gate
   schematic diagram

e. NEMA AND logic function diagram

FIGURE 12–1 Various AND logic symbols

The OR function can be seen in Figure 12–2 in relay logic form. The relay coil C will energize if either A or B closes. Note that the switches are electrically in parallel. Closing one of the switches allows current to flow, completing the circuit. The equivalent semiconductor gate is shown in Figure 12–2b. As the truth table shows, a high at either A or B will give a high output at C. The output is low only when both inputs are low. The IEEE symbol for the OR function is shown in Figure 12–2d, while the NEMA symbol is found in Figure 12–2e.

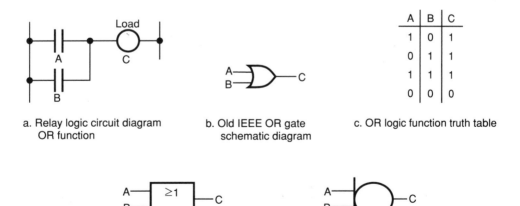

A	B	C
1	0	1
0	1	1
1	1	1
0	0	0

a. Relay logic circuit diagram
   OR function

b. Old IEEE OR gate
   schematic diagram

c. OR logic function truth table

d. New IEEE OR function diagram

e. NEMA OR logic function diagram

FIGURE 12–2 Various OR logic symbols

The final logic expression needed to complete the three basic logic classifications is the *NOT* circuit. The relay logic NOT circuit appears in Figure 12–3a. Recall that the NOT operation is a complement operation in Boolean algebra. This means that the output of this circuit will be the complement of the input. Note that the *A* coil contacts are normally closed. This means that the *A* coil (not shown) is NOT energized when the *B* coil is energized. Conversely, when the *A* coil is energized, the *A* contacts will open and the current will not flow through *B*. We can see the corresponding NOT semiconductor function in 12–3b with the truth table in 12–3c. The semiconductor NOT gate is also called an inverter. The IEEE symbol for the NOT function is shown in Figure 12–3d, while the NEMA symbol is shown in Figure 12–3e.

Most of the logic functions can be made with these three functions. For example, the *NOT* function can be combined with the *AND* function to give the *NAND* function. The relay logic equivalent of the *NAND* function can be seen in Figure 12–4a. The top part of the relay circuit you will recognize as the *AND* part. The bottom part represents the *NOT*. The combination of these two relay circuits is the *NAND*. When both switches *A* and *B* are closed, current flows through *C*. Energizing coil *C* opens the *IC* contact, removing current from *D*, which then deenergizes. The semiconductor version of the *NAND* circuit is shown in Figure 12–4b with the accompanying truth table in Figure 12–4c. Note the circle at the output of the *AND* symbol. This circle at the output (or input) indicates a complementary function or inversion. The IEEE symbol for the *NAND* function is shown in Figure 12–4d, while the NEMA symbol is found in Figure 12–4e.

The *NOR* function shown in Figure 12–5a,b,c is the combination of the *OR* and *NOT* digital functions. As in the *NAND* function, the relay logic

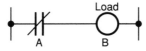

A	B
1	0
0	1

a. Relay logic circuit diagram
NOT function

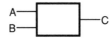

b. Old IEEE NOT logic function
schematic diagram (Inverter)

c. NOT logic function truth table

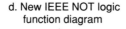

d. New IEEE NOT logic
function diagram

e. NEMA NOT logic function diagram

FIGURE 12–3 Various NOT logic diagrams

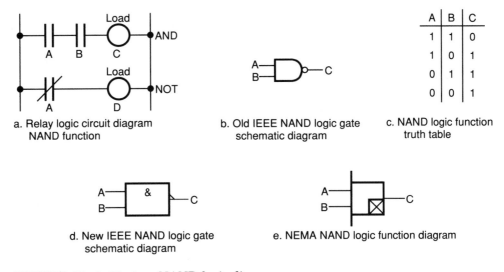

a. Relay logic circuit diagram
NAND function

b. Old IEEE NAND logic gate
schematic diagram

c. NAND logic function
truth table

d. New IEEE NAND logic gate
schematic diagram

e. NEMA NAND logic function diagram

FIGURE 12–4 Various NAND logic diagrams

equivalent of the *NOR* function is the combination of the *OR* and *NOT* functions. The semiconductor logic function of this relay logic circuit is shown in Figure 12–5b. Note that this circuit is equivalent to the diagram shown in Figure 12–5d. The IEEE symbol for the NOR function is shown in Figure 12–5e, while the NEMA symbol is found in Figure 12–5f.

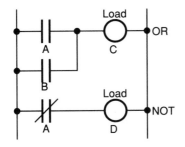

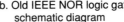

A	B	C
1	1	0
1	0	0
0	1	0
0	0	1

a. Relay logic circuit diagram
NOR function

b. Old IEEE NOR logic gate
schematic diagram

c. NOR logic function truth table

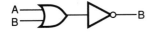

d. IEEE NOR equivalent circuit

e. New IEEE NOR logic
function diagram

f. NEMA NOR logic
function diagram

FIGURE 12–5 Various NOR logic diagrams

12–2.2 Relay Logic Motor Control Example

A familiar electrical motor control circuit appears in Figure 12–6. Pushing the *START* momentary contact pushbutton allows current to flow through the *M* relay coil. Energizing the *M* coil changes the state of the three *N/O M* contacts. AC power is then applied to the three-phase motor. Current also flows through *1M* contacts when the relay picks up. When the *START* button is released, current continues to flow through its own contacts to the *M* coil. You will recognize this as a latching relay configuration. The relay coil *M* will deenergize under three conditions:

    1) if AC power is lost
    2) if an overload relay trips
    3) if the STOP button is pushed.

We can simulate this relay logic circuit with semiconductor digital logic circuits, as shown in Figure 12–7. Device *A* is an *OR* gate, while devices *B* and *C* are inverters. Device *D* is a three-input *AND* gate. If the *START* switch is changed, a high is applied to the lower input to the *OR* gate. Since both inverters have a low input, three highs are present at the input to the *AND* gate. Three highs at the inputs to the *AND* gate will cause the output to go high, turning on the three SSRs at the output. The three SSRs apply three-phase power to the motor as well as providing isolation. Changing the state of either the *STOP* or *O/L* switches will cause the output of the *AND* gate *D* to go low, removing the power from the motor. Note also that a high is fed back from the output of the *AND* gate to the input of the *OR* gate. If the *START* switch changes state, one high will still be present at the input to the *OR* gate from

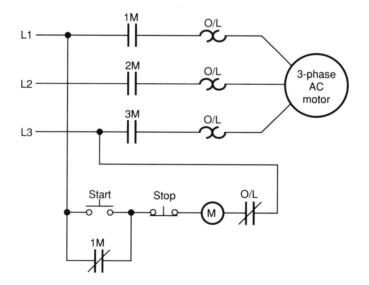

FIGURE 12–6 Three-phase AC motor starting circuit using relay logic symbols

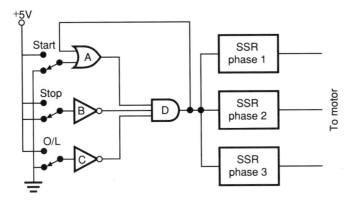

FIGURE 12–7 TTL digital logic motor starter

the *AND* output. The output of the *AND* gate is then kept high by its own output.

This example shows how semiconductor digital logic can do the same function as relay logic. Semiconductor digital logic has the advantage of being more reliable, being cheaper to build and operate, and taking up less space. These advantages come from the inherent differences between electromechanical relays and semiconductor digital logic devices. Both digital and relay logic systems have the same disadvantage in that they are both difficult to modify. When the task changes, the circuit that performs that task must be changed. Modification means rewiring and troubleshooting the new circuit. Circuit redesign and troubleshooting add great expense and is a time-consuming job. A microprocessor-based control system can do the same job as the relay or semiconductor digital logic circuit. Furthermore, it has the advantage of being able to change the tasks it performs easily in software.

12–3 MICROCOMPUTER MOTOR CONTROL

12–3.1 Introduction

Before we consider how a microcomputer controls a motor, let's review what we know about microcomputers. The block diagram of a microcomputer in Figure 12–8 shows a typical bus-oriented microcomputer structure. The central unit is called the central processing unit (CPU). The CPU controls the flow of information between itself and the memory and the *Input/Output (I/O)* sections. The CPU communicates with the memory and I/O sections through control lines that connect each unit. The control lines control the reading and writing of information. Together, the control lines make up the system's **control bus.** In addition to the control bus, the *systems data bus* carries information to and from the memory and I/O sections. The third set of connections is called the

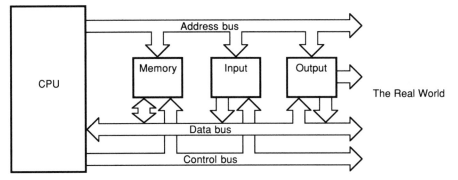

FIGURE 12–8 Block diagram of a microcomputer system

address bus. The address bus points to a specific memory location or to a unique I/O device.

The main purpose of the I/O block is to allow the microprocessor to communicate directly with people or with another computer or machine. The most familiar types of I/O equipment are CRT terminals and printers. Other forms include AC and DC motors, indicator lamps, switches, solenoids, relays, and disk systems. Other typical microprocessor I/O equipment includes analog-to-digital converters (A/D) and digital-to-analog converters (D/A). Any device that responds to an electrical signal or produces an electrical signal can be used as an I/O device in a microprocessor-based system.

Recall that microprocessor systems use two different types of I/O, memory-mapped and isolated I/O. *Memory-mapped I/O* treats the I/O device as if it were a location in memory. *Isolated I/O* treats the I/O device separately from memory. Both schemes are used in microprocessor applications, since it is rare that the entire memory is used for program and data storage. Intel and Zilog microprocessors (the 8085 and the Z-80) can use either type of I/O, but the Motorola 6800 can use only the memory-mapped I/O scheme.

Many microprocessors use a special interface device called a peripheral interface adapter or PIA. A PIA is a special IC chip that contains two or three parallel I/O ports or locations that can be programmed to handle input or output data. Some of these interfaces contain RAM (random-access memory) or ROM (read-only memory). These devices allow almost any TTL compatible input or output device to be connected to the microprocessor. They contain all the special input-output circuitry needed to interface the microprocessor to the outside world.

12–3.2 Motor Control Example

Recall that we simulated a simple relay logic motor control with TTL logic gates in Figure 12–7. The motor control consisted of START and STOP pushbuttons and a motor overload relay. The diagram in Figure 12–9 shows a

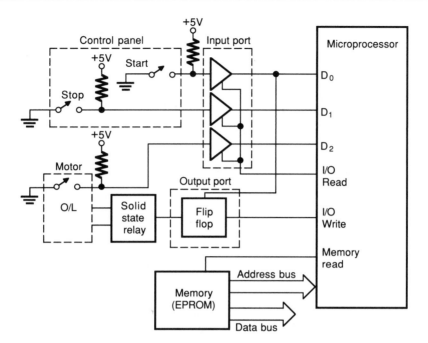

FIGURE 12-9 Block diagram of a microcomputer-based motor control

simple method to simulate this function with a microcomputer. The basic system has four parts. First, the microprocessor controls the system. Later in our example, we will use an 8085-based system to accomplish our control plan. The second part of the system is the memory, which contains the control program. Third, the input port allows the computer to look at the state of the three switches controlling the motor. Last, the output port allows the computer to turn the motor on and off.

The microprocessor has four buses associated with it. The power bus, generally not shown, provides all the DC voltages needed to run the microcomputer. The remaining buses in this system are the data bus, address bus, and the control bus. This design uses three data lines from the microprocessor's data bus, lines *D0, D1,* and *D2.* Three control lines are used: the I/O read, I/O write, and memory read. The program is stored in an EPROM and determines the logic with which the system operates. The erasable programmable read-only memory (EPROM) is a device in which the user can store digital data permanently. When the power is shut off, the data remains in the EPROM.

The input port allows the condition of the switches to be placed in the computer via the data bus. The computer can then perform operations on this data, determining what the motor state should be. The input is connected to a tri-state bus driver, which places the information about the state of the switches on the data bus only when the I/O READ control line enables the bus driver. Since the data from the program stored in memory must also use the data bus, a conflict would occur if the memory and the input port used the data bus at the same time. Thus, all data bus inputs use the bus at different times. The control lines make sure that this conflict does not occur.

Note that the output uses the *DO* line on the data bus. In our system, when the microprocessor is not reading memory or the state of the inputs, the data bus can be used (line *DO*) to turn the motor on or off. The data line *DO* controls a flip-flop that, in turn, controls the motor through a solid-state relay. The overload switch, connected to the motor, turns on when the motor overheats. The state of the overload switch makes up one of the three inputs to the computer system.

The diagram in Figure 12–10 shows an actual 8085-based motor control system. The actual control lines to this microprocessor are $\overline{WR}$, $\overline{RD}$, and $I/O\,\overline{M}$. (The bars over the letters represent an active low. This means that the line must go low to cause that function to occur.) If the $I/O\,\overline{M}$ line is high, the processor will be doing an *I/O* function, reading or writing from or to the ports. If the $I/O\,\overline{M}$ line is low, the processor will be reading or writing to the memory. The input lines are connected to a 4066 CMOS switch, which has an active high enable on the control line. Note the gate connected to the 4066 control line is controlled by inputs from the processor, $\overline{WR}$ and $I/O\,\overline{M}$. The output of

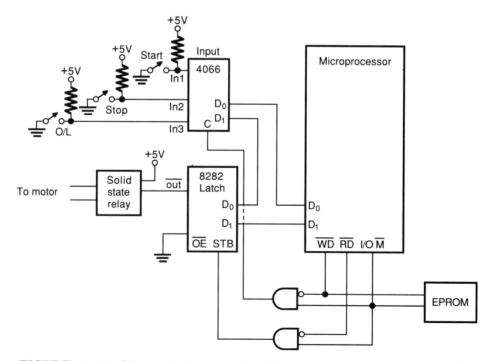

FIGURE 12–10 Schematic diagram of an 8085 microcomputer motor control

this gate will go high when I/O $\overline{M}$ goes high and $\overline{WR}$ goes low. The gate will drive the control line on the 4066 high, allowing the switches to be connected to the data bus. The 4066 control line uses an active high enable. A high control voltage closes the switch. The second gate connects to the 8-bit 8282 latch. The strobe input is an active high enable and connects internally to the clock pin of the internal flip-flop. When the strobe input (STB) goes low, the data from the DO data line is latched from the data bus. Note that the 8282 output enable pin is held low at all times. The $\overline{OE}$ pin requires an active low enable. When pulled low, any data that has been latched in will appear at the output. In our case, we have a solid-state switch connected to one output. When the output goes low, the solid-state relay will turn on, turning on the motor.

A flow chart of the control program can be seen in Figure 12–11. The program starts by turning on the motor. Since we have no way of determining the state of the motor in this simple system, we must start with it off. We then examine the overload switch, which is the most important switch in the system. We must never turn the motor on if this switch is on. To do so would possibly damage the motor. If the overload switch is on, we turn off the motor. If the motor is already off, no harm will be done. If the motor is on, however, we must turn if off immediately. The program branches back to the point marked A and starts the process all over again.

FIGURE 12–11 Flow chart motor starter program

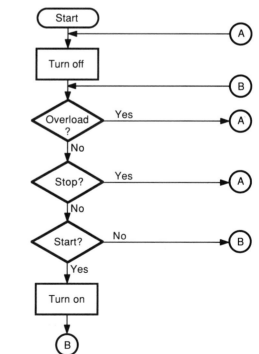

The program will stay in this loop until the overload switch opens, signaling that it is safe to turn on the motor. We next determine whether we have turned on the motor. If we have turned the motor on, we examine the *STOP* switch. If the *STOP* is open (not on), we loop back to the beginning of the program. If the motor is not on, we examine the *START* switch. If the *START* switch is not on, we loop back to the beginning of the program. If the *START* switch has been pressed, we turn the motor on and loop back to the beginning of the program.

A program using the 8085 microprocessor to run the program just described is shown in Table 12–1.

TABLE 12–1   8085 Microprocessor Motor Control

1	START:	MVI A 01H	; Set the output port high to be sure
2		OUT 02H	; the motor is off upon power up
3	CTN1:	IN 01H	; Read the state of the switches and mask
4		ANI 04H	; out all but the desired overload switch bit.
			; 0=no overload, 1=overload
			; (switch will normally be closed indicating
			; no overload.)
5		JNZ START	; loop back to start and turn off motor
6		IN 01H	; read the state of the switches and
7		ANI 02H	; mask out all but the STOP switch bit.
			; 0=no stop, 1=stop
			; switch will normally be closed indicating
			; no stop
8		JNZ START	; loop back to start and turn off motor
9		IN 01H	; read the status of the switches
10		ANI 01H	; and mask out all but the START switch bit.
			; 0=no start, 1=start (switch will be closed
			; to run motor)
11		JZ START	; loop back to start and turn off motor
12		MVI A 00H	; turn on motor
13		OUT 02H	;
14		JMP CTN1	; loop back to beginning, motor should be ON

This simple program demonstrates that a simple motor control system can be written on a microprocessor. The advantage of the microcomputer system is obvious; it can be changed easily by simply rewriting the program. No rewiring will be necessary for this change.

Simple microprocessor programs or those that need to run very quickly are written in assembly language, as is the 8085 program in Table 12–1. (Since it is difficult for most people to write computer programs in binary machine language, computer manufacturers have devised English-like words to represent the binary machine codes. A program using these codes is called an assembly language program.) Assembly language programs suffer the disadvantage of not being transportable. A program written for the 8085 cannot be used for the 6800. Higher level languages, such as PASCAL, BASIC, and C, are often used because they can be moved to other systems with little or no change and because long programs are easier to maintain in high-level languages. Higher level languages tend to run more slowly than assembly language programs.

Table 12–2 shows a program written in the BASIC computer language. It accomplishes the same result as the 8085 program listed in Table 12–1. The

TABLE 12–2 Microcomputer Motor Control Program Written in Applesoft™ BASIC

```
100 REM    THIS PROGRAM CHECKS THE STATE OF AN AUTOMATED
110 REM    DC MOTOR SYSTEM AND TURNS THE MOTOR ON OR OFF
120 REM    BY MICROCOMPUTER CONTROL
130 REM
140 REM    SET INITIAL VALUES FOR INPUT AND OUTPUT SLOTS
150 INSLOT = -16256 + 5 * 16
160 OUTSLOT = -16384 + 4 * 256
170 REM    INITIALIZE BEGINNING STATE AND TURN OFF
180 STATE$ = "OFF"
185 REM    READY TO OUTPUT
190 POKE -16368, 0
195 REM    OUTPUT BINARY 1'S TO 8 OUTPUT CHANNELS
200 POKE OUTSLOT, 255
210 REM    BEGINNING OF LOOP READ INPUTS
220 REM    SET INPUT READY FOR READING
230 POKE INSLOT, 256
240 REM    RETRIEVE INPUT VALUES FROM MEMORY
250 C = PEEK (INSLOT + 1) * 256 + PEEK (INSLOT)
260 REM    SET OVERLOAD FLAG BY CHECKING CHANNEL VALUE FOR 1
270 OVER$ = "NO"
280 IF C    AND 1 THEN OVER$ = "YES"
290 REM    IF OVERLOAD THEN TURN MOTOR OFF
300 IF OVER$ = "YES" GOTO 180
310 REM    SET ON AND OFF INPUTS
320 LET TURNON$ = "NO"
340 IF C    AND 2 THEN TURNON$ = "YES"
350 LET TURNOFF$ = "NO"
360 IF C    AND 4 THEN TURNOFF$ = "YES"
370 REM    CHECK TO TURN OFF
380 IF TURNOFF$ = "YES" GOTO 180
390 REM    CHECK TO TURN ON IF NO, GOTO LOOP BEGINNING 230
400 IF TURNON$ = "NO" GOTO 230
410 REM    OTHERWISE TURN ON
420 REM    SET TO TURN ON
430 POKE -16368,0
440 REM    TURN IT ON
450 POKE OUTSLOT, 0
460 REM GOTO THE BEGINNING OF THE LOOP
470 GOTO 230
999 END
```

REM or REmark statements serve to document the program in case future changes need to be made.

12–3.3 Microprocessor Speed Control

Up to this point, we have considered only ON-OFF control of motors. For many applications, this kind of control is adequate. Microprocessors are capable of executing complex sequential control plans. Since changes are done in software, no rewiring needs to be done. Microprocessors are also able to control the speed of a motor.

As you can see from the simplified block diagram in Figure 12–12, the microprocessor is part of a closed-loop motor speed control system. As with most closed-loop speed control systems, the microprocessor provides the control over the motor speed and also continuously samples the motor speed. Based on the actual motor speed, the microprocessor decides whether or not to adjust the speed to bring it back into specified limits. In this circuit, the microprocessor adjusts the speed of a DC motor by varying the average voltage applied to the motor. The average voltage is varied by pulse-width modulation. If the motor speed needs to be increased, the *on* pulse is lengthened with respect to the period. Another way of saying this is that to increase the speed of the motor, the microprocessor increases the duty cycle at one of its output ports.

A basic speed control system in block diagram form appears in Figure 12–13. The microprocessor, a Motorola 6800, uses port *B* of the 6820 PIA as an output port. Line *PBO* drives the LED of an optoisolator. An optoisolator is a device that contains an LED and a photosensor. The only connection between them is the light that travels from the LED to the photosensor. The output of the optoisolator drives a power MOSFET, which directly controls the current flow through the DC motor. The microcomputer turns the LED inside the optoisolator on and off by the state of the logic bit on PBO. If the logic bit is 1, or 5 V, the LED is off. If the logic bit is 0 or O V, the LED is on. The program directs the microprocessor to change the duty cycle of the output wave form, resulting in power control by pulse-width modulation. The microprocessor provides a square wave signal with a variable duty cycle under software control.

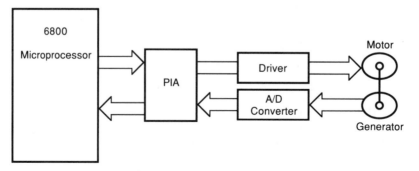

FIGURE 12–12 Block diagram microprocessor speed control

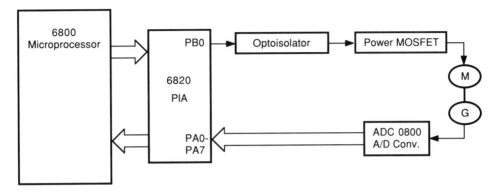

FIGURE 12–13 Detailed block diagram microprocessor speed control of a DC motor

The longer the duration of the LED on pulse (logic 0), the greater the amount of average current in the motor and the higher its speed. The duty cycle is varied by changing the delay byte in the program below.

An *analog-to-digital converter* is necessary to adjust the motor speed, as seen in Figure 12–13. The DC motor used in this system is a servomotor with a separate set of armature windings. The second set of windings is used as a DC generator. As the motor increases speed, the generator output voltage will also increase proportionally. The generator output voltage is filtered and connected to the input of an A/D converter. A National Semiconductor ADC 0800 successive approximation A/D converter was used in this circuit. The A/D converter generates an 8-bit code that represents the analog voltage at the input. The microprocessor reads the 8-bit code through part A of the PIA. The user enters into memory an 8-bit code that represents the desired motor speed. The program directs the microprocessor to compare the code of desired speed with the code representing the actual motor speed. If the actual speed of the motor is above or below the desired speed, the microprocessor changes the speed of the motor until actual speed and desired speed are the same. The microprocessor changes the speed by adjusting the duty cycle of the output square wave.

The program listed in Table 12–3 will allow the microprocessor to control the speed of the motor.

TABLE 12–3 Microcomputer Speed Control Program

1	0000 C6	LDAB	Load accumulator B W/00
2	01 00		
3	02 F7	STAB	Store accumulator B at 8007
4	03 80	CRB	
5	04 07		
6	05 C6	LDAB	Load accumulator B w/01
7	06 01		
8	07 F7	STAB	Store accumulator B at 8007

9	08 80	DDR	
10	09 06		
11	0A C6	LDAB	Load accumulator B w/04
12	0B 04		
13	0C F7	STAB	Store accumulator B at 8007
14	0D 80	CRB	
15	0E 07		
16	0F 86	LDAA	Load accumulator A w/00
17	10 00		
18	11 B7	STAA	Store contents of A at 8005
19	12 80	CRA	
20	13 05		
21	14 86	LDAA	Load accumulator A w/00
22	15 00		
23	16 B7	STAA	Store contents of accumulator at 8004
24	17 80	DDR	
25	18 04		
26	19 86	LDAA	Load accumulator A w/04
27	1A 04		
28	1B B7	STAA	Store accumulator A at 8005
29	1C 80	CRA	
30	1D 05		
31	1E 86	LDAA	Load accumulator A with delay byte
32	1F —		Delay byte-checking A/D input
33	20 97	STAA	Store at 0080
34	21 80		
35	22 C6	LDAB	Load accumulator B w/01
36	23 01		
37	24 F7	STAB	Store at 8006
38	25 80		DR Port B output 01
39	26 06		
40	27 CE	LDX	Load index register w/high delay byte
41	28 —		High delay byte
42	29 —		High delay byte
43	2A FD		
44	2B 09	DEX	Decrement index register X
45	2C 26	BNE	Branch if not equal to
46	2D C6	LDAB	Load accumulator B w/00
47	2E 00		
48	2F F7	STAB	Store at 8006
49	30 80		Port B output
50	31 06		
51	32 CE	LDX	Load index register
52	33 —		Low delay byte
53	34 —		Low delay byte
54	35 FD		
55	36 09	DEX	Decrement index register

56	37 26	BNE	
57	38 6A	DEC	Decrement delay byte for checking A/D
58	39 00		
59	3A 80		
60	3B 26	BNE	Branch if not equal to zero
61	3C E5		
62	3D B6	LDAA	Load accumulator A from 8004
63	3E 80		
64	3F 04		
65	40 43	COM	Complement
66	41 81	CMP	Compare
67	42 —		Speed code byte
68	43 2D	BLT	Branch if less than
69	44 4C		
70	45 2E	BGT	Branch if greater than
71	46 3A		
72	47 7E	JMP	Jump to 001E

SUBROUTINE FOR SPEED > CODED SPEED BYTE

73	81 DE	LDX	Load index register w/low delay byte
74	82 33	PULB	Pull data from B
75	83 09	DEX	Decrement low delay
76	84 DF	STX	Store index register
77	85 33		
78	86 DE	LDX	Load index register w/high delay bytes
79	87 28		
80	88 08	INX	Increment register X —high delay
81	89 DF	STX	Store index register
82	8A 28		
83	8B 7E	JMP	Jump to 001E
84	8C 00		
85	8D 1E		

SUBROUTINE FOR SPEED < CODED SPEED BYTE

86	91 DE	LDX	Load index register w/low delay bytes
87	92 33	PULB	Pull data from B
88	93 08	INX	Increment low delay
89	94 DF	STX	Return low delay bytes
90	95 33		
91	96 DE	LDX	Load index register w/high delay bytes
92	97 28		
93	98 09	DEX	Decrement high delay byte
94	99 DF	STX	Return high delay bytes
95	9A 28		
96	9B 7E	JMP	Jump to 001E
97	9C 00		
98	9D 1E		

12–3.4 Microprocessor Stepper Motor Control

Introduction The microprocessor often controls the stepper motor, especially in computer peripheral devices. Currents through the motor coils generate fields that position the rotor. The motor in Figure 12–14 has four coils that control the rotor position. A current driver provides enough current to create the field in the stator poles. The stepper coil driving circuit (Figure 12–14) generates a large current flow when the step voltage is applied. When the *STEP* voltage is applied, 24 V is applied to the stator coil. The *COIL ON* voltage allows current to flow through the coil. The *COIL ON* voltage also remains after the step to hold the coil in position. Moving the rotor requires more current than holding it at a constant position.

The output ports of a computer, usually *CMOS* or *TTL,* do not normally provide enough current to drive a stepper motor stator. The output port of the computer is usually connected to a current driver, a transistor, Darlington, or power MOSFET. The power MOSFET is ideal, for it not only gives more driving current but also isolates the power circuit from the low voltage microprocessor output port. Without this isolation, it is possible to damage the output ports with high voltages and currents.

Stepper Motor Example The diagram in Figure 12–15 shows a stepper motor driven from a TTL compatible microcomputer interface. The pattern of current flow in the coils determines the direction of rotation. In our example, we will have four coils to control, and, therefore, four drivers. Two coils energized at a time produce one step of 1.8°. A complete revolution will require 200 steps. By sending the binary pattern in Table 12–4 from a PIA, the direction and speed of the rotor can easily be controlled.

FIGURE 12–14 Stepper motor coil driving circuit

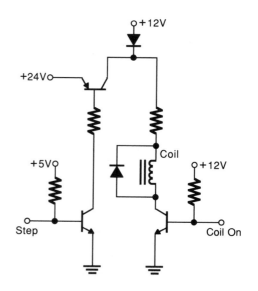

TABLE 12–4 Stepper Motor Bit Control Patterns

Step Number	Counterclockwise Rotation	Clockwise Rotation
1	0 0 1 1	1 1 0 0
2	0 1 1 0	0 1 1 0
3	1 1 0 0	0 0 1 1
4	1 0 0 1	1 0 0 1

This particular bit pattern energizes two coils at the same time. Note that the CCW pattern shifts the high bits left in each successive step. The CW pattern shifts the high bits right one position. The faster the pattern is moved, the faster the rotor turns.

This computer's software controls the direction of rotation and the number of steps taken. The software must have the ability to remember the position of the rotor. The flowchart for the motor control operation is shown in

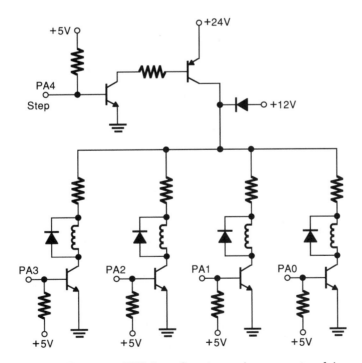

FIGURE 12–15 TTL interface to a microcomputer-driven stepper motor

Figure 12–16. From this flowchart, software routines control the motor with a 8085 microprocessor and a 6800 microprocessor. In both software routines, the memory location LOC indicates the current bit pattern of the previous data sent to the stepper motor. It is this bit pattern that the software program modifies to rotate the motor in the right direction. In both programs, the *B* register contains information on the direction of rotation and how many steps to be taken. If the *B* register contains a positive number, the rotation is CW. If the B register contains a negative number, the rotation is CCW.

The software programs, written for the 8085 and the 6800 microprocessors, are listed in the following programs.

TABLE 12–5 Stepper Motor Control Program—8085 Microprocessor

```
1   ;8085A ASSEMBLY LANGUAGE PROGRAM
2   ;SUBROUTINE TO CONTROL A STEPPER MOTOR
3   ;B=NUMBER OF STEPS AND DIRECTION
4   ;CW = + AND CCW = −
5   ;
6   ;TEST FOR DIRECTION
7   STEP:      MOV    A,B      ;GET DIRECTION
8              ADD    A        ;DIRECTION TO CARRY
9              LDA    LOC      ;GET PRIOR LOCATION
10             JC     STEP2    ;COUNTERCLOCKWISE?
11             RRC             ;SETUP NEW POSITION
12  ;ROTATE    STEPPER MOTOR
13  STEP1:     STA    LOC      ;SAVE NEW POSITION
14             ORI    10H      ;SETUP STEP BIT
15             OUT    PORTA    ;START ROTATION
16             CALL   DELAY    ;WAIT 1 MSEC.
17             MOV    A,B      ;GET COUNT
18             ANI    7FH      ;STRIP OFF DIRECTION
19             DCR    A        ;DECREMENT COUNT
20             JZ     STEP3    ;DONE?
21             DCR    B        ;DECREMENT COUNT
22             JMP    STEP     ;MOVE ANOTHER STEP
23  ;SETUP FOR COUNTERCLOCKWISE
24  STEP2:     RLC             ;SETUP NEW POSITION
25             JMP    STEP1    ;GO STEP MOTOR
26  ;STOP ROTATION AND END SUBROUTINE
27  STEP3:     LDA    LOC      ;GET LOCATION
28             ANI    15H      ;CLEAR STEP BIT
29             OUT    PORTA    ;STOP ROTATION
30             RET             ;RETURN FROM SUBROUTINE
31  LOC:       DB     33G      ;INITIAL POSITION
```

TABLE 12–6 Stepper Motor Control Program—6800 Microprocessor

```
1   *6800 ASSEMBLY LANGUAGE PROGRAM
2   *SUBROUTINE TO CONTROL A STEPPER MOTOR
3   *ACC B=NUMBER OF STEPS AND DIRECTION
4   *CW = + AND CCW = −
5   *
6   *TEST FOR DIRECTION
7   STEP    LDAA    LOC      GET PRIOR POSITION
8           TSTB             CHECK FOR DIRECTION
9           BMI     STEP2    COUNTERCLOCKWISE?
10          LSRA             SETUP NEW POSITION
11          BCC     STEP1    NO CARRY?
12          ADDA    #$80     ADJUST RESULT
13  *START ROTATION
14  STEP1   STAA    LOC      SAVE NEW POSITION
15          ORAA    #$10     SETUP STEP BIT
16          STAA    PORTA    START ROTATION
17          JSR     DELAY    WAIT FOR 1 MSEC.
18          TBA              GET COUNT
19          ANDA    #$7F     STRIP DIRECTION BIT
20          DECA             DECREMENT COUNT
21          BEQ     STEP3    COUNT ZERO?
22          DECB             DECREMENT COUNT
23          BRA     STEP     MOVE ANOTHER STEP
24  *IF COUNTERCLOCKWISE
25  STEP2   ASLA             SETUP NEW POSITION
26          BCC     STEP1    NO CARRY?
27          INCA             ADJUST POSITION
28          BRA     STEP1    GO ROTATE
29  *FINISH SUBROUTINE
30  STEP3   LDAA    LOC      GET POSITION
31          ANDA    #$OF     CLEAR STEP BIT
32          STAA    PORTA    STOP ROTATION
33          RTS              RETURN FROM SUBROUTINE
34  LOC     FCB     #$33     INITIAL VALUE
```

Stepper Motor Position Feedback Some applications require that the controller know the position of the stepper motor. A common use of stepper motors today is in the line feed mechanism in the computer printer. In this application, the stepper is pulsed a number of times to move the paper up one line. Half the number of pulses will move the paper up one-half of a line. In this application, the system does not need to keep track of the stepper rotor position.

FIGURE 12–16 Stepper motor rotation flow chart

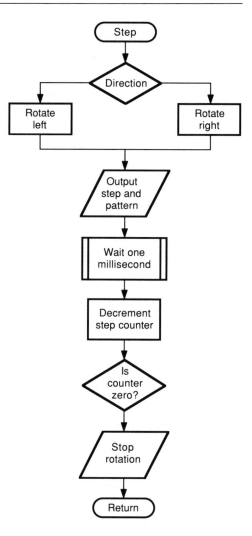

Another application of a stepper motor in a dot matrix printer is in the print head driver. Unlike the line feed mechanism, the computer does need to keep track of the position of the print head. If the system did not keep track of the print head, it would be impossible to print information on a page with any reliability. One popular method of keeping track of the print head is to start it out at a known position. This position is called the home position. In most printers, this position is at the left side of the paper. The sensing of the home position is done with a LED-phototransistor, as shown in Figure 12–17. The head driver motor receives a home signal, moving it to the left side of the paper. When the head arrives at the left side of the paper, the light path from the LED to the phototransistor is broken. The broken beam is the signal to the

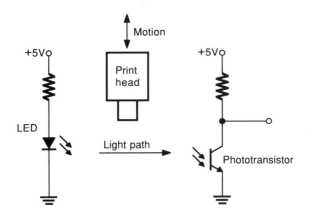

FIGURE 12–17 LED-phototransistor printer home sensor

computer that the head has reached the home position. Once the home position is known, the software can tell the head stepper motor the correct number of steps to any point on the paper without further feedback.

12–4 PROGRAMMABLE CONTROLLERS

12–4.1 Introduction

We have seen how the relays, digital ICs, and microprocessors have been used to control a motor. We noted that the hardware-based relay logic was effective but unreliable and slow. Slow speed results from the mechanical action of the relay contacts opening and closing. The same mechanical action makes relay logic unreliable. Digital logic ICs are more reliable and faster than relay logic. Both digital and relay logic, however, suffer from a common disadvantage; they are difficult to change. Any change in logic function requires a rewiring of the circuit, an expensive and time-consuming operation.

We noted that the microprocessor has all the advantages of the digital logic circuit. Microprocessor circuits are reliable, fast, and do not consume much power when compared to the similar relay logic circuit. The microprocessor has an advantage that the relay and digital logic circuits do not have. Microprocessors do not require rewiring to change the logic function. Logic functions are changed by changing the program, the computer software.

Although changing the program is easier and less time-consuming than rewiring the hardware, software changes must be done carefully. First, a programmer must be available to change the software to meet the new logic functions. The program must then be tested to work out any problems, or bugs. This process may take some time, even with experienced programmers. Furthermore, the users of the equipment will not usually be the ones who will make the changes to the program. The disadvantages of the hardwired relay and semiconductor logic led to the development of the programmable controller (PC).

The first PCs were developed by the automobile industry which needed a control system that was easy to use, maintain, and repair. It also needed to be small, rugged, relatively inexpensive, and able to communicate with other computer equipment. The PC filled all of these requirements. Furthermore, PCs could be easily reprogrammed by users. No knowledge of traditional computer programming was required. As a result of these advantages, PCs rapidly replaced relays in the automotive industry. This introduction was not without some difficulty. Maintenance personnel were accustomed to the sound of the clattering relays. In fact, the relay noise was used to tell if the process was functioning correctly. So important were these sounds that PC manufacturers actually considered adding sound circuitry to the PC to make the system behave more like relay circuitry.

PCs were first used as event-based *sequential controllers.* If a given event happens and a certain action is taken, this is called *event-based control.* For example, an event may be the pressing of a motor *START* button. The action taken may be to apply power to the motor. This is event-based control. Sequential control carries with it the idea of a series of events happening in sequence. Sequential control happens when switches, motors, lights, solenoids, or instruments turn on and off in preselected sequences or intervals. For example, first motor #1 is turned on. Then, after a 10-second delay, motor #2 is turned on. Events occur in a predetermined sequence. PCs in modern industry do far more than event-based, sequential control. They are also far more reliable and flexible. PCs are still used primarily for sequential control but can now be programmed to do comparisons and combinational logic. PCs even replaced some computer control applications, when plant engineers found that they could do the same jobs faster, cheaper, and with greater reliability. Although PCs started out in the auto industry, they are found today in almost every type of manufacturing application.

What actually is a PC? After four years of work, in 1978 NEMA released a standard for programmable controllers, NEMA Standard ICS3-1978, part ICS3-304. NEMA defines a PC as "a digitally operated electronic apparatus that uses a programmable memory for the internal storage of instructions for implementing specific functions, such as logic, sequencing, timing, counting, and arithmetic, to control machines or processes through digital or analog input or output modules, various types of machines or processes. A digital computer which is used to perform the functions of a programmable controller is considered to be within this scope. Excluded are drum and similar mechanical type sequencing controllers." This definition is very wide. It could apply to any digitally based process control circuit, even a microcomputer. Since NEMA is an organization including manufacturers of PC equipment, we should not be surprised to see a definition wide enough to include everyone's equipment. Estimates by industry observers state that more than 50 products would fit this definition of a PC.

12–4.2 PC Basics

Regardless of the manufacturer, all PCs have certain things in common. All PCs have input and output interfaces, memory, a method of programming, a central processor, and a power supply. These functions are shown in the functional block diagram in Figure 12–18. The PC first examines the inputs to see if an action needs to be done at the output. The combinations of input and output states are called logic states. Switches and motors, for example, may be on or off. The logic combinations are carried out according to a control plan or program. The program may have instructions to turn on a motor if the START button is depressed or to stop the motor if the STOP button is pushed. The program is stored in memory using the programming device. All the instructions (logic combinations) stored in memory are evaluated or scanned at regular intervals and compared to the inputs and outputs.

PC Central Processing Unit (CPU) The CPU is the brain of the PC since it organizes all the PC activity. Data comes from the input section and, based on the stored program, the CPU performs logical decisions and drives outputs. The CPU first evaluates the inputs in a predetermined sequence stored in memory. A number (or number-letter combination) refers to data locations known as addresses. The processor uses these addresses to sort and fetch I/O and memory data. The address is similar to a pigeonhole in a small post office. Each hole has a code that refers to its specific position. Data (mail) is placed in

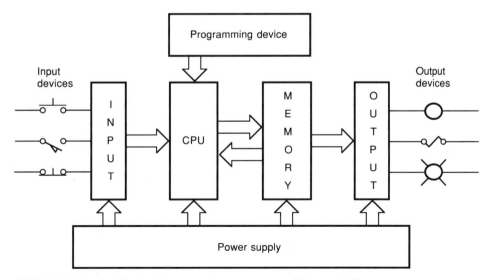

FIGURE 12–18 Block diagram of a programmable controller

a particular hole to be processed or retrieved later. The user selects the I/O address by assigning an output or input circuit to a specific physical device. The CPU does more than logical functions. It can time, count, perform arithmetic calculations, and do comparisons. It replaces the electromagnetic counters, timers, and relays found in hardwired relay logic circuitry.

The time it takes the CPU to complete one operating cycle is called the scan time. During the scan time, the following things are done:

1) all inputs are scanned
2) all logical comparisons are made and solved
3) all arithmetic calculations are made
4) all outputs are updated.

Scan times in PCs run from 5 to 200 ms. Scan times depend on the size and type of machine and the size of the control program. Based on the program and the status of inputs, one or more control actions can be started. After one scan is completed, another is started. The scan repeats so that any changes in the status of inputs can be acted on.

PC Memory   The PC memory system provides a way to hold the control plan in the PC and references to input and output devices. Memory in a PC is divided into two different areas: the executive program area and application program area. The executive program is provided by the PC manufacturer to direct the activity of the CPU. It cannot be changed by the user. The user, however, enters the application program into memory with the programming device. The application program contains specific instructions about the control process. The complexity of the program determines how much memory is needed. Specifically, the amount of inputs and outputs and the length of the program determine this memory size requirement.

As in microcomputers, the PC stores information in pieces called bits. A grouping of 8 bits is called a word, or byte. Memory size is specified in thousands of bytes, or K. Each one thousand bytes (actually 1024 bytes) is called a K byte. PC memory capacity depends on the machine and task. PCs may have memory ranging from 100 bits to 512 K bytes. Since each manufacturer's machine is different, an accurate estimate of how much memory is needed for a specific task is difficult. A general estimate can be found by writing out the program and counting the number of instructions used. The total memory can be estimated by multiplying the number of instructions by the number of words used in each instruction.

The executive program is typically stored in a read-only memory chip (ROM chip). The ROM is usually a programmable ROM, or PROM. Some PROMs are programmed by the manufacturer at the factory by burning open fusible links. These types of PROMs cannot be changed by any means. Some types of PROMs, however, can be changed. Ultraviolet light erases the erasable PROM (EPROM), which is programmed electrically. The electrically-erasable PROM (EEPROM), as its name suggests, is erased and programmed by an electrical signal. All of these types of memory retain their program information when the PCs are shut off.

Application programs are usually stored in a different type of memory called random-access memory (RAM). RAM memory is made up of integrated transistors or MOSFETs. The most common type is the NMOS memory. NMOS memory is available in two forms: the dynamic form called dynamic memory, or DRAM, and the static form called SRAM. Dynamic memory requires periodic refreshing, while SRAM hold its data as long as power is applied. Since DRAM requires additional circuitry to refresh the memory, it is normally used only on larger systems. Another type of RAM, called complementary metal oxide semiconductor (CMOS) RAM, is now popular in PC memory. CMOS RAM requires far less power than NMOS RAM. In some PCs, DRAM and SRAM memory sections have batteries that retain the contents of memory in the case of power loss. A new type of RAM, called non-volatile RAM (NOVRAM), is now being seen in PCs. NOVRAM retains its memory contents when power is removed. It does not, therefore, need a battery as other types of RAM do.

PC Input/Output　The input/output (I/O) section of the PC consists of input and output cards or modules mounted in a cabinet. The I/O sections are called modular, because they can easily be plugged in the system. The I/O modules accept input signals from machines or pilot devices, such as limit switches, pushbuttons, thumbwheel switches, pressure switches, proximity switches, transducers, and analog/digital solid-state devices. In general, input devices provide information about the condition or status of an operation. Inputs usually tell something about a machine or workpiece status or an operator's demands. For example, in a drilling operation, the part must be correctly positioned before the drill motor is turned on. An input sensor tells the system when the part (called the workpiece) is correctly positioned. Another example is a motor START/STOP switch, a device that reflects an operator demand to turn a motor on or off.

The input module takes these inputs and converts them into a form that can be evaluated by the CPU. The output part of the I/O module converts the CPU instructions into the right kind of signal to control external output devices. Output devices would include solenoids, relays, motor starters, alarms, indicators, and annunciators. Most PC I/O modules accept digital logic (including binary-coded decimal) as well as analog inputs and can produce the same type of output signals.

Each I/O module can have up to 32 different input or output circuits. The I/O racks can be placed close to the CPU/memory unit or as far away as 15,000 ft. I/O modules are connected to the CPU/memory unit by either twisted pair or fiber-optic cables. I/O modules are usually found in racks, such as the one shown in the photo in Figure 12–19. Each rack contains one or more input and output modules. The wires from the input/output devices are attached to

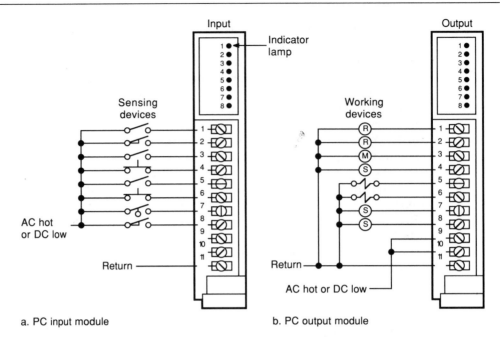

a. PC input module   b. PC output module

FIGURE 12–19 Programmable controller I/O modules

the modules by terminal strips with screw attachments. Lights are often added to I/O modules to show the ON or OFF status of each I/O circuit. Each I/O circuit is usually fused to prevent damage to the module. Some PCs add a blown fuse indicator to make troubleshooting easier.

PCs are now classified by size according to the number of I/O ports. The names of the PCs according to size are as follows:

TABLE 12–7 Size Classifications of Programmable Controllers

Name	Number of I/O Ports
Mini	0–64
Small	1–12B
Medium	128–512
Large	512–up

PC Power Supplies The PC power supply provides all the voltages needed to operate the PC. The power supply converts 120 or 240 VAC into the DC voltage required by the CPU, memory, and I/O modules. The power supply may be part of the CPU/memory unit or located at some remote station.

PC Programming Devices Before the PC can do anything useful, the user must enter instructions into memory in the form of a program. The programming device allows the user to do this program entry. Programming units may be small, hand-held units, such as the one shown in Figure 12–20. The hand-held programmer is about the size of a desktop calculator. It has its own power cord and non-volatile memory. It allows users to program from remote locations, such as in an office or at home. Some PCs have CRTs with graphics capabilities. The photo in Figure 12–21 shows a CRT displaying a program. This particular programmer has a large membrane keyboard to facilitate program entry. Many PCs are now programmable by personal computers. Programming by personal computers does not require purchase of a special programming unit.

12–4.3 PC Languages

All programs must be entered into memory using a special language. The PC language, like all languages, has grammar, syntax, and vocabulary that allow the user to write a program stating what the CPU is to do. All PC manufacturers use a slightly different language to do this. All PC languages are designed to instruct the CPU in the carrying out of the control plan.

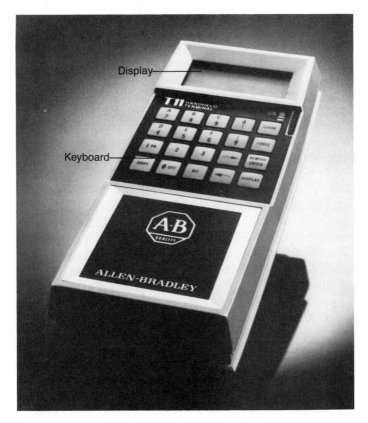

FIGURE 12–20 Allen-Bradley hand-held programmer

FIGURE 12–21   CRT displaying a program

Ladder Logic Programming   Several programming languages are used with PCs. First, and most popular, is the relay logic diagram. Its popularity dates to the introduction of the PC in the automotive industry in the early 1970s. The automotive industry, at that time, needed a system that could be modified easily to reflect yearly model changes. Prior to the introduction of the PC, all control operations were carried out by relays. Relay circuits were laid out in ladder diagram form. Plant personnel, by virtue of training and experience, were (and still are) well acquainted with relay logic in the form of ladder diagrams. The PC manufacturers were faced with the problem of how to introduce something new and overcome the natural resistance to the introduction of high technology equipment. It would have been unrealistic to expect plant personnel to learn a totally new language (like Boolean algebra) to work their machines. The PC manufacturers made the new equipment programmable in ladder diagram form. This was a wise decision for two reasons. First, the equipment could be installed and personnel trained in a short time. Second, it helped overcome the natural resistance to new technology by relating it to old concepts.

     Ladder diagrams do have some disadvantages. First, long programs using ladder diagrams tend to be slow because all inputs must be evaluated every program cycle, whether they have changed or not. In most systems, many of the inputs will not change frequently. The system is, therefore, slowed down considerably with irrelevant logic. Second, ladder logic, although well suited to sequential process control, is not well suited to continuous process control since it is not easily modified. The ladder diagram programming language is said to lack extensibility. The language also makes it difficult to give diagnostics for errors.

As we have seen in previous chapters, ladder diagrams use symbols that represent relay coils, normally-open and normally-closed contacts. The coils and contacts can be arranged in series, parallel, or series parallel connections, as shown in Figure 12–22. The diagram is made up of a series of rungs. Each rung contains one or more inputs and the output or outputs are directly controlled by the inputs. The example shown in Figure 12–22 is a reciprocating motion problem. When the start button is pressed, 1CR is energized, causing the table to move forward. When the table touches 2LS, the motor is reversed and the table moves in the opposite direction. When the table touches 1LS, the motor reverses again and the process repeats. In the PC version of this circuit, the CPU scans each rung, evaluating the inputs and deciding what to do with each output. The PC performs the same function as the relays in Figure 12–22. A PC relay logic diagram equivalent is shown in Figure 12–23.

Another more complicated PC ladder logic diagram is shown in Figure 12–24. This diagram shows a motor controlled by two limit switches, LS1 and LS2, two pressure switches, P1 and P2, a timer, pump switch, run switch, and emergency stop switch.

Boolean Language Programming Relay logic diagrams are expected to be a popular method of programming PCs for many years to come. Another language sometimes used in PC programming is the Boolean statement. The relay logic circuit can be represented by logic functions. For example, a string of N/O series contacts can be represented by an AND gate. Each one of these logic states can be transformed into a statement in Boolean algebra. These Boolean

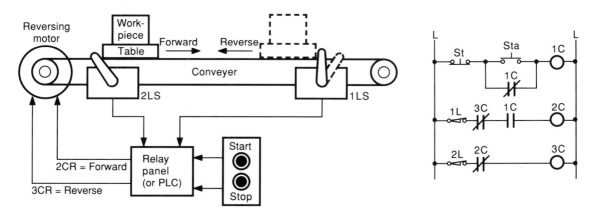

FIGURE 12–22 Example ladder diagram—Reciprocating motion

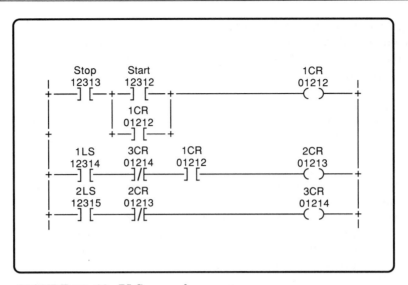

FIGURE 12–23 PLC example program

statements are known as a statement or instruction list. Usually found in small to medium sized PCs, Boolean statements can be executed very quickly. This makes the Boolean language run faster than a comparable statement in ladder logic. A standards organization called the International Electrotechnical Commission (ILC) is working on a standard for this language.

A Boolean statement representing the relay ladder logic diagram of Figure 12–22 is as follows:

$$* \text{STOP} + (\text{START} * \text{ICR}) ** \text{ILS} ** 3\text{CR} = 2 \text{ CR}$$

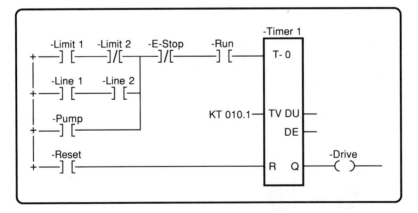

FIGURE 12–24 Programmable logic controller program representation using modified relay logic

An example of a PC Boolean program is shown in Figure 12–25. This Boolean statement is the equivalent of the program shown in Figure 12–24. Note the correspondence between the ladder logic and the Boolean expression of that logic.

Boolean expressions may be more familiar to systems designers. Boolean expressions do, however, have some drawbacks. Translation of the control plan into Boolean expressions, especially for complicated programs, is a difficult mental exercise. The difficulty is compounded if those responsible for maintaining and using the program are unfamiliar with Boolean expressions. Boolean expressions, because of their flexibility, are also more difficult to test and debug. This increases the time needed to make program changes, something a PC should be able to permit quickly.

Flowchart Programming Language You are probably familiar with the concept of a flowchart if you have programmed a computer. A flowchart is a pictoral language that shows the interconnections of variables within a process. This type of language, also called a flow block and a control system flowchart, is more popular in Europe than in the United States. This diagram

: A	-LIMIT 1	OVERRIDE LIMIT SWITCH
: AN	-LIMIT 2	CONVEYOR LIMIT SWITCH
: O		
: A	-LINE 1	LINE 1 READY
: A	-LINE 2	LINE 2 READY
: O	-PUMP	
:]		
: AN	-E-STOP	EMERGENCY STOP
: A	-RUN	RUN ENABLE
: L	KT010.1	TIMERSETTING
: SR	-TIMER 1	
: A	-RESET	RESET THE TIMER
: R	-TIMER 1	TIMER RESET INPUT
: L	-TIMER 1	LOAD TIMER VALUE TO ACC1
: T	DW1	TRANSFER VALUE TO DATA WORD 1
: LD	-TIMER 1	LOAD BCD VALUE TO ACC1
: T	DW2	TRANSFER VALUE TO DATA WORD 2
: A	-TIMER 1	DELAY TIMER
: =	-DRIVE	MAIN DRIVE MOTOR

FIGURE 12–25 Example of a Boolean language program on a programmable controller

closely resembles the type of diagram used by process engineers when designing a process control system. The flowchart language gives PC programmers the familiarity of logic design with the flexibility of a block diagram. The diagram in Figure 12–26 shows a PC flowchart of the process of Figure 12–24. Compare this diagram to the ladder and Boolean expressions of the previous figures.

PC Graphics Language   Another language gaining popularity is the graphics language, an example of which is shown in Figure 12–27. This diagram shows the functions of the automatic control system as a series of steps and transitions. A step is an action that a machine performs on command. A transition is a binary state that allows the next step to take place. Graphic programming was designed for sequential control problems where the steps are in a sequence and may be time-dependent. Typical applications are container filling applications, rotary machinery control, and drilling and punching operations. The advantage of the graphics language is in the overview that it gives of the entire process. This makes it easier to understand and, therefore, to program and troubleshoot the process.

Traditional Computer Languages   A code or mnemonic language with vocabulary similar to logical functions is used to program some PCs. This language is similar to a microprocessor assembly language. The following program, written in mnemonic code, represents the program for the relay logic,

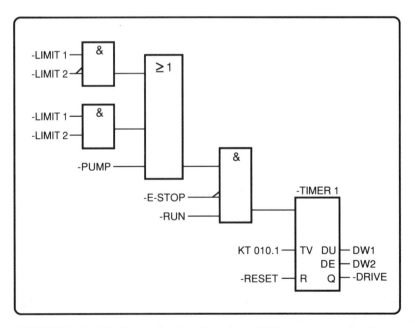

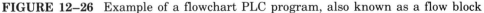

FIGURE 12–26   Example of a flowchart PLC program, also known as a flow block

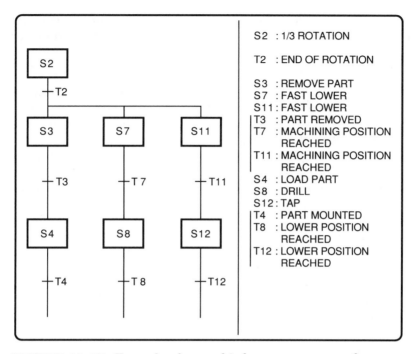

FIGURE 12–27 Example of a graphic language program for a programmable controller

ladder diagram of Figure 12–22:

```
LOAD    1PB
AND     2CR
OR      3LS
AND     4CR
CAND    5CR
STORE   SOLA
```

All programming languages have advantages and disadvantages. The relay logic diagram seems to be the most popular. Some PC manufacturers feel that the number of people on the factory floor will decrease in the future, thus decreasing the pressure to have a language understood by plant electricians. PCs may also need a programming language more suited to mathematics, control, and data communications.

As PCs became more popular and acceptable to industry, PC manufacturers began to add more features, increasing the PC's capabilities. Some PC manufacturers have tried to make their products more versatile by adding the capability to program in the BASIC computer language. The user may then use the BASIC programming language for those tasks that are more efficiently done in that language. Although the ladder diagram is convenient for sequential control, BASIC is well suited to data acquisition and process control.

12–4.4 PC Programming Example

Up to this point we have discussed the programmable controller in general terms only. It may now be helpful to discuss a particular programmable controller. The Allen-Bradley Company was the first company to produce a programmable controller for the automobile industry. Since that time, it has produced the most popular line of programmable controllers in industry. We will use an Allen-Bradley programmable controller in our discussion.

Like most programmable controllers in modern industry, the Allen-Bradley line of controllers uses the basic ladder diagram language for programming.

To gain familiarity with the Allen-Bradley method of programming, we will use the simple example of a motor control relay 1CR in a latching configuration, a start-stop station, and a set of motor overload relay contacts (Figure 12–28a). The programmable-controller representation of this circuit is shown in Figure 12–28b.

In our example, let us use the basic START-STOP station controlling a motor controller with an overload relay, as shown in Figure 12–28. The Allen-Bradley format for this control circuit is as follows:

Note first the numbering system used to identify the devices. Allen-Bradley uses a 5-digit system. The first number tells whether the device is an input or output device. A *1* indicates an input, and a *0* indicates an output. The last four digits identify the input/output (I/O) rack, module, and terminal. This 5-digit number tells the controller where it needs to look for specific information. For example, using the START switch, the controller looks for the START switch in the second output rack, terminal number 13 in the third output module.

We can immediately see a difference between the relay logic diagram and the PC representation of it. In the relay logic diagram, both the overload switch and the STOP switch are normally closed. This is correct, since we want

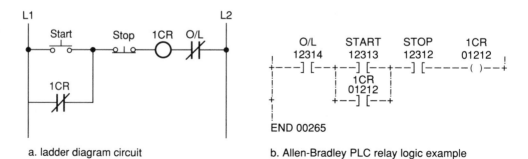

a. ladder diagram circuit   b. Allen-Bradley PLC relay logic example

FIGURE 12–28 Motor starter—two formats

current to flow through the rung, unless the motor is overloaded or we want to stop it by pressing the STOP switch. In the PC ladder diagram, we see the STOP and O/L switch appear to be normally open. Here we encounter a difference between hardwired ladder logic and PC ladder logic.

In the hardwired relay logic, current flows if electrical continuity is present. In the PC, current flows through contacts only when there is logical continuity. The normally open symbol - - -] [- - - in PC terminology is called an *examine ON* instruction. The CPU sees this instruction as a request to check the address in the input register to see if a device has been turned on. When an input device is turned on, the CPU detects this condition, setting a bit in the correct input address register to *1*. The *1* indicates that the device has been turned on. In an examine ON condition, the CPU will examine that register, looking for an ON condition. If the CPU finds an ON condition, a logically true condition results. Current may then flow through the contacts. If the CPU examines the input register and finds the device in an off condition, a logically false condition occurs. No current flows through contacts associated with a logically false condition.

We can see now what happens in Figure 12–28b more clearly. Since the normally-closed overload switch is on, the CPU will detect this condition and set the appropriate input register. When the CPU scans the examine ON instruction for the overload switch and finds it on, the switch condition is logically true. The same condition results for the STOP switch, since it will be a normally-closed switch also. The entire rung is false until the START button is depressed. When the operator presses the START button, the CPU gets the START examine ON instruction and, seeing the switch on, declares the switch condition true. The entire rung is true and current can flow through 1CR, as shown in Figure 12–29.

The normally closed contact - - -]/[- - - is called the *examine OFF* instruction. The CPU sees the examine OFF instruction as a request to examine an input device register for an off condition. If the CPU finds an OFF condition, it declares the condition true. Current can then flow through the associated set of contacts. If the CPU finds an ON condition during an examine OFF instruction, the condition will be false and no current can flow. Only an OFF condition will allow current to flow under an examine OFF instruction.

FIGURE 12–29 Allen-Bradley PLC relay logic example with entire rung true

```
                              Current flow through rung
             -----------------------------------------------> 
              TRUE       TRUE       TRUE
              O/L        START      STOP
              12314      12313      12312      1CR
                                               01212
             +---] [--+--] [--+--] [------( )--+
             |        |  1CR   |
             |        |  01212 |
             +        +--] [--+
             |
             END 00265
```

The actual keyboard entry of the program with all its steps is shown in Figure 12–30.

Note in the program in Figure 12–30 that instruction number generates an END (of program) display on the screen. As shown in Figure 12–28b, the message END appears just after the last instruction on the last rung. The number following END shows the amount of memory used as expressed in words. Each instruction of the program uses one word of memory space. If the first 256 words of the PC memory are set aside as a data table, the user program will start at word 257. In this example, therefore, the program with its nine instructions produces the number 00265, which represents the total actual words used in the PC memory.

Recent developments have given the PC capabilities far exceeding the simple ON/OFF control demonstrated here. Many PC manufacturers offer modules that control the speed of motors and even the motor's acceleration and deceleration. Most of these modules are part of a larger group of PC modules called intelligent I/O modules. The regulation of speed, acceleration, and deceleration are left to the capabilities of the module itself, freeing the CPU in the main PC for other tasks. Most of the intelligent I/O modules are used to drive low power DC servo motors and stepper motors. The DC motor control modules are often called axis positioning modules or APMs. APMs get their name from the fact that a motor may be controlling one particular axis on a machine. Recent developments in APMs allow control of more than one axis at a time.

Step	Program Instruction	Explanation
1. [- -] → [-][-]	[- -] [1] [2] [3] [1] [4]	Examine STOP input Determine if closed
2. [ ] → [-ᴛ-]	[ ]	Start leg of branch
3. [- -] → [-][-]	[- -] [1] [2] [3] [1] [3]	Examine O/L input Determine if closed
4. [ ] → [-•-]	[ ]	Start new leg of branch
5. [- -] → [-][-]	[- -] [0] [1] [2] [1] [2]	Examine 1CR output Determine if energized
6. [ ] → [-ᴛ-]	[ ]	End branch legs
7. [- -] → [-][-]	[- -] [1] [2] [2] [1] [2]	Examine stop input Determine if closed
8. OK	[-( )-] [0] [1] [2] [1] [2]	Energize 1CR output if answer to examine instruction of step #1 and step #7 and to either step #3 or #5 is yes
9. [ ] → [end]	[ ] [3] [7] [7] [1] [7]	End of Program

FIGURE 12–30 Keyboard entry of a PC program on an Allen-Bradley PC

This simple example demonstrates the power of the PC. We have seen the PC used to control a simple motor circuit. The operator does not need to know how to program in BASIC or PASCAL or even how a computer works. Programs can easily be written and changed by people who know relay logic well. Because of its versatility, the PC has been called the industrial revolution of the 1970s. Certainly those who have worked with both hardwired relay logic and the PC will echo this sentiment.

CHAPTER SUMMARY

- Many industrial control diagrams are written in relay logic.
- The basic logic functions are AND, OR, NOT, NOR, and the inverter.
- Semiconductor digital logic circuits can replace the more bulky EMR versions. Both semiconductor and EMR relay logic circuits are difficult to change.
- The microcomputer can replace both the semiconductor and EMR logic circuitry. It is easier to modify the function of the circuit.
- The microcomputer logic is changed through the computer program (software).
- The programmable controller is a special microcomputer system used for industrial control.
- PCs contain input and output interfaces, memory, a programming method, a CPU, and a power supply.
- Most PCs in the USA are programmed in relay logic.
- PCs are designed to exist in the harsh industrial environment. Most computers, including microcomputers, cannot stand up as well.

QUESTIONS AND PROBLEMS

1. Draw the relay logic circuits for the following digital logic functions:
   a. AND
   b. OR
   c. NOT
   d. NOR
2. Draw the schematic diagram of the circuit in Figure 12–31 using the following notations: (note: the English equivalent for this circuit is— Either A and B, or A and not C and D, is equal to E.)
   a. TTL logic
   b. Allen-Bradley PC
   c. Boolean equation

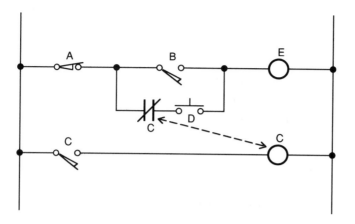

FIGURE 12–31 Relay logic circuit

3. List the one advantage and one disadvantage of the following types of motor control systems:
   a. hard-wired relay logic
   b. TTL logic
   c. microcomputer
   d. programmable controller

4. List and explain the basic parts of a programmable controller.

5. Draw a basic functional block diagram of a DC feedback motor controller, explaining each functional block.

6. What was the reason for the creation of the first programmable controller?

7. Why is a STOP switch, usually represented by a normally-closed contact, programmed as a set of normally-open contacts on a PC?

8. Explain the meaning of examine ON and examine OFF.

9. State the function of the input and output modules of a PC.

13

AC AND DC GENERATORS

At the end of this chapter, you should be able to

- state how field strength can be varied in a DC generator.
- list the types of armatures used in DC generators.
- state the three types of self-excited generators and describe their characteristics.
- explain the concept of voltage regulation.
- calculate the percentage of voltage regulation, given no-load and full-load voltages.
- describe the differences between the two basic types of AC generators.
- list one advantage and one disadvantage of each of the two basic types of AC generators.
- describe exciter generators within alternators and discuss their construction and purpose.
- explain the factors that determine the maximum power output of a generator.

13–1 INTRODUCTION

The electrical energy for the operation of most electronics equipment depends on a machine, called a generator, that converts mechanical energy into electrical energy by electromagnetic induction. A generator that produces alternating current is called an AC generator, or alternator. A generator that produces direct current is called a DC generator. Both types of generators operate by the induction of voltage into coils across which magnetic flux is passed. Regardless of the size, all generators work by the same principle. A magnetic field cuts through one or more conductors, or one or more conductors cut a magnetic field.

Most generators, therefore, have two sets of conductors. One set of conductors produces the voltage while the other generates the field by electromagnetism. The conductors that generate the voltage are called the armature windings. The conductors that generate the electromagnetic field are called the field windings.

Electromagnetic generation of electric current also requires relative motion between the armature and field. To provide relative motion in an AC generator, the machine is divided into two basic parts: the rotor and the stator. The rotor rotates inside the stator, driven by any one of a number of commonly used power sources, such as gas or hydraulic turbines, electric motors, steam, or internal combustion engines. As its name implies, the stator is the stationary part of the generator.

13–2 THE PRACTICAL DC GENERATOR

The actual construction and operation of a practical DC generator differs somewhat from the elementary generator discussed in Chapter 1. The differences lie in the armature construction, the way the armature is wound, and the method of developing the main field.

A generator that has only one or two armature loops has high ripple voltage, as seen in Figure 1–14 in Chapter 1. The high ripple voltage lowers the average voltage produced by the generator. The lowered voltage produces too little current to be of any practical use. To increase the amount of current output, a number of loops of wire are used. These additional loops do away with most of the ripple. The loops of wire, called windings, are evenly spaced around the armature so that the distance between each winding is the same.

13–2.1 Commutation

Recall that commutation in a DC generator is the reversing of current in the individual armature coils to make sure that only DC flows in the load.

The commutator in a practical generator differs from the elementary generator. It has several segments instead of two or four as in our elementary generators. The number of segments must equal twice the number of armature coils, one for each end of the coil.

We can see the commutation process in Figure 13–1. Note that commutation occurs at the same time in the two coils that are being short-circuited by the brushes. Coil B is being short-circuited by the negative brush and coil E by the positive brush. As in the wound-field DC motor, armature reaction causes a shifting of the total field. The brushes are then placed on the commutator in a position that short-circuits the coils when they are moving through the neutral plane. The coils moving through the neutral plane generate no voltage, causing no sparking when they make contact with the brushes.

FIGURE 13–1 Commutation in a DC generator

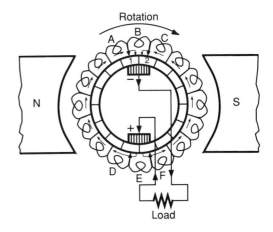

Current through the armature takes two paths. If 100 amperes of load current flow, each path will carry 50 amperes. Each coil on the left side will carry 50 amperes in a given direction, and each coil on the right side will carry 50 amperes in the opposite direction. Current reverses in a given coil when the coil is connected to a brush. In Figure 13–1, as coil A approaches the negative brush, it carries the full 50 amperes which flows through commutator segment *1* and the left half of the negative brush where it joins the 50 amperes from coil *C*.

Note in Figure 13–1 that the negative brush covers half of segment *1* and half of segment *2*. Coil *B* moves parallel to the field, generating no voltage, so no current flows through it. As rotation continues in a clockwise direction, the negative brush covers more of segment *1* and less of segment *2*. Soon the negative brush will cover only segment *1*. At that time, the current in segment *1* increases from 50 to 100 amperes. Current in segment *2* decreases from 50 amperes to zero. When segment *2* leaves the brush, no current flows from segment *2* to the brush and commutation is completed.

The reversal of current in the coils takes place very quickly. In a four-pole generator, for example, each current in coil reverses at a rate of several thousand times per minute. It is very important that commutation be accomplished without sparking, as sparking causes wear on the commutator. As in the DC motor, interpoles and brush position reduce sparking to a minimum.

13–2.2 Armature Construction

Two major types of armatures are used in generators in industry today, the gramme-ring and the drum-type.

Gramme-ring Armatures The diagram of a gramme-ring armature is shown in Figure 13–2.

The gramme-ring winding is formed by winding insulated wire around a hollow iron ring. Each coil is tapped at regular intervals and connected to two

FIGURE 13–2 End view—gramme-ring armature

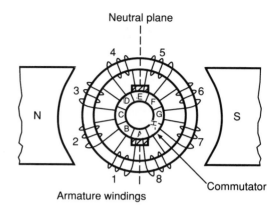

FIGURE 13–3 Composite view—gramme-
ring armature

commutator segments as shown. One end of coil *1* goes to segment *A*; the other
end of coil *1* goes to segment *B*. One end of coil *2* goes to segment *C*; the other
end of coil *2* goes to segment *B*. The rest of the coils are connected in the same
way, in series, around the armature. To complete the series arrangement, coil
8 connects to segment *A*. Each coil connects, therefore, in series with every
other coil.

Figure 13–3 shows a composite view of a gramme-ring armature. It illustrates more graphically the physical relationship of the coils and commutator
locations.

The windings of a gramme-ring armature are placed on an iron ring. A
disadvantage of this arrangement is that the windings located on the inner
side of the iron ring cut few lines of force. Therefore, they have little, if any,
voltage induced in them. For this reason, the gramme-ring armature is not
widely used.

Drum-type Armature A drum-type armature is shown in Figure 13–4. The
armature windings are placed in slots cut in a drum-shaped iron core. Each
winding completely surrounds the core so that the entire length of the conductor cuts the main magnetic field. Therefore, the total voltage induced in the

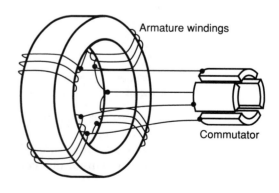

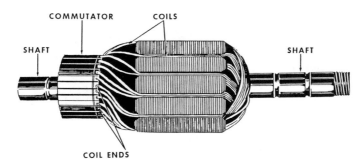

FIGURE 13–4 DC generator armature

armature is greater than in the gramme-ring. You can see that the drum-type armature is much more efficient than the gramme-ring. This accounts for the almost universal use of the drum-type armature in modern DC generators.

Drum-type armatures are wound with either of two types of windings—the lap winding or the wave winding.

The lap winding, shown in Figure 13–5a, is used in DC generators designed for high current applications. The windings are connected to provide several parallel paths for current in the armature. Lap-wound armatures used in DC generators require several pairs of poles and brushes.

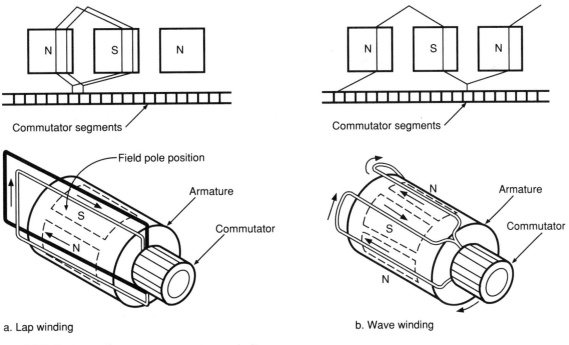

a. Lap winding

b. Wave winding

FIGURE 13–5 Drum-type armature windings

The wave winding, shown in Figure 13–5b, is used in DC generators employed in high voltage applications. Note that the two ends of each coil are connected to commutator segments separated by the distance between poles. This configuration allows the series addition of the voltages in all the windings between brushes. This type of winding requires only one pair of brushes. In practice, a practical generator may have several pairs to improve commutation.

13–2.3   Field Excitation

When a DC voltage is applied to the field windings of a DC generator, current flows through the windings and sets up a steady magnetic field. This is called field excitation.

This excitation voltage can be produced by the generator itself or it can be supplied by an outside source such as a battery or power supply. A generator that supplies its own field excitation is called a **self-excited generator.** Self-excitation is possible only if the field pole pieces have retained a slight amount of permanent magnetism, called residual magnetism. When the generator starts rotating, the weak residual magnetism causes a small voltage to be generated in the armature. This small voltage applied to the field coils causes a small field current. Although small, this field current strengthens the magnetic field and allows the armature to generate a higher voltage. The higher voltage increases the field strength and so on. This process continues until the output voltage reaches the rated output of the generator. This process is called building.

13–2.4   Classification of Generators

Self-excited generators are classed according to the type of field connection they use. There are three general types of field connections—series-wound, shunt-wound (parallel), and compound-wound. Compound generators are further classified as cumulative-compound and differential-compound. These last two classifications are not discussed in this text.

Series-Wound Generator   In the series-wound generator, shown in Figure 13–6, the field windings are connected in series with the armature. Current that flows in the armature flows through the external circuit and through the field windings. The external circuit connected to the generator is called the load circuit.

A series-wound generator uses very low resistance field coils consisting of a few turns of large diameter wire. Resistance to current flow must be low, since load current must flow through the armature.

The voltage output increases as the load circuit starts drawing more current. Under low-load current conditions, the current that flows in the load and through the generator is small. Since small current means that a small magnetic field is set up by the field poles, only a small voltage is induced in the

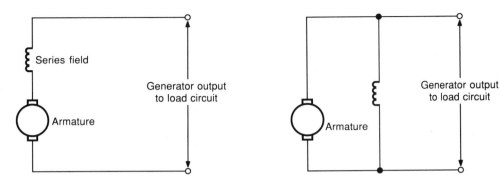

FIGURE 13–6 Series-wound DC generator

FIGURE 13–7 Shunt-wound DC generator

armature. If the resistance of the load decreases, the load current increases. Under this condition, more current flows through the field. This increases the magnetic field and increases the output voltage. A series-wound DC generator has the characteristic that the output voltage varies with load current. This is undesirable in most applications. For this reason, this type of generator is rarely used in daily practice.

Shunt-Wound Generators In a shunt-wound generator like the one shown in Figure 13–7 the field coils consist of many turns of small wire. They are connected in parallel with the load. In other words, they are connected across the output voltage of the armature. The armature current is equal to the sum of the field current and the load current. Since the field resistance is usually high, the field current is small compared to the load current.

Current in the field windings of a shunt-wound generator is independent of the load current since currents in parallel branches are independent of each other. The field current in the shunt generator, and therefore field strength, is not changed by load current. The output voltage remains more nearly constant than does the output voltage of the series-wound generator. The armature current, however, varies directly with the load current. As the load draws more current from the generator, the armature current increases.

In actual use, the output voltage in a DC shunt-wound generator varies inversely as load current varies. The output voltage decreases as load current increases because the voltage drop across the armature resistance increases ($E = IR$). Since the armature resistance is small, the decrease in output voltage is small.

Buildup in the Shunt Generator After the prime mover, which is a source of mechanical energy, brings the generator up to its proper speed, the generator output voltage must build up to the rated voltage. This building process must occur before connecting any load to the generator. A schematic diagram of the shunt generator is shown in Figure 13–8a without a load and with only a few turns of wire in the field.

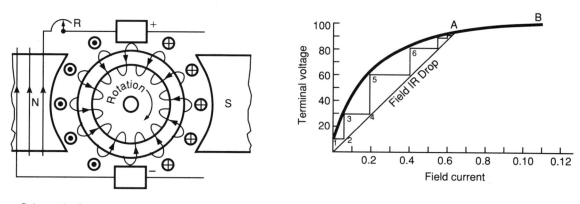

a. Schematic diagram

b. Field saturation curve

FIGURE 13–8 Buildup of the voltage of a DC shunt generator

The diagram in Figure 13–8b shows the field saturation curve for the generator. The strength of the field depends on a number of physical factors, including the permeability of the field pole pieces. As the current through the field windings increases, the field flux and the output voltage also increase. The field strength is, however, not linear with changes in field current, as Figure 13–8b shows. The field saturation curves bend toward the right at high values of field current. The decrease in output voltage is caused by the tendency of the field iron to saturate.

Line *OA* in Figure 13–8b represents the relationship between voltage and current in the field coil circuit for one value of field circuit resistance. Line *OA* is called an IR-drop curve for the field circuit. The IR drop is assumed to be a straight line in constant-temperature operation. The field rheostat allows adjustment of the field current over a wide range of values. The rheostat resistance is designed to be at least equal to the field winding resistance.

At start, the prime mover drives the generator to rated speed. The load is not connected. The armature conductors cut the small residual magnetic field present in the pole pieces, generating about 10 V across the brushes (point *1* on the curve). The left-hand rule for generators tells us that the generated voltage is applied to the field winding in a direction that adds to the strength of the residual magnetic field. Ten volts across the field will cause about 0.07 amperes of current to flow through the coils (see point *2* in Figure 13–8b).

This current strengthens the field further, and the armature voltage increases to 30 V, as shown at point *3* on the curve. During the process of voltage buildup, the effect of the armature circuit resistance is neglected, since the field current is only a small fraction of an ampere. A generated voltage of 30 V applied to the field causes a field current of 0.2 amperes, as shown in point *4* on the IR drop curve in Figure 13–8b. The 0.2 ampere current flow causes a generated voltage of 60 V, as seen at point *5*. The 60 V potential causes 0.4 amperes of current to flow, increasing the terminal voltage to 80 V (point *6*).

The rise in voltage continues until the field current increases to 0.6 amperes and the terminal voltage levels off at 90 V (point *A*). No further increase in generated voltage occurs for the given value of field circuit resistance. The amount of field saturation flux produced is enough to generate the 90 volts needed to circulate 0.6 amperes of current through the field coils. The field resistance needed to circulate this current is

$$\text{Rf} = \frac{V}{I} = \frac{90 \text{ V}}{0.6 \text{A}} = 150 \text{ } \Omega$$

Field saturation limits the generated voltage to this value as determined by the setting of the field rheostat.

To increase the shunt generator terminal voltage further, decrease the field rheostat. For example, if the terminal voltage is to be increased to 110 volts, the corresponding field current from the saturation curve is 1.2 amperes (point *B*). The field resistance would be

$$\text{Rf} = \frac{V}{I} = \frac{110 \text{ V}}{1.2 \text{ A}} = 91.8 \text{ } \Omega$$

If we decrease the field resistance from 150 Ω to 91.8 Ω, the terminal voltage will increase from 90 V to 110 V.

A common fault of DC generators is the failure to build. Four circumstances can cause failure to build. First, the pole pieces may not contain any residual magnetism. Physical shocks and heat commonly cause loss of the residual magnetic field. The residual magnetic field can be restored by connecting a DC voltage across the field for a short time. Make sure that the polarity of the applied voltage matches the voltage that will eventually be applied to the field. This procedure is called flashing the field. The second cause of failure to build is improper connection of the field windings. If the field windings are improperly connected, the generated current will oppose the residual magnetic field. The remedy for this malfunction is simply to reverse the field connections. Third, the field resistance may be too high or the field winding may be open. In this case, a check of the field resistance with an ohmmeter should reveal the problem. Finally, the armature winding may have an abnormally high resistance or may even be open. Again, an ohmmeter check will show this fault.

Shunt Generator Output Voltage Characteristics In a series-wound generator, output voltage varies directly with load current. In the shunt-wound generator, output voltage varies inversely with load current. We can see this output voltage behavior in Figure 13–9. Note that the terminal voltage of this shunt generator decreases slightly as the load current increases from no-load to rated full-load. The figure also shows that if a heavy overload occurs, the terminal output voltage falls more rapidly. Between the rated full-load and the

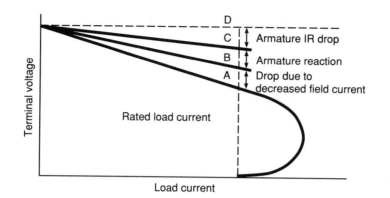

FIGURE 13–9 DC shunt generator external characteristics

breakdown point, the shunt field current reduces and so does the magnetization of the field. The dotted portion of curve *A* shows the way the terminal voltage falls below the breakdown point. In large generators, the breakdown point occurs at several times the rated load current. Generators are not designed to operate at these large values of load current. Most generators will overheat at twice the value of full-load current.

Curve *B* shows the external voltage of a shunt generator that has a constant field current with varying loads. Curve *C* represents the external voltage for the same range of loads without the effects of armature reaction and with a constant field current. Curve *D* shows the effects of the IR drop across the armature resistance. In summary, the output voltage of the shunt generator is well regulated, compared with the series generator. Designers keep the shunt generator output voltage from varying by keeping armature resistance and reaction low.

A combination of the two types of generators (series and shunt) can overcome the disadvantages of both. This combination of windings is called the compound-wound DC generator.

Compound-Wound Generator Compound-wound generators have a series-field winding in addition to a shunt-field winding as shown in Figure 13–10. The shunt and series windings are wound on the same pole pieces.

FIGURE 13–10 Compound generator schematic diagram

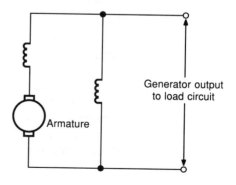

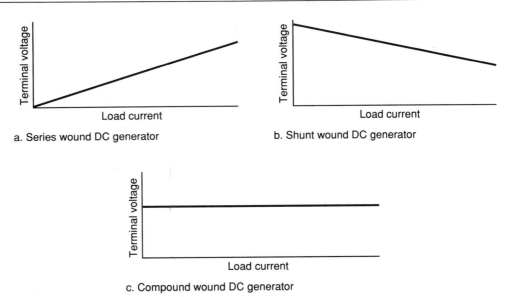

a. Series wound DC generator   b. Shunt wound DC generator

c. Compound wound DC generator

FIGURE 13–11 Output voltage characteristic of various generators

In the compound-wound generator, when load current increases, the armature voltage decreases, just as in the shunt-wound generator. This causes the voltage applied to the shunt-field winding to decrease, which results in a decrease in the magnetic field. This same increase in load current, since it flows through the series winding, causes an increase in the magnetic field produced by that winding.

By proportioning the two fields so that the decrease in the shunt field is just compensated by the increase in the series field, the output voltage remains constant. This is shown in Figure 13–11, which shows the output voltage characteristics of the series, shunt, and compound-wound generators. As you can see, by adjusting the effects of the two fields (series and shunt), a compound-wound generator provides a constant output voltage under varying load conditions. Actual curves are seldom, if ever, as perfect as shown.

Generator Construction Figure 13–12 shows an entire DC generator with the component parts installed. The cutaway drawing helps you to see the physical relationship of the components to each other. Note that the construction of the wound-field DC generator shown is almost identical to the wound-field DC motors studied in earlier chapters.

13–2.5 Voltage Regulation

The regulation of a generator refers to the voltage change that takes place when the load changes. It is usually expressed as the change in voltage from a no-load condition to a full-load condition, expressed as a percentage of full-load. Expressed as a formula,

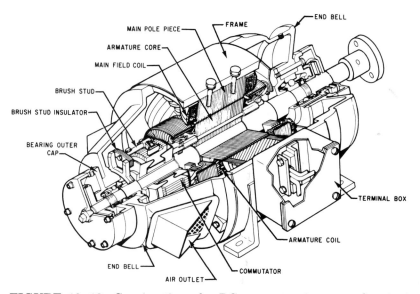

FRAME
MAIN POLE PIECE
ARMATURE CORE
MAIN FIELD COIL
BRUSH STUD
BRUSH STUD INSULATOR
BEARING OUTER CAP
END BELL
END BELL
AIR OUTLET
COMMUTATOR
ARMATURE COIL
TERMINAL BOX

FIGURE 13–12 Construction of a DC generator (cutaway drawing)

$$\frac{E_{nl} - E_{fl}}{E_{fl}} \times 100 = \text{percent of regulation}$$

where E_{nl} is the no-load terminal voltage and E_{fl} is the full-load terminal voltage of the generator. For example, to calculate the percent of regulation of a generator with a no-load voltage of 462 V and a full-load voltage of 440 V

No-load voltage = 462 V

Full-load voltage = 440 V

$$\text{Percent of regulation} = \frac{E_{nl} - E_{fl}}{E_{fl}} \times 100$$

$$= \frac{462 \text{ V} - 440 \text{ V}}{440 \text{ V}} \times 100$$

$$= \frac{22 \text{ V}}{440 \text{ V}} \times 100$$

$$= .05 \times 100$$

Percent of regulation = 5%

Note that the lower the percent of regulation, the better the generator. In the above example, the 5% regulation represented a 22-V change from no load to full load. A 1% change would represent a change of 4.4 volts which, of course, would be better.

13-2.6 DC Generator Voltage Control

Since a generator produces an output voltage, some control needs to be present to adjust the output voltage. Voltage control in generators is either manual or automatic. In most cases the voltage control process involves changing the resistance of the field circuit. By changing the field circuit resistance, the field current is controlled. Controlling the field current permits control of the output voltage. The major difference between the various voltage control systems is merely the way in which the field circuit resistance and the current are controlled.

The concept of voltage regulation should not be confused with voltage control. As described previously, voltage regulation is an internal action occurring within the generator whenever the load changes. Voltage control is an imposed action, usually through an external adjustment, for the purpose of increasing or decreasing the generator output terminal voltage.

Manual Voltage Control The hand-operated field rheostat shown in Figure 13-13 is a typical example of manual voltage control. The field rheostat is connected in series with the shunt field circuit. This provides the simplest method of controlling the terminal voltage of a DC generator.

This type of field rheostat contains tapped resistors with leads to a multiterminal switch. The arm of the switch may be rotated to make contact with the various resistor taps. This varies the amount of resistance in the field circuit. Rotating the arm in the direction of the lower arrow (counterclockwise) increases the resistance and lowers the output voltage. Rotating the arm in the direction of the raised arrow (clockwise) decreases the resistance and increases the output voltage.

Most field rheostats for generators use resistors of alloy wire. They have a high specific resistance and a low temperature coefficient. These alloys include

FIGURE 13-13 DC generator manual voltage control

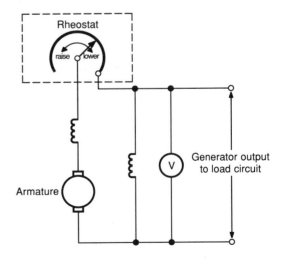

copper, nickel, manganese, and chromium. They are marked under trade names such as Nichrome, Advance, Manganin, and so forth. Some very large generators use cast-iron grids in place of rheostats and motor-operated switching mechanisms to provide voltage control.

13–2.7 Automatic Voltage Control

Automatic voltage control can be used where load current variations exceed the built-in ability of the generator to regulate itself. An automatic voltage control device senses changes in output voltage and causes a change in field resistance to keep output voltage constant.

The actual circuitry involved in automatic voltage control will not be covered in this text. Whatever control method is used, the range over which voltage can be changed is a design characteristic of the generator. The voltage can be controlled only within the design limits.

13–2.8 Parallel Operation of Generators

When two or more generators are supplying a common load, they are said to be operating in parallel. The purpose of connecting generators in parallel is simply to provide more current than a single generator is capable of providing. The generators may be physically located quite a distance apart. However, they are connected to the common load through the power distribution system.

There are several reasons for operating generators in parallel. The number of generators used may be selected according to the load demand. By operating each generator as near as possible to its rated capacity, maximum efficiency is achieved. A disabled or faulty generator can be taken off the line and replaced without interrupting normal operations.

13–2.9 Safety Precautions

You must always observe safety precautions when working around electrical equipment to avoid injury to personnel and damage to the equipment. Electrical equipment frequently has accessories that require separate sources of power. Lighting fixtures, heaters, externally powered temperature detectors, and alarm systems are examples of accessories whose terminals must be deenergized. When working on DC generators, you must check to ensure that all such separate circuits are deenergized and tagged before you attempt any maintenance or repair work. You must use the greatest care when working on or near the output terminals of DC generators.

13–2.10 Amplidyne

An amplidyne is a special purpose DC generator. This type of DC generator supplies large, precisely controlled currents to large DC motors that drive heavy physical loads. As in any generator, the amplidyne must have a prime mover coupled to the generator. The generator is connected to behave as a high-gain amplifier. In the amplidyne, a small DC voltage controls a large current in the output. In a normal separately-excited DC generator, a small DC voltage applied to the field winding controls the output voltage of the generator. An increase in the voltage in the separately-excited field causes an increase in the output voltage of the generator. In a typical generator, a change in the field voltage from 0 V to 3 V DC may cause the generator output voltage to vary from 0 V DC to 300 V DC. If we consider the voltage applied to the field as an input and the 300 V taken across the brushes as an output, the gain will be 100.

$$\text{Gain} = \frac{V_{out}}{V_{in}} = \frac{300 \text{ V}}{3 \text{ V}} = 100 \qquad \textbf{(eq. 13–1)}$$

This equation states that the output voltage will be 100 times greater than the input voltage. In the amplidyne generator, gains can increase to greater than 10,000.

The diagram in Figure 13–14 shows a separately-excited DC generator. The 10 V potential across the field causes 1 ampere of field current to flow, drawing 100 W of power from the supply. Let us assume that this generator will produce 87 amperes of current at 115 V. This represents an output power of 10,000 W. The power gain of this generator is 100. One hundred watts of input power controls 10,000 watts of output power.

An amplidyne is a special type of DC generator. The following changes will convert the separately-excited generator we have just discussed into an amplidyne.

The first step is to short the brushes together, as shown in Figure 13–15a. This removes almost all of the resistance in the armature circuit. Because of the very low resistance in the armature circuit, a much smaller amount of field flux will produce full-load current in the armature (about 87 amperes in our example). The smaller control field now requires a control voltage of only one volt to produce the full-load current at the output.

The next step is to add another set of brushes. This new set of brushes becomes the output brushes of the amplidyne. The new set of brushes is placed

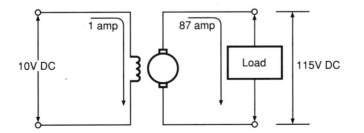

FIGURE 13–14 Shunt generator—input 100 W, output 10,000 W

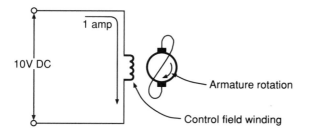

a. Separately excited generator with shorted armature

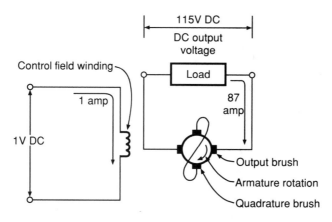

b. Separately excited generator with shorted armature and an
additional set of brushes

FIGURE 13–15 Developing the amplidyne from a separately-excited DC generator

against the commutator in a position perpendicular to the original brushes, as
shown in Figure 13–15b. The previously shorted brushes are now called the
quadrature brushes. This is because they are connected in quadrature (perpendicular to the output brushes). The output brushes are in line with the armature flux. The output brushes, therefore, pick off the voltage induced in the
armature windings at this point. The voltage at the output of the generator
will be the same, 115 V DC.

As we have seen, the original separately-excited generator in Figure
13–14 produced a 10,000-W output with a 100-W input. The amplidyne of
Figure 13–15 produces the same 10,000-W output with only a 1-W input. A
device that produces a 10,000-W output with an input of 1 W has a power gain
of 10,000. The power gain of the original generator has been greatly increased.

The amplidyne is sometimes used to excite the fields of large synchronous
motors and generators. Another use of the amplidyne is in the positioning of
heavy loads using DC motors. Let us assume that a large turning force is

needed to rotate a heavy object, such as an antenna, to a very precise position. If we have a low power, small voltage that represents the amount of antenna motion we need, we can use that voltage to control the field winding of an amplidyne. Because of the amplidyne's ability to amplify, its output can be used to drive a powerful DC motor, which turns the heavy object (in this case, the antenna). When the source of the input voltage senses the correct movement of the object, the input voltage drops to zero. The field is no longer strong enough to allow an output voltage to be developed. The motor then ceases to drive the object.

13–2.11   Rototrol

Since the development of the amplidyne, several manufacturers have introduced similar products. The amplidyne is based on a separately-excited DC generator, while the rototrol, developed by Westinghouse, is based on a series DC generator. In the series generator, the field is in series with the current flowing through the armature. The series DC generator is also a self-excited generator. This means that under a no-load condition, the field is made up of residual magnetism only. As the load increases, the field strength increases, as does the generator DC output voltage. To increase or decrease the output voltage, we must be able to increase and decrease the flux in the field.

The rototrol has additional field windings added to the generator. These additional windings allow the field flux to be varied, which, in turn, changes the series generator output voltage. Figure 13–16a shows a simplified schematic diagram of a rototrol. The tuning resistor allows the output voltage curve to be adjusted to a linear portion of the characteristic curve. Note the two extra fields shown in the diagram. The pattern or reference field normally has a constant voltage applied to it. The pilot or control field is normally connected to a variable DC potential. When the output voltage is at the desired value, the flux from the pilot field is equal to and opposite to the flux from the pattern field.

The operation of the rototrol is seen from the diagram in Figure 13–16b. The tuning resistor is adjusted until Vo is at the desired voltage. A decreased output load on the rototrol will increase the output voltage. The pilot-field flux will then exceed the pattern-field flux. The resultant flux in the two fields will oppose the flux in the self-energizing series field, reducing the output voltage back to a value close to the original.

13–2.12   Regulex

The regulex, developed by Allis-Chalmers, is very similar to the rototrol, as can be seen from Figure 13–17. The main difference between these two is in the field winding. In the rototrol, the field is in series with the armature. In the regulex, the field is a shunt field, in parallel with the armature. The tuning resistor allows regulex adjustment for linear characteristics. The regulex has

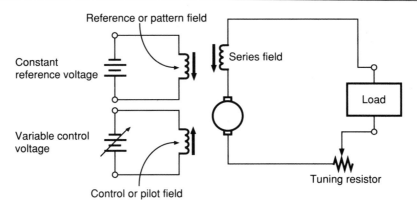

a. Basic rototrol circuit

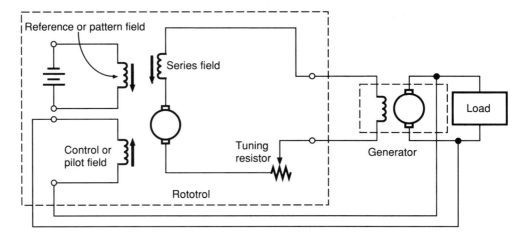

b. Rototrol controlling the output voltage of a separately-excited generator

FIGURE 13–16   Rototrol construction and operation

two fields, a reference or pattern field and a control field. The purpose of these two fields is identical to the purposes of the similarly named field in the rototrol.

13–3   PRACTICAL ALTERNATORS

The AC generator described in Chapter 1 was very simple to help us understand the basic AC generator principle. The rest of this chapter will describe the practical alternator commonly used in industry today. We will cover such concepts as prime movers, field excitation, armature characteristics and limi-

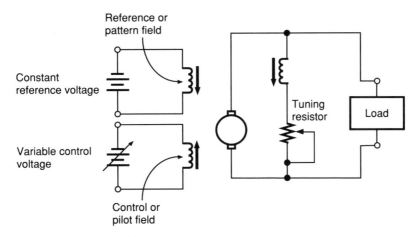

FIGURE 13–17 Basic regulex circuit

tations, single-phase and polyphase alternators, controls, regulation, and parallel operation.

13–3.1 Types of AC Generators

Industry uses several different types of AC generators or alternators, as they are sometimes called. All types of AC generators, however, perform the same function of changing mechanical motion to alternating current electrical energy. The two basic types of AC generators are the rotating field and the rotating armature.

Rotating Armature Alternators In the rotating armature AC generator, an electromagnet provides a stationary magnetic field called the stator. The armature (usually called the rotor) revolves in the stator field. As the rotor revolves, its windings cut lines of force in the field, producing AC voltage. In this type of generator, the AC output voltage is taken from slip rings.

The rotating armature AC generator is not widely used in industry for several reasons. A major reason is that this type of generator has limited output power. The amount of power output is restricted by the low current-carrying ability of the slip-rings and brushes.

The rotating-armature alternator is similar in construction to the DC generator in that the armature rotates in a stationary magnetic field as shown in Figure 13–18. In the DC generator, the EMF generated in the armature windings is converted from AC to DC by means of the commutator. In the alternator, the generated AC is brought to the load unchanged, by means of slip rings. The rotating armature is found only in alternators of low power rating and generally is not used to supply electric power in large quantities.

Rotating-Field Alternators The rotating-field alternator has a stationary armature winding and a rotating-field winding as shown in Figure 13–19. The

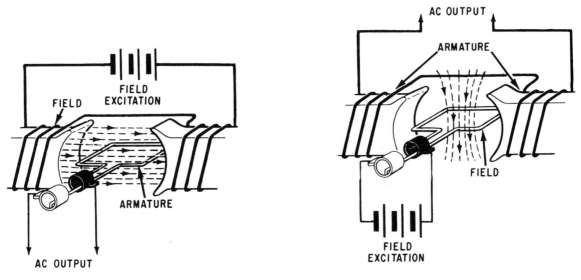

FIGURE 13-18 Rotating-armature alternator   **FIGURE 13-19** Rotating-field alternator

advantage of having a stationary armature winding is that the generated voltage can be connected directly to the load.

A rotating armature requires slip rings and brushes to conduct the current from the armature to the load. The armature, brushes, and slip rings are difficult to insulate, and arc-overs and short circuits can result at high voltages. For this reason, high-voltage alternators are usually rotating-field alternators. Since the voltage applied to the rotating field is low voltage DC, the problem of high voltage arc-over at the slip rings does not exist.

The stationary armature, or stator, of this type of alternator holds the windings that are cut by the rotating magnetic field. The voltage generated in the armature as a result of this cutting action is the AC power that will be applied to the load.

The stators of all rotating-field alternators are similar. The stator consists of a laminated iron core with the armature windings embedded in this core as shown in Figure 13-19. Note that the core is attached to the stator frame.

13-3.2 Functions of Alternator Components

A typical rotating-field AC generator consists of an alternator and a smaller DC generator built into a single unit. The output of the alternator section supplies alternating voltage to the load. The only purpose for the DC exciter generator is to supply the direct current required to maintain the alternator field. This DC generator is referred to as the *exciter*. Figure 13-20 is a simplified schematic of the AC generator, including the exciter.

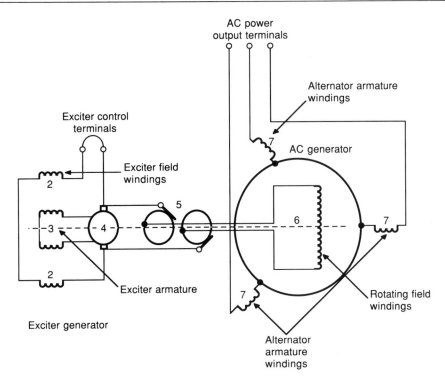

FIGURE 13–20 AC generator schematic diagram

The exciter is a DC, shunt-wound, self-excited generator. The exciter shunt field, shown in Figure 13–20, creates an area of intense magnetic flux between its poles. When the exciter armature rotates in the exciter-field flux, voltage is induced in the exciter armature windings. The output from the exciter commutator is connected through brushes and slip-rings to the alternator field. Since this is direct current, already converted by the exciter commutator, the current always flows in one direction through the alternator field. Thus, a fixed-polarity magnetic field is maintained at all times in the alternator field windings. When the alternator field is rotated, its magnetic flux is passed through and across the alternator armature windings.

The same types of exciters used in the synchronous motor are used to excite the synchronous alternator. The illustration described in Figure 13–20 uses a DC generator with brushes and a commutator. Both brushless excitation and static excitation using voltage regulators provide a method of excitation with lower maintenance costs. In larger alternators, the amplidyne principle is sometimes used to provide control over the excitation current.

The armature is wound for a three-phase output, which will be covered later in this chapter. Remember, a voltage is induced in a conductor if it is stationary and a magnetic field is passed across the conductor. A voltage is produced in the same way if the field is stationary and the conductor is moved. The alternating voltage in the AC generator armature windings is connected through fixed terminals to the AC load.

13–3.3 Prime Movers

All generators, large and small, AC and DC, need a source of mechanical power to turn their rotors. This source of mechanical energy is called a prime mover.

Prime movers are divided into two classes for generators: high-speed and low-speed. Steam and gas turbines are high-speed prime movers, while geared turbines, internal combustion engines, water, and electric motors are considered low-speed prime movers.

The type of prime mover plays an important part in the design of alternators since the speed at which the rotor turns determines certain characteristics of alternator construction and operation.

13–3.4 Alternator Rotors

Two types of rotors are used in rotating-field alternators: turbine-driven and salient-pole rotors.

The turbine-driven rotor (also called the smooth rotor) shown in Figure 13–21 is used when the prime mover is a high-speed turbine.
The windings in the turbine-driven rotor are arranged to form two or four distinct poles. The windings are firmly embedded in slots to withstand the tremendous centrifugal forces encountered at high speeds.

The salient-pole rotor (often called a wound rotor) shown in Figure 13–22 is used in slow-speed alternators. The salient-pole rotor often consists of several separately wound pole pieces, bolted to the frame of the rotor. You will

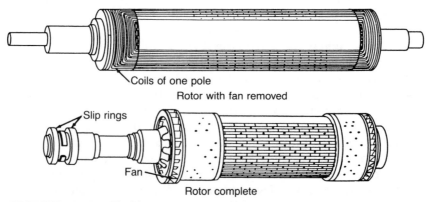

FIGURE 13–21 Turbine rotor construction

FIGURE 13–22 Salient-pole rotor

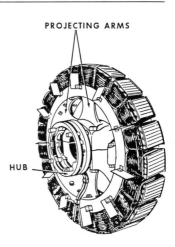

PROJECTING ARMS

HUB

note the similarities of construction between this rotor and the salient-pole rotor of the synchronous motor studied in an earlier chapter.

If you could compare the physical size of the two types of rotors with the same electrical characteristics, you would see that the salient-pole rotor has a greater diameter. At the same number of revolutions per minute, it has a greater centrifugal force than does the turbine-driven rotor. To reduce this force to a safe level so that the windings will not be thrown out of the machine, the salient pole is used only in low-speed designs, typically under 1200 rpm. The high-speed turbine driven rotor sees speeds in excess of 1200 rpm.

13–3.5 Alternator Characteristics and Limitations

Alternators are rated according to the voltage they are designed to produce and the maximum current (or load) they are capable of providing. Normal load ratings specify the amount of current that the generator can handle continuously. Its overload rating is the above-normal load it can carry for a specified length of time only. The load rating of a particular generator indicates the amount of heat the generator can stand. Since current flow causes the heat, the generator's overload rating identifies very closely with its current-carrying capacity.

The maximum current that can be supplied by an alternator depends upon the maximum heating loss that can be sustained in the armature. This heating loss (which is an I^2R power loss) acts to heat the conductors and, if excessive, destroys the insulation. Alternators are rated in terms of this current and in terms of the voltage output. The alternator rating in small units is in volt-amperes. In large units it is termed kilovolt-amperes. In either case, the specified armature current and voltage are specified at a particular frequency and power factor. The power factor normally specified is 80% lagging. For example, a single-phase alternator designed to give 100 amperes at 1,000 V is rated at 100 kVA. This machine can supply a 100-kW load at unity power factor or an 80-kW load at 80% power factor.

The AC generator usually also specifies the maximum heating loss that can be sustained by the field. In the example above, if the generator supplied a 100-kVA load at 20% power factor, the current needed to keep the desired terminal voltage would cause excessive heating in the field.

When an alternator leaves the factory, it is designed to do a very specific job. The speed at which it is designed to rotate, the voltage it will produce, the current limits, and other operating characteristics are built in. This information is usually stamped on a nameplate on the case so that the user will know the limitations of the generator.

Single-Phase Alternators A generator that produces a single, continuously alternating voltage is known as a single-phase alternator. All of the alternators discussed so far fit this definition. The stator (armature) windings are connected in series. The individual voltages add to produce a single-phase AC voltage. Figure 13–23 shows a basic alternator with its single-phase output voltage.

The definition of phase learned in studying AC circuits may not be helpful in studying single-phase alternators. "Out of phase" meant "out of time" when applied to AC circuits. It may now be easier to think of the word phase as meaning voltage, as in single-voltage. The need for a modified definition of phase in this usage will become easier to recognize as we continue.

Single-phase alternators are found in many applications. They are most often used where the loads being driven are relatively light. The reason for this will be more apparent as we get into multiphase alternators (also called polyphase). Power used in homes, farms, shops, light industry, and ships to operate portable tools and small appliances is single-phase power. Single-phase alternators always generate single-phase power. However, all single-phase power does not come from single-phase alternators.

Three-Phase Alternators As discussed in Chapter 1, the three-phase generator has three single-phase windings. Each winding is spaced so that each induced voltage is 120° out-of-phase with the voltage in the other two wind-

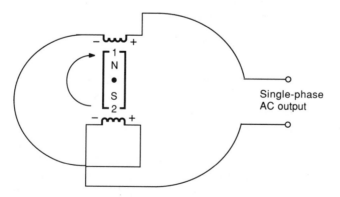

FIGURE 13–23 Single-phase alternator

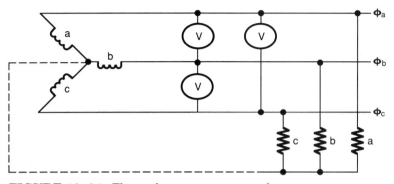

FIGURE 13–24 Three-phase, wye-connected system

ings. Three-phase generators can be connected either in delta or wye configurations. The wye configuration is shown in Figure 13–24. This three-phase generator supplies three separate loads. When the loads are unbalanced, drawing unequal currents, the neutral (seen as a dotted line) is added. Note that the neutral serves as a common return circuit for all three phases. The neutral also maintains a voltage balance across the loads. No current flows in the neutral unless the loads become unbalanced. This system is referred to as a 3-phase, 4-wire system and is widely used in industry and for aircraft AC-power systems.

A three-phase stator can also be connected in delta configuration, as shown in Figure 13–25. In the delta-connected alternator, the start end of one phase winding is connected to the finish end of the next. The three junction points are connected to the line wires as seen in the figure. This system supplies a three-phase, delta-connected load at the right-hand end of the three-phase line. Because the phases are connected directly across the line wires, the phase voltage is equal to the line voltage. When the generator phases are properly connected in delta under no-load conditions, little current flows in the delta. If, however, any one of the phases is reversed, a short-circuit occurs that will probably damage the stator windings.

To avoid connecting a phase in reverse we must be able to test the circuit before closing the delta. One way to do this is to connect a voltmeter to mea-

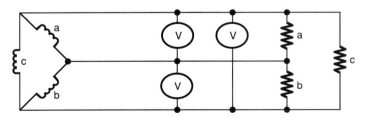

FIGURE 13–25 Three-phase, delta-connected system

FIGURE 13–26 Measuring the open spot in the delta with a voltmeter

sure the open spot in the delta, as shown in Figure 13–26. If the voltmeter reads a significant voltage, the delta should not be closed. For example, if the voltmeter reading is twice the phase voltage, then one of the coils is reversed.

13–3.6   Power Measurement in Three-Phase Alternators

The wattmeter connections for measuring true power are shown in Figure 13–27 for both delta and wye-connected generators. The method shown in Figure 13–27a uses three wattmeters with their current coils inserted in series

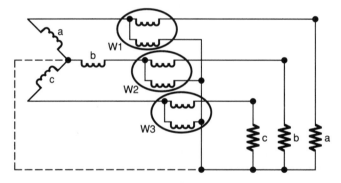

(a) Three-wattmeter method wye connection

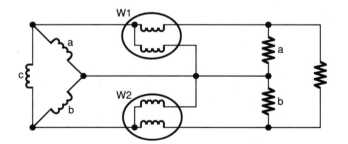

(b) Two-wattmeter method delta connection

FIGURE 13–27   Wattmeter connection for power measurement

with the line wires. The potential coils are connected between the line and neutral wires. The total true power is equal to the arithmetic sum of the reading on the three wattmeters.

The method shown in Figure 13–27b uses two wattmeters with their current coils connected in series. Their potential coils connect between the two line wires with the current connections and the third wire that does not have a current coil. The total true power is equal to the algebraic sum of the two wattmeter readings. If one meter reads backwards, its potential coil connections need to be reversed. After reversal, the total true power is then equal to the difference between the two wattmeter readings.

13–3.7 Frequency of the Generated Voltage

The frequency of the AC generator voltage depends on the speed of the rotor and the number of poles. The faster the prime mover turns the rotor, the higher the frequency. The more poles on the rotor, the higher the frequency of the generated voltage at a given rotor speed. When a rotor has rotated so that a north and south pole have passed by one winding, the voltage induced will have gone through one complete cycle. For a given output frequency, the greater the number of poles, the lower the rotor speed must be. For example, a two-pole generator must rotate at twice the speed of a four-pole generator to produce the same frequency in the generated voltage. The frequency of the generator in Hz (cycles per second) is related to the number of poles and the speed,

$$f = \frac{PN}{120}$$

where P is the number of poles and N the speed in rpm. For example, a two-pole 3,600 rpm generator has a frequency of

$$f = \frac{2 \times 3,600}{120} = 60 \text{ Hz}$$

A four-pole 1,800 rpm generator has the same frequency, 60 Hz. A six-pole 500 rpm generator has a frequency of 25 Hz.

Parallel Operation of Alternators Three situations demand that alternators be connected in parallel. First, paralleling alternators increases the amount of load current available. Recall from your study of power supplies that we can increase the amount of current from a power supply by connecting two supplies in electrical parallel. The same reasoning holds true for the alternator, which is an AC power supply. Second, paralleling permits a power reserve. In times where an unpredicted load demand occurs, a reserve alternator can supply the necessary extra current. Third, paralleling allows shutting down an alternator without interrupting service. Paralleling alternators can be thought of as a meshing of gears in the mechanical world. For gears to mesh properly, both must be synchronized and running at the same speed. If gears are meshed when running at different speeds, the gears will be damaged. Likewise, alternators running at different frequencies and voltages can be damaged when connected in parallel. In small generators, incorrect paralleling will usually result in a blown circuit breaker. In large generators, stators can be torn off their mounts with great explosive force. At the very least, equipment will be damaged. Any personnel in the vicinity may be seriously injured. NEVER connect two or more generators in parallel without instructions from your supervisor or instructor.

To avoid damage, alternators must be synchronized as closely as possible before connecting them. In this process of synchronization, one alternator is placed on line or on the bus. This alternator is known as the bus generator. The other alternator (known as the incoming generator) is then synchronized to the bus generator prior to connecting it to the bus.

Three conditions must be met before the generators can be connected in parallel safely. First, the generators must have equal terminal voltages. We can test for this condition by measuring the terminal voltage of the incoming generator. If the voltage is not equal to the other generators on line, the field voltage may be increased to make the voltages equal. Second, the frequencies of the incoming generator must be equal to the others. Again, frequency can be measured by a frequency meter or a frequency counter. If the frequency of the incoming generator is not equal to the bus frequency, the incoming generator prime mover speed must be adjusted. Third, the incoming generator must be in phase with the bus and must have the proper phase sequence. Two alternators may have the same frequency but be out of phase with each other. The incoming alternator may, for example, be lagging behind the bus by 45°. Phases must also be in the same sequence. The bus may have a phase sequence of ABC, while the incoming generator has a sequence of ACB. Before the incoming generator can come on line, the B and C phases must be reversed. If the incoming generator has the sequence ACB and the bus generator has a sequence of ABC, connecting the incoming generator will connect phase C to phase B on one line and B to C on the other. Since the voltages in the two lines are out of phase, heavy short-circuit currents will flow between them, causing heavy torque fluctuations. The torque fluctuations are caused by the reversal of the stator CEMFs, which add to the line CEMFs rather than opposing them. A phase sequence meter can be used to determine the phase sequence of the bus generator. If the incoming generator has the wrong phase sequence, it may be corrected by reversing the connection of any two of the three phase leads.

In actual industrial practice, the incoming machine runs at a voltage slightly higher than the bus and at a frequency slightly faster than the bus frequency. This "high and fast" mode insures that the incoming machine does not drag down the bus when it is put on line.

13–3.8 Synchronizing Alternators

One way to synchronize alternators is to use synchronizing lamps, as shown in Figure 13–28. Note that the lamps are connected directly between the incoming generator's output and the bus. In this way, the two AC sources can be synchronized before the incoming generator's main power contactor is closed. At the instant the two generators are synchronized, lamps L2 and L3 will glow with maximum brightness and L1 will be dark.

Let us assume that the incoming generator is lagging. All three lamps will glow steadily since the voltage across them is the difference in the frequency of the two generators. As the lagging generator is accelerated, the frequency of the potential applied to the lamps decreases until the light flickers visibly. The flickering will have a rotating sequence if connections are correct.

At a point near synchronism, lamp L1 will be dark because it is connected between like phases. The two-phase voltage will be so nearly synchronized that the difference in voltage will not be able to light the lamp. Even so, there may be sufficient difference to damage the system if connected at this time. The lamps L2 and L3 prevent this. Under perfect synchronization, the phase voltages across L2 and L3 are both 120° apart because of their cross-connection. At perfect synchronization, both lamps should be equally bright. If the generators are even slightly out of phase and nearly synchronized, one of the lamps will be increasing in voltage as the other is decreasing. This action would cause a visible difference in the intensity between L2 and L3. The incoming generator may then be adjusted to perfect synchronism by adjustments

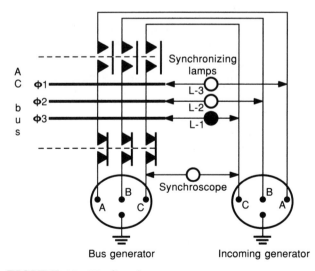

FIGURE 13–28 Synchroscope

that make both *L2* and *L3* equal in intensity while *L1* is off. This procedure is based on paralleling two 120-VAC alternators, such as those found in school laboratories. A more practical method of synchronizing higher voltage alternators uses the synchroscope.

Closed synchronization of alternators can be achieved by using a synchroscope. A drawing of a synchroscope is shown in Figure 13–29. Inside the synchroscope is a small induction motor attached to a shaft that is free to move in a complete circle. The shaft is attached to the pointer on the face of the synchroscope. The voltage from the two alternators powers the induction motor. A difference in frequencies between the two alternators causes the pointer to turn. If the incoming generator is too slow, the pointer will be turning in the slow direction, as seen in Figure 13–29. As the speed of the incoming generator is increased, the speed of the pointer's rotation will decrease. The pointer will stop moving when the incoming generator's frequency and the bus are equal. They will not be in phase, however, until the pointer points straight upwards. At that time, the incoming alternator is synchronized and may be put on the bus.

13–3.9   Principles of Alternator Voltage Control

In an alternator, the amount of voltage produced depends on three factors: the number of conductors in series per winding, the speed of rotation, and the strength of the magnetic field. Any of these three factors can be used to adjust the alternator output voltage. The number of windings is normally fixed by the manufacturer. Since the frequency of the alternator must be held constant under normal operation, it is not feasible to change the speed of the prime mover. The only practical method of output voltage adjustment is the strength of the magnetic field. Since the field is produced by an electromagnet, the field strength can be varied by changing the current through the coil. Adjusting the field voltage causes a change in the field current. Variation in the exciter armature DC output voltage changes the AC generator field strength. The alternator output voltage depends directly on the value of the exciter output voltage. This relationship allows a large AC potential to be controlled by varying a small DC potential.

FIGURE 13–29   Diagram of a synchroscope

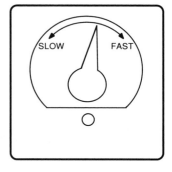

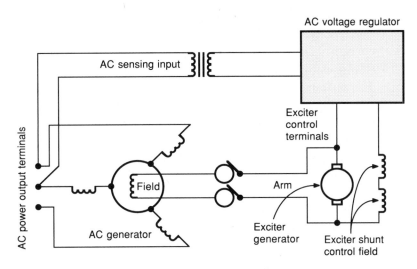

FIGURE 13–30 Automatic control of alternator output voltage using a voltage regulator

A voltage regulator provides voltage control of the DC exciter, which is usually a shunt DC generator. The purpose of the alternator voltage regulator is to keep the alternator output voltage constant under changing load conditions. We can see how this is done by examining Figure 13–30. If the load increases, pulling the alternator output voltage down, the voltage regulator senses this through the AC sensing input. The voltage regulator increases the voltage to the shunt field, producing more field current. The alternator field increases, causing the alternator output voltage to increase to a value close to its original value.

CHAPTER SUMMARY

- A generator produces electricity by electromagnetic induction.
- Commutation is the process used to get current from a DC generator. The coil connections must be reversed as the coil passes through the neutral plane.
- There are two basic classifications of DC generators: self-excited and separately-excited.
- The separately-excited generator receives current for the field coils from an outside source.
- The self-excited generator uses its own output voltage to supply the field coils.
- The three basic connections of self-excited generators are the series, shunt, and compound field connections.

- Series-wound DC generators have field windings and armature windings connected in series. The output voltages vary greatly with load current.

- Shunt-wound DC generators have field windings connected in parallel. The output voltage varies inversely with load current.

- The compound-wound DC generator has both series and shunt-wound fields. Its performance is somewhere between the pure series and pure shunt-wound generator performance.

- The AC generator exists in two basic forms: the rotating field and rotating armature types.

- In the rotating-field generator, two basic forms of rotors are normally used: the turbine-driven and the salient-pole.

- The device that drives the rotor of a generator is called the prime mover.

- Most AC generators, or alternators as they are sometimes called, produce three-phase AC.

- The three-phase AC can be connected in wye or delta, depending on the application.

- When placing more than one generator on line, all generators must be synchronized.

QUESTIONS AND PROBLEMS

1. What component causes a generator to produce DC rather than AC at its output terminals?

2. How can field strength be varied in a practical DC generator?

3. Why is the drum-type armature preferred over the gramme-ring?

4. What are the three classifications of self-excited DC generators?

5. A DC generator's no-load voltage is 100 V. Under full-load conditions, the voltage drops to 95 V. Calculate the voltage regulation.

6. Most AC generators have a small DC generator built into them. What is the name of this generator? What is its purpose?

7. What type of prime mover requires a specially designed high-speed alternator?

8. In a three-phase alternator, what is the phase relationship between the individual output voltages?

9. What two factors determine the frequency of the output voltage of an alternator?

10. How is output voltage controlled in practical alternators?

11. What generator characteristics must be considered when alternators are synchronized for parallel operation?

APPENDIX A

Section 12.53 of NEMA MG 1–1978, entitled "Motors and Generators," states that the nominal efficiency figure on the motor nameplate should be the average efficiency determined by testing a large group of the same type of motors. NEMA further sets the minimum efficiency that can be associated with a given nominal efficiency. Older motors (those constructed before 1982) were permitted to have a letter designating the efficiency of the motor. Since 1982, NEMA requires efficiencies to be expressed in a numerical value, rather than by an index letter. The nominal and minimum efficiencies are given below.

TABLE A–1 Motor Nameplate Efficiencies

Nominal Efficiency	Minimum Efficiency	Nominal Efficiency	Minimum Efficiency
95	94.1	80.0	77.0
94.5	93.6	78.5	75.5
94.1	93	77.0	74.0
93.6	92.4	75.5	72.0
93	91.7	74.0	72.0
92.4	91.0	72.0	68.0
91.7	90.2	70.0	66.0
91	89.5	68.0	64.0
90.2	88.5	66.0	62.0
89.5	87.5	64.0	59.5
88.5	86.5	62.0	57.5
87.5	85.5	59.5	57.5
86.5	84	57.5	52.5
85.5	82.5	55.0	50.5
84.0	81.5	52.5	48.0
82.5	80.0	50.5	46.0
81.5	78.5		

TABLE A–2   Selected Metric—English/English—Metric Conversions

Quantity	SI Unit	English to SI	SI to English
Force	newton (N)	1 oz. = 0.278 N 1 lb. = 4.448 N 1 kilopound = 9.807 N	1 N = 3.597 oz. = 0.225 lb. = 0.102 kp
Energy (work)	joule (J)	1 Btu = 1055.06 J 1 kwh = 3.6×10^6 J 1 Ws = 1 J 1 kcal = 4186.8 J	1 J = 9.478×10^{-4} Btu = 2.778×10^{-7} kwh = 1 Ws = 2.389×10^{-4} kcal
Power	watt (W)	1 HP = 746 W	1 W = 1.341×10^{-3} HP
Torque	newton-meter (N·m)	1 lb. ft. = 1.356 N·m 1 oz. in. = 7.062×10^{-3} N·m 1 kilopoundmeter = 9.807 N·m 1 lb. in. = 0.113 N·m	1 N·m = 0.737 lb. ft. = 8.851 lb. in. = 0.102 kpm = 141.61 oz. in.
Temperature	degree Celsius (°C)	$F = (C \times \frac{9}{5}) + 32$	$C = (F - 32) \times \frac{5}{9}$

TABLE A–3   Comparing AC Motors based on two-pole, three-phase machines at 10,000 rpm 2 HP with no cooling fans

Parameter	Type of Motor				
	Synchronous Reluctance	Synchronous PM (ferrite)	Synchronous PM (rare earth)	Synchronous Separately excited	Induction Motor
Efficiency (full-load)	70%	72%	82%	74%	75%
Power Factor	0.63	0.64	0.85	0.8	0.85
Starting Current (times full load current)	10X	—	9X	8X	7X
Internal Heating	Fair/High	Fair	Low	Fair	Fair
Pull-out Torque (% of FLT)	160%	160%	160%	140%	N/A
External Excitation	No	No	No	Yes 60 Watts	No

APPENDIX B: CHECKING SEMICONDUCTORS WITH AN OHMMETER

One of the most common ways for checking any semiconductor device is to use an ohmmeter to measure relative resistance between the device's different terminals. Ohmmeter checks are usually used to determine if the device has any gross defects (short, open, very high leakage current, etc.). In addition, the ohmmeter can often be used to determine the lead configuration of the device if it is not known; for example, to establish which lead is the emitter, which is the base and which is the collector of a transistor.

A typical ohmmeter applies a voltage through a series resistor to any device connected to its terminals. The amount of current which flows as a result is an indication of the resistance of the device. The ohmmeter scale is calibrated directly in ohms to give a reading of the device resistance. Since most semiconductor devices are polarity-sensitive, it is important to note how the ohmmeter is connected to the device. The ohmmeter terminal labeled *common*, or *ground*, is the *negative* terminal; the other one is the *positive* terminal.

The following table will describe the common ohmmeter checks for the various devices. These should be studied and understood thoroughly since they are an important part of a technician's laboratory skills.

Device	Positive ohmmeter lead tied to	Negative lead tied to	Expected results
P-N diodes (also zener diodes, photodiodes, and P-N junctions of any device)	anode (forward bias)	cathode	Low resistance; usually 10–1000 Ω depending on type of diode and ohmmeter range used. Reading should be smaller on lower ohmmeter ranges.
	cathode (reverse bias)	anode	Very high resistance; typically 1 MΩ or greater for germanium and 10 MΩ or greater for silicon.
Tunnel diodes	anode (forward bias) cathode (reverse bias)	cathode anode	Very low resistance in either direction. Resistance is usually slightly lower when anode is negative and cathode positive.

Device	Positive ohmmeter lead tied to	Negative lead tied to	Expected results
Photoconductive cells	either end	either end	Ohmmeter reading should be the same in either direction and depends on cell sensitivity and amount of ambient light. Cell resistance should increase considerably when cell is darkened.
Photodiodes, photovoltaic cells, and LEDs	Same as for P-N diodes		
NPN transistor	emitter	base	High resistance (reverse-biased junction) unless ohmmeter voltage exceeds E-B breakdown voltage, BV_{EBO}.
	base	emitter	Low resistance (forward-biased junction).
	collector	base	High resistance.
	base	collector	Low resistance.
	emitter	collector	High resistance in both directions. Reading is also usually higher when emitter is negative and collector positive.
	collector	emitter	
PNP transistor	Same as for NPN, except that all polarities are reversed		
Four-layer diode and silicon unilateral switch (SUS)	anode (forward bias)	cathode	High resistance; $> 1 \text{ M}\Omega$.
	cathode (reverse bias)	anode	High resistance; usually greater than in other direction but may be impossible to detect with some ohmmeters.
SBS and DIAC	either end	either end	$> 1 \text{ M}\Omega$ in either direction.

Device	Positive ohmmeter lead tied to	Negative lead tied to	Expected results
Silicon controlled rectifier (SCR), light-activated SCR, gate-controlled switch (GCS)	anode (forward bias)	cathode	$> 1\,M\Omega$ (might be less for very high current SCRs).
	cathode (reverse bias)	anode	$> 1\,M\Omega$ but usually greater than in forward direction.
	gate cathode	cathode gate	Similar to P-N diode with low resistance when gate is positive and high resistance when gate is negative.
	gate anode	anode gate	$> 1\,M\Omega$ in either direction.
TRIAC	either anode 1 or 2	either anode 1 or 2	Very high resistance; $> 1\,M\Omega$ but may be less in very high current TRIACs.
	gate anode 1	anode 1 gate	Low resistance in both directions.
	gate anode 2	anode 2 gate	High resistance in both directions.
Unijunction transistor	base 1 base 2	base 2 base 1	Same resistance in either direction; typically $4\,k\Omega$–$10\,k\Omega$.
	emitter (forward bias)	base 1	Moderate resistance; usually in the $3\,k\Omega$–$15\,k\Omega$ range.
	base 1	emitter	Very high resistance; $> 1\,M\Omega$.
	emitter (forward bias)	base 2	Moderate resistance; usually in the $2\,k\Omega$–$10\,k\Omega$ range. Usually less than the emitter-base 1 forward resistance.
	base 2	emitter	Very high resistance; $> 1\,M\Omega$.

Device	Positive ohmmeter lead tied to	Negative lead tied to	Expected results
Programmable UJT (PUT)	anode	cathode	High resistance; > 1 MΩ.
	cathode	anode	
	anode	gate	Low resistance (forward bias).
	gate	anode	High resistance.
	gate	cathode	High resistance in either direction.
	cathode	gate	
N-channel JFET	drain	source	Same resistance in either direction; typically 500 Ω –5 kΩ.
	source	drain	
	gate	drain or source	Low resistance (forward-biased P-N junction).
	drain or source	gate	High resistance; > 10 MΩ unless ohmmeter battery exceeds JFET breakdown voltage BV_{GDO}.
P-channel JFET	Same as for N-channel, except that all polarities are reversed		
E-MOSFET	drain	source	Very high resistance; > 10 MΩ.
	source	drain	
	gate	drain or source	Very high resistance; > 100 MΩ for either direction.
DE-MOSFET	drain	source	Moderate resistance; in 500 Ω–5 kΩ range.
	source	drain	
	gate	drain or source	Very high resistance; > 100 MΩ for either direction.
	drain or source	gate	

Source: *Fundamentals of Electronic Devices,* third edition, by Ronald Tocci (Columbus: Merrill, 1982). Reprinted by permission.

GLOSSARY

A—D converter A circuit that converts an analog or continuous voltage or current value into an output digital code.

Accelerating torque In a synchronous motor, the torque available at any time during the acceleration period, over and above the torque requirements of the load at that speed.

Accessory A device that controls the operation of a magnetic motor control.

Across-the-line starter See full-voltage starter.

Address bus In a microprocessor system, a group of lines used to identify a peripheral device or a memory location.

Air gap The space between the rotating and stationary members in an electric motor.

Alternator A generator that produces AC (alternating current).

Ambient temperature The temperature of the medium (such as air or water) into which the heat of the object is dissipated.

Amortisseur windings See damper windings.

Ampere-turn The magnetomotive force produced by the current of one ampere in a coil of one turn.

Amplidyne A special DC generator in which a small DC voltage applied to field windings controls a large output voltage from the generator.

Analog-to-digital converter A circuit that takes an analog input and converts it to a corresponding digital output.

Anode In a semiconductor device, the terminal from which electrons flow.

ANSI The acronym for the American National Standards Institute, an organization that sets standards in many industrial areas.

Apparent Power The product of the rms voltage and current in a circuit.

Armature The laminated iron core with wire wound around it in which EMF is produced by magnetic induction in a motor or generator. Usually the rotating part, but in AC machines, sometimes stationary.

Armature reaction The current that flows in the armature winding of a DC motor and pro-duces magnetic flux, in addition to that produced by the field current.

Autotransformer A transformer where one winding serves as both primary and secondary.

Base speed The lowest rated speed obtained at rated load and voltage at the temperature rise specified in the rating.

Bidirectional mode In a stepper motor, the mode of operation where the stepper motor may be operated in either direction.

Bifilar In a stepper motor, a type of winding made up of two overlapping wires wound on the same pole.

Bipolar A type of motor drive in which current flows through the stator windings in both directions.

Brake An electromechanical friction device used to stop and hold a load.

Breakdown torque The maximum torque a motor will develop at rated voltage without an abrupt drop or loss in speed.

Brush A piece of current-carrying material (usually carbon or graphite) that rides directly on the commutator of a commutated machine and conducts current from the power supply to the armature windings.

Brushless DC motor A DC motor whose commutation is done by electronic devices rather than with a mechanical commutator and brushes.

Bushing A cylindrical metal sleeve inserted into a machine part to reduce friction.

Cathode In a semiconductor device, the terminal to which electrons flow.

Centrifugal switch A switch that opens to disconnect the start winding when the rotor speed reaches a certain preset speed and reconnects the start winding when the speed falls below a preset value.

Circuit breaker A device for interrupting a circuit between separable contacts under normal and abnormal conditions.

Closed loop A broadly applied term referring to any system where the output is measured and compared to the input. The output is then adjusted to a desired level. In motion control systems, the term is used to describe a system where velocity or position is controlled.

Cogging The jerking rotation of a motor armature caused by the tendency of the armature to prefer certain positions.

Commutator In a DC motor or generator, a split ring, each segment of which is connected to an end of a corresponding armature coil.

Compound motor A DC motor with both shunt- and series-connected fields.

Constant-current source A source that provides a constant current to a load regardless of the demands of the load.

Constant–torque region A region on the operating curve of a motor where the torque produced by the motor is constant.

Contact A current-conducting member of a device that is designed to complete or interrupt power to a circuit.

Contact bounce The bouncing movement that occurs when two switch contacts come together.

Contactor A device used to repeatedly establish or interrupt an electrical power circuit.

Continuous armature current An operating condition in a DC motor drive where the armature current never decreases to zero.

Continuous duty A motor designed to run constantly with a duty cycle of 100%.

Control bus In a microprocessor system, the lines that control the reading and writing of information.

Converter In motor control circuitry, a circuit that converts a fixed-frequency AC input to a variable-frequency AC output voltage.

Copper loss The loss in electrical power caused by current flowing through coils of wire.

Core loss See iron loss.

Corona An electrical discharge in a gas that takes place whenever a potential is greater than the dielectric strength of the gas, usually air.

Counter electromotive force (CEMF) The voltage induced into the armature of a DC motor or the stator of an AC motor, caused by the armature coils cutting the field magnetic flux. Also called back EMF.

Countertorque That force between two magnetic fields of like polarity that reduces or decreases the applied or generated torque. Also called motor reaction.

Coupling coefficient An expression of the extent to which two inductors are coupled by magnetic lines of force. The coefficient of coupling is expressed as a decimal or a percentage of the maximum possible coupling and is represented by the letter K.

CPU Acronym for central processing unit. A group of registers and logic that form the arithmetic logic unit, plus another group of registers and the associated decoding logic that form the control unit.

Critical damping In a stepper motor system, when the response to a step change is achieved in the shortest possible time.

CSI See current-source inverter.

Cumulative compound motor A compound DC motor in which the two fields aid each other in polarity.

Current-source inverter An inverter that takes an adjustable DC input voltage and converts it to a variable-frequency AC output current.

Cycloconverter A type of converter used in motor controls that converts a fixed-frequency AC input to a variable-frequency AC output voltage at a lower frequency.

Damper windings Squirrel cage windings embedded in the rotor of a synchronous motor that allow it to start as an induction motor. Also called amortisseur windings.

Damping In a stepper motor system, the indication of the rate of decrease of a signal to its steady-state value. In viscous damping, a blade moving in a fluid provides the damping. In slip-clutch damping, damping is provided by the friction between two metal plates rubbing together.

Data bus The group of lines carrying information to and from the memory and I/O sections.

DC drive A broad term used to describe the circuitry that controls a DC motor.

Dead center In á DC motor or generator, that position of the armature coil that does not produce a voltage or torque.

Delta A type of three-phase electrical connection where the terminals are connected in a closed ring. The instantaneous voltage around the ring is equal to zero. Used mostly in three-phase transformers and AC motors.

Detent torque In a PM stepper motor system, the maximum torque that can be applied to the shaft externally with the stator not energized. Also called residual torque.

Diac A bidirectional thyristor that is triggered by breakover.

Dielectric strength The highest voltage that can be applied to an insulating material without breaking down the material.

Differential compound motor A compound DC motor in which the two fields oppose each other in polarity.

Disconnect A device or group of devices that removes conductors of a circuit from the source of supply.

Discontinuous armature current An operating condition in a DC motor drive where the armature current decreases to zero at some point.

Distributed capacitance In transformers, a measure of the capacitance within a winding resulting from the coils and the dielectric material between the coils.

Drive A broad term used to describe the circuitry that controls a motor.

Drop-out voltage In a relay, the voltage at which the armature unseats from the magnetic assembly.

Duty cycle In a repeating wave form, the ratio of time on (or pulse width) to the total cycle time.

Dynamic braking A system of electrical braking in which the motor, when used as a generator, converts the mechanical energy of the load into electrical energy and, in doing so, exerts a slowing force on the load.

Dynamo A generalized electrical machine that either converts mechanical energy in the form of torque to electrical energy, or electrical energy to torque.

ECKO Acronym for eddy current killed oscillator A transducer operated on the principle of changing the reactance of the tank circuit of an oscillator enough to stop the oscillations and trigger the output.

Eddy currents Those currents induced in the body of a conducting mass or coil by a rate of change in the magnetic flux.

Efficiency In a motor, the ratio of mechanical power expressed in HP or Ws out to electrical power in. Expressed in a percentage or a decimal value.

Electromotive force (EMF) A synonym for voltage, usually restricted to generated voltage.

Electronic commutation In a DC motor, when the switching of current through the armature is done by semiconductor devices rather than by a conventional mechanical commutator.

Enclosure A housing that provides some degree of mechanical, electrical, and environmental protection for control devices.

End bell That part of the motor housing that supports the bearing and acts as a protective guard to the electrical and rotating parts of the motor.

EPROM An acronym for erasable programmable read only memory.

Equipment record A record kept by maintenance personnel of the maintenance done on a piece of electrical equipment.

Error voltage A voltage that reflects the difference between the desired value of a variable and the actual value of the variable.

Eutectic alloy A special type of solder that melts rapidly when a specific temperature is reached. Used in thermal overload relays.

Event-based control A type of sequential control in which the occurrence of one event causes a certain action to take place.

Examine-off instruction In a programmable controller, the instruction that returns a true condition if the status indicator is on and the normally-closed contacts pass power.

Examine-on instruction In a programmable controller, the instruction that returns a true condition if the status indicator is on and the normally-open contacts pass power.

Excitation The creation of a magnetic field by passing current through a coil.

Excitation current In transformers, that current drawn by the transformer when the secondary is unloaded. In DC motors, that current in a shunt motor resulting from the voltage applied across the field.

Exciter In a synchronous AC motor or generator, a DC generator that provides the field in the rotor.

Factor-of-merit An experimentally derived number used to predict the performance of a servo motor.

Field A term commonly used to describe the stator of a DC motor.

Field weakening A method of increasing the speed of a motor by reducing the stator magnetic field intensity.

Float switch A switch that is actuated by a liquid level.

Flux The magnetic field created around an energized conductor or permanent magnet.

Force The quantity that produces or tends to produce motion.

Fractional HP motor A motor with a continuous rating of less than 1 HP, open construction, at 1700–1800 r/min.

Frame size A NEMA designation made up of letters and numbers. All motors of the same frame size have the same dimensions.

Freewheeling diode A diode that is connected across the armature of a DC motor.

Full-load current The armature current of a motor operated at its full-load torque.

Full-load speed The speed attained by the output shaft of the motor with rated load connected and delivering its rated output.

Full-load torque The torque necessary to produce rated power output at full-load speed.

Full-voltage controller See full-voltage starter.

Full-voltage starter A motor starter that applies full voltage to the motor at starting. Also called across-the-line starter.

Fuse An overcurrent protective device with a part that opens when heated by an overcurrent.

Ground fault A current imbalance between the hot and neutral leads of a line supplying power.

Hall-effect device A sensor that produces a voltage output when magnetic flux impinges on it.

Heat sink A device used to draw heat away from an object.

Holding current The anode current needed to keep a thyristor conducting.

Holding torque In a stepper motor system, the maximum external force or torque that can be applied to a stationary, energized motor without causing the rotor to turn.

Horsepower (HP) The rate at which a motor can perform work.

Hysteresis The time lag of the magnetic flux in a magnetic material behind the magnetizing force producing it. Hysteresis losses are produced by the molecular friction of the molecules trying to align themselves with the magnetizing force applied to the material.

Hysteresis loss The resistance offered by materials to becoming magnetized, resulting in expended energy.

Inch A control function that provides for the momentary operation of a drive for the purpose of accomplishing a small movement of the driven machine. Also called inching.

Induction motor A type of AC motor in which the rotor gets its power through electromagnetic induction.

Inertia A measure of an object's resistance to a change in velocity.

Inrush current In a solenoid or relay coil, the current through the coil the first instant the coil is energized.

Insulation class A letter designation that identifies the ability of the insulation to withstand temperature.

Integral HP motor A motor built into a frame having a continuous rating of one HP or more, open construction at 1700-1800 rpm. Usually greater than 9″ in diameter.

Interlock A safety device that removes power from a circuit or system when the possibility of danger to personnel exists.

Intermittent duty A motor designed to run at a duty cycle less than 100%.

Inverter In logic circuitry, a circuit whose output is always in the opposite state or phase from the input. Also called a NOT circuit. In motor control circuitry, a device that takes a DC input and converts it to a variable-frequency AC output voltage.

I/O Acronym for input/output. Any circuit that introduces data into or extracts data from a data communications system.

I^2R In a circuit, the power dissipated by the current flowing through an electrical resistance, also called true power.

Iron loss The losses caused by the iron in a motor rotor. Consists of eddy current and hysteresis losses. In a transformer, called core loss.

Isolated I/O In a microprocessor system, a system that treats an I/O device separately from memory.

JEDEC An acronym for the Joint Electronic Device Engineering Council, an organization that sets standards for electronic devices in industry.

Kilowatt-hour The amount of power (in kilowatts) consumed in one hour.

Large-apparatus AC motor Integral HP machines with power ratings from 200 to 100,000 HP

Latching The behavior of a switch in which the switch stays on after the actuation is withdrawn.

Limit switch A switch that is actuated by the mechanical motion of an object.

Line-commutation A condition that occurs when the line voltage forces the current through a thyristor to such a low value that the device shuts off. Also called self-commutation.

Locked-rotor current The steady-state current taken from the line with the rotor at a standstill with rated voltage and frequency applied to the motor.

Locked-rotor torque The minimum torque that a motor will develop at rest for all angular positions of the rotor with rated voltage applied to the motor at rated frequency. Also called stall torque.

Magnetic controllers A motor controller that controls power to the motor by relay action.

Magnetomotive force (MMF) The magnetic energy supplied to establish the flux between the poles of a magnet.

Maintained contact A switch in which the contacts remain open or closed after actuation.

Manual controllers Motor controllers operated by hand, usually with pushbuttons.

Microstepping In a stepper motor system, an electronic control technique that adjusts the current in the stator windings to give additional steps between poles.

Megger See megohmmeter.

Megohmmeter A meter that measures very large values of electrical resistance, commonly used to check the insulation breakdown in conductors.

Memory-mapped I/O In a microprocessor system, the treatment of an I/O device as if it were a memory location.

Mode In a stepper motor system, a sequence of excitation that causes the rotor to rotate in the desired manner.

Momentary contact A switch in which contacts remain open or closed only as long as the switch is actuated.

Monofilar In a stepper motor, a type of winding made up of one winding wound on each pole.

Motor controller A control circuit that starts, stops, reverses, changes the speed of a motor, and protects it from overheating.

Motoring A condition in which a dynamo operates as a motor.

Motor starter A control circuit that starts a motor. An electrical controller for accelerating a motor from rest to normal speed.

Mutual flux In a transformer, the flux common to both primary and secondary.

Mutual inductance A circuit property existing when the relative position of two inductors causes the magnetic lines of force from one to link with the turns of the other, represented by the letter M.

Nameplate A plate fixed on the outside of the motor that specifies parameters and operating characteristics of the motor.

Negative feedback A feedback signal in a direction that reduces the variable represented by the feedback.

NEMA An acronym for the National Electrical Manufacturer's Association, an organization that sets standards for the electrical industry.

No-load The state of a machine rotating at a normal speed under rated conditions, but where no load is required of it.

No-load speed The speed attained by the output shaft of the drive motor with no external load connected and with the drive adjusted to deliver rated output.

OEM An acronym for original equipment manufacturer.

Off-delay In a time-delay relay, that condition begun by the deenergizing of the coil.

On-delay In a time-delay relay, that condition begun by the energizing of the coil.

One-phase excitation See wave drive.

One-two phase excitation In a stepper motor system, a type of excitation in which two phases are energized at the same time. Also called the half-step excitation mode.

Open loop Refers to a motion control system where no external sensors are used to give velocity or position control.

Opto-isolator A device that allows a signal to be sent from one place to another without the usual requirement of a common-ground potential.

Overcurrent relay A relay that operates when the current through it is equal to or greater than a preset level.

Overload The overcurrent to which a device or system is subjected in the course of its normal operating conditions.

Overload relay An overcurrent relay that functions at a predetermined value of overcurrent, causing the load to be disconnected from the power supply.

Overshoot In a stepper motor system, the distance a motor shaft can rotate beyond the step angle before it comes to rest at a step-angle position.

Permanent-magnet motor A DC motor whose field is created by a permanent magnet.

Permeability The relative ease with which the magnetic domains in a magnetic core can be made to line up to create a magnetic field.

Phase A term that indicates the space relationship of the windings and the changing values of the AC voltage and currents applied to those windings.

Phase control The process of varying the point within the cycle at which forward conduction is permitted to begin.

Photoswitch A switch actuated by light.

Pick-up voltage In a relay, the minimum coil

voltage that causes the armature to move.

Pilot device A switch that is actuated by a nonelectrical means, as in a limit switch.

Plugging A control function that provides braking by reversing the motor line voltage polarity or phase sequence so that the motor develops countertorque.

Plunger The movable part of the solenoid core.

Poles The elements that develop one or more sets of magnetic flux fields in a motor. In the induction motor, the number of poles is determined by the configuration of the stator winding. In DC machines, the poles are either permanent or wound magnets fixed in the motor frame. Synchronous motors have AC poles established by the stator winding and DC poles built into the rotor.

Pole saturation The point at which the magnetic pole of a motor or generator will take on no additional magnetic flux.

Pony brake A simple mechanical device, normally a piece of wood with an adjustable strap, used to test the torque output of a motor.

Pony motor An external motor used to start the synchronous motor and drive it to near synchronous speed.

Positive feedback A feedback signal in a direction that increases the variable represented by the feedback.

Power The rate of doing work, measured in watts or horsepower. 1 HP = 746 W.

Power factor The ratio of total watts of true power to the total rms voltamperes apparent power. The cosine of the phase angle between current and voltage in an AC circuit.

Power factor relay A relay that monitors the power factor of the synchronous motor system and protects the motor from damage after pull-out.

Prime mover That source of power used to turn an AC or DC generator.

Programmable controller A solid-state, Boolean logic system that stores a predetermined control sequence in memory. The controller, which is triggered by various inputs, activates certain outputs in response to the predetermined logic.

Pull-in Describes the situation where the synchronous AC motor goes into synchronous operation.

Pull-in torque In a stepper motor system, the maximum torque at which an energized motor can start and run in synchronism. In a synchronous motor, the maximum constant torque that will accelerate the motor into synchronism at rated voltage and frequency.

Pull-out Describes the situation where the synchronous AC motor breaks out of synchronous operation.

Pull-out torque The maximum torque that can be applied to the rotor without loss of synchronism. In a synchronous AC motor, the maximum torque developed by the motor for one minute before it pulls out of step due to an overload. Also called breakdown torque in an induction motor.

Pull-up torque In a synchronous motor, the minimum torque developed by the motor between standstill and pull-in.

Pulse rate In a stepper motor system, the frequency of the step pulses applied to a stepper motor.

Pushbutton A switch having a manually operated plunger, rocker, or button for activating the switch.

PWM inverter An inverter that uses a pulse-width modulation technique to control the frequency applied to a polyphase AC motor.

Ramping The acceleration and deceleration of a motor.

Rated torque The torque production capability of a motor turning at a specified speed. The maximum torque a motor can deliver to a load.

Rectifier A circuit that converts an AC voltage to a DC voltage.

Reduced-voltage controller See reduced voltage starter.

Reduced-voltage starter A motor starter that applies less than the full-line voltage to the motor at starting.

Regenerative braking A form of dynamic braking in which the energy of the motor is returned to the power supply.

Relay logic A type of logic circuit that uses the electromagnetic relay as a logic element.

Reluctance That characteristic of a magnetic material that resists the flow of magnetic flux through it.

Resonance In a stepper motor system, the inability of a rotor to follow the step input command. Occurs in VR steppers where the natural frequency of the rotor is the same as the frequency of the input pulses. Can cause missteps, oscillation, and reversing.

Rest mode In a stepper motor, the mode of operation where the rotor is at rest and the stator windings are not energized.

Reverse-blocking triode thyristor A thyristor with three terminals that can conduct in only one direction.

Reversing starter A motor starter that allows the motor to start in either direction.

Rotor The rotating member of a machine. In DC machines with stationary field poles and universal motors, it is commonly called the armature.

Salient pole A stator or field pole constructed so that the poles are concentrated into confined arcs and the winding is wrapped around them.

Saturable core reactor A type of inductor used to control the amount of AC power delivered to a load.

Sealed current In a solenoid or relay, the steady-state current through the coil after the armature has completely closed.

Self-excited generator A DC generator that provides its own DC field excitation.

Separately-excited generator A DC generator that receives its field excitation from an external source.

Sequential controller A controller that initiates a series of actions in a predetermined sequence until the task is accomplished.

Service factor A parameter, normally found on a motor nameplate, that indicates how much above the nameplate rating a motor can be driven without overheating.

Servo motor An AC or DC motor used in a remote control application that has identical characteristics in both directions of rotation.

Set point A voltage reference that reflects the desired value of the controlled variable.

Settling time In a stepper motor, the time

taken from the application of the last pulse command until the point where the motor comes to rest after the damping oscillations.

Slew In a stepper motor system, the portion of a rotor movement made at a constant speed. Normally a high speed area of operation where the motor can run in one direction but cannot start, stop, or reverse direction on command.

Slewing mode See slew.

Slip The difference between the synchronous speed and the actual speed of the motor to the synchronous speed, expressed as a ratio or a percentage.

Slip-clutch damping In a stepper motor system, a type of mechanical damping that uses a heavy inertial wheel sliding between two collars.

Slip ring A conductive band mounted on the armature of an alternator or synchronous motor and insulated from the armature.

Snap-action The rapid motion of contacts from one position to another. Normally used to refer to switches.

Solenoid An electromagnetic device that converts electrical energy into mechanical energy, usually linear motion.

Solid-state relay A relay with isolated input and output whose functions are achieved by means of electronic components and without the use of moving parts.

Speed The linear or rotational velocity of a motor shaft or other object in motion.

Speed regulation The motor speed change between minimum-load and full-load torque, expressed as a percentage of the full-load motor speed.

Spider The core of a synchronous motor or generator rotor.

Squirrel-cage rotor A type of AC motor rotor in which heavy conductors are embedded in the rotor body.

Stall mode In a stepper motor system, the mode of operation where the stator windings are energized but the rotor is not moving.

Stall torque The amount of torque applied to a rotor, which will cause it to stop rotation.

Starting current The amount of current drawn when a motor is energized.

Starting torque The torque delivered by a motor when energized.

Start-stop region In a stepper motor system, the region of the operating curve where a motor can stop, start, and reverse in synchronism with the external pulse signal.

Step angle In a stepper motor system, the angle through which the rotor turns in response to one pulse signal. Depends on the structure of the motor and the excitation mode.

Step angle accuracy In a stepper motor system, the per-step deviation from the nominal step angle of the motor. Can be expressed as a percent or an angle.

Step-down transformer A transformer that produces a lower voltage at the secondary than is applied to the primary.

Step-up transformer A transformer that produces a higher voltage at the secondary than is applied to the primary.

Stepper motor A motor in which the rotor moves in discrete increments when power is applied to the stator.

Stepping rate In a stepper motor system, the number of steps the rotor will produce in a specified amount of time.

Stiff A term used to describe a voltage source that has good voltage regulation.

Synchrogram A graph comparing the time behavior of two or more electrical or mechanical signals.

Synchronous condenser A synchronous motor used to correct power factor.

Synchronous speed The speed of rotation of the magnetic flux of a motor, produced by or linking the primary winding.

Synchronous torque In a synchronous AC motor, the amount of torque produced when the motor is synchronized.

Systems data bus See data bus.

Thermal overload relay An overload relay that functions or trips by means of a thermally responsive system.

Three-wire control A type of motor control where three wires are connected to the starter.

Thyristor A semiconductor switch with at least three junctions and four layers.

Time-delay relay A control relay in which the contacts open or close at a preset time interval after the coil is energized or deenergized.

Torque That force that causes a body to rotate.

Torque-to-inertia ratio In a stepper motor system, the holding torque divided by the inertia of its rotor. The higher the ratio, the better a motor's capability to accelerate a load will be.

Torque motor A motor designed to produce torque at either a low speed or at zero rotor speed.

TTL (Transistor-transistor logic) Describes a common form of digital logic device family used in modern digital equipment.

Transient voltage A voltage present for only a short period of time.

Turns ratio In a transformer, the ratio of the voltage induced in the primary to the voltage induced in the secondary.

Two-phase excitation In a stepper motor system, a four-step switching sequence.

Two-wire control A motor control circuit where only two wires are required to connect the pilot device to the motor starter.

Undervoltage protection The effect of a device, operating on the reduction or failure of a voltage, to cause and maintain the interruption of power to the main circuit. Also called low-voltage protection.

Unipolar drive A type of motor drive in which current flows through the stator windings in one direction.

Universal motor A motor that can be run on either AC or DC power.

V-curve Synchronous motor curves that indicate the changes in current for a constant load

and varied rotor excitation.

Variable-voltage inverter An inverter that takes an adjustable DC input voltage and converts it to a variable-frequency AC output voltage.

Variable-reluctance stepper motor A stepper motor whose rotor is constructed of soft-iron. The rotor is positioned by seeking a minimum reluctance position around the stator.

Varistor An electronic component in which the resistance is inversely proportional to the voltage applied across it.

VCO See voltage-controlled oscillator.

Viscous damping In a stepper motor system, a type of mechanical damping that uses an inertial body moving in a viscous liquid.

Voltage-controlled oscillator An oscillator in which the frequency of oscillation is determined by the input voltage to the oscillator.

VVI See variable voltage inverter.

Wave drive In a stepper motor system, a type of excitation in which only one phase is energized at a time. Also called the one-phase drive.

Windage The power loss in a rotating machine caused by the friction between the rotor and the air surrounding it.

Work The product of displacement of an object and the force that caused the displacement. Measured in foot-pounds or joules.

Wound-field DC motor A type of DC motor that has an armature made from coils of wire, a commutator, and brushes.

Wye A type of three-phase electrical connection where all terminals with the same instantaneous polarity are connected at a neutral junction. Also called the star connection.

INDEX

WE VALUE YOUR OPINION—PLEASE SHARE IT WITH US

Merrill Publishing and our authors are most interested in your reactions to this textbook. Did it serve you well in the course? If it did, what aspects of the text were most helpful? If not, what didn't you like about it? Your comments will help us to write and develop better textbooks. We value your opinions and thank you for your help.

Text Title _____ Edition _____

Author(s) _____

Your Name (optional) _____

Address _____

City _____ State _____ Zip _____

School _____

Course Title _____

Instructor's Name _____

Your Major _____

Your Class Rank _____ Freshman _____ Sophomore _____ Junior _____ Senior

_____ Graduate Student

Were you required to take this course? _____ Required _____ Elective

Length of Course? _____ Quarter _____ Semester

1. Overall, how does this text compare to other texts you've used?

   _____ Superior _____ Better Than Most _____ Average _____ Poor

2. Please rate the text in the following areas:

	Superior	Better Than Most	Average	Poor
Author's Writing Style	_____	_____	_____	_____
Readability	_____	_____	_____	_____
Organization	_____	_____	_____	_____
Accuracy	_____	_____	_____	_____
Layout and Design	_____	_____	_____	_____
Illustrations/Photos/Tables	_____	_____	_____	_____
Examples	_____	_____	_____	_____
Problems/Exercises	_____	_____	_____	_____
Topic Selection	_____	_____	_____	_____
Currentness of Coverage	_____	_____	_____	_____
Explanation of Difficult Concepts	_____	_____	_____	_____
Match-up with Course Coverage	_____	_____	_____	_____
Applications to Real Life	_____	_____	_____	_____

3. Circle those chapters you especially liked:
   1   2   3   4   5   6   7   8   9   10   11   12   13   14   15   16   17   18   19   20
   What was your favorite chapter? _____
   Comments:

4. Circle those chapters you liked least:
   1   2   3   4   5   6   7   8   9   10   11   12   13   14   15   16   17   18   19   20
   What was your least favorite chapter? _____
   Comments:

5. List any chapters your instructor did not assign. _____

6. What topics did your instructor discuss that were not covered in the text?_____

   _____

7. Were you required to buy this book? _____ Yes _____ No

   Did you buy this book new or used? _____ New _____ Used

   If used, how much did you pay? _____

   Do you plan to keep or sell this book? _____ Keep _____ Sell

   If you plan to sell the book, how much do you expect to receive? _____

   Should the instructor continue to assign this book? _____ Yes _____ No

8. Please list any other learning materials you purchased to help you in this course (e.g., study guide, lab manual).

   _____

9. What did you like most about this text? _____

   _____

10. What did you like least about this text? _____

   _____

11. General comments:

   May we quote you in our advertising? _____ Yes _____ No

   Please mail to: Boyd Lane
                     College Division, Research Department
                     Box 508
                     1300 Alum Creek Drive
                     Columbus, Ohio 43216

   Thank you!